Energy Storage and Release through the Solar Activity Cycle

Christophe Marqué • Alexander Nindos
Editors

Energy Storage and Release through the Solar Activity Cycle

Models Meet Radio Observations

Previously published in *Solar Physics* Volume 273,
Issue 2, 2011

Springer

Editors
Christophe Marqué
Royal Observatory of Belgium
Brussels, Belgium

Alexander Nindos
University of Ioannina
Ioannina, Greece

ISBN 978-1-4614-4402-2
Springer New York Heidelberg Dordrecht London

Library of Congress Control Number: 2012940773

Cover illustration: A solar eruption as observed at 164 MHz by the Nançay Radioheliograph on 4 May 2000. The left hand figure shows the quiet corona prior to the eruption (the blue circle represents the solar photospheric limb) and the next two figures display two steps of the ongoing eruptive event, respectively 10 and 30 minutes later. Credits: The Nançay Radioheliograph data are provided by the Observatoire de Paris.

Printed on acid-free paper

Springer is part of Springer Science+Business Media (www.springer.com)

Contents

Solar Phys (2011) 273:307–308
DOI 10.1007/s11207-011-9877-4

Preface

C. Marqué · A. Nindos · L. van Driel-Gesztelyi · C.H. Mandrini

Received: 10 October 2011 / Accepted: 10 October 2011 / Published online: 25 October 2011

The Community of European Solar Radio Astronomers (CESRA) organizes triennial workshops on investigations of the solar atmosphere using observations at radio and other wavelengths. Although special emphasis is given to radio diagnostics, the workshop topics are of interest to a large community of solar physicists. The 2010 workshop, "Energy Storage and Release through the Solar Activity Cycle – Models Meet Radio Observations", addressed

Energy Storage and Release through the Solar Activity Cycle – Models Meet Radio Observations
Guest Editors: Christophe Marqué and Alexander Nindos

C. Marqué
Solar Terrestrial Center of Excellence-SIDC, Royal Observatory of Belgium, Av. Circulaire 3, 1180 Brussels, Belgium
e-mail: cmarque@oma.be

A. Nindos
Section of Astrogeophysics, Physics Department, University of Ioannina, Ioannina 45110, Greece
e-mail: anindos@cc.uoi.gr

L. van Driel-Gesztelyi
Mullard Space Science Laboratory, University College London, Holmbury St. Mary, Dorking, Surrey RH5 6NT, UK

L. van Driel-Gesztelyi (✉)
Observatoire de Paris, LESIA, CNRS, UPMC, Université Paris Diderot, 5 Place Jules Janssen, 92195 Meudon Cédex, France
e-mail: Lidia.vanDriel@obspm.fr

L. van Driel-Gesztelyi
Konkoly Observatory of Hungarian Academy of Sciences, Budapest, Hungary

C.H. Mandrini
Instituto de Astronomía y Física del Espacio, IAFE, UBA-CONICET, CC. 64 Suc. 28, 1428 Buenos Aires, Argentina
e-mail: mandrini@iafe.uba.ar

explosive energy conversion, large-scale instabilities, and nonthermal processes in the active and quiet solar atmosphere. Many of the articles included in this Topical Issue were presented in the CESRA 2010 workshop, but the issue was also made open for general submission on the core topics.

The recent long declining phase of Solar Cycle 23 gave us the opportunity to study the solar atmosphere at different levels of activity and its impact on the heliosphere. Radio diagnostics contributes important information, notably because it covers the whole range from the low atmosphere to 1 AU and because of its sensitivity to nonthermal electron populations. Radio observations of the quiet Sun have confirmed that the "quiet" emission may not be that quiet after all – it can be intense and show considerable time variation. Studying the quiet Sun at radio wavelengths has the advantages that it can be done from the ground and that the physics of the processes involved can be understood without the need to worry about some well-known burdens (*e.g.* non-LTE effects, excitation and ionization equilibria, abundances) that one needs to face at other spectral ranges.

At microwaves, the emission from non-flaring active regions comes from gyro-resonance radiation above sunspots and free–free radiation from hot plasma trapped in active region loops. The gyro-resonance emission mechanism is unique to radio data and is a useful tool to measure the magnetic field above sunspots. Radio observations are valuable in providing information on temporal variations of the emission from active regions. Pulsations of the radio emission show a large variety of temporal and spatial signatures, periods, bandwidths, and amplitudes. Study of these phenomena is a powerful tool for coronal plasma and magnetic field diagnostics.

Radio observations of flares and coronal mass ejections (CMEs) help us address questions about energy release and its properties, the configuration of the flare-CME source regions, particle acceleration and transport, and the origin of solar energetic particles. Furthermore, both flares and CMEs can ignite shock waves, and radio observations offer the most robust tool to study them. All the above aspects of radio emission, both from the flaring and non-flaring solar atmosphere, are reflected in the present Topical Issue, which includes one invited review paper.

The CESRA 2010 workshop was held in La Roche-en-Ardenne (Belgium). The members of the Scientific Organizing Committee were H. Aurass (Germany), K.-L. Klein (France), A. MacKinnon (UK), C. Marqué (Belgium; co-chair), V. Melnikov (Russia), A. Nindos (Greece; co-chair), S. Pohjolainen (Finland), and S. Poedts (Belgium) as president of the Solar Physics Section within the Joint Astrophysics Division of the European Physical Society and the European Astronomical Society.

The members of the Local Organizing Committee were O. Boulvin, E. D'Huys, C. Marqué, (co-chair), A. Vandersyppe (co-chair) and P. Vanlommel, with the support of the personnel of the Royal Observatory of Belgium and its director, R. Van der Linden.

The workshop received financial support from the Royal Observatory of Belgium, the Solar Terrestrial Center of Excellence, the Belgian Federal Science Policy Office, the European Space Agency, the Observatoire de Paris, the European Physical Society, and from the Community of European Solar Astronomers.

The editors would like to thank both the authors and the referees of the articles in this volume.

Solar Phys (2011) 273:309–337
DOI 10.1007/s11207-011-9788-4

ENERGY STORAGE AND RELEASE – MODELS MEET RADIO OBSERVATIONS

Radio Emission of the Quiet Sun and Active Regions (Invited Review)

K. Shibasaki · C.E. Alissandrakis · S. Pohjolainen

Received: 9 February 2011 / Accepted: 8 May 2011 / Published online: 15 June 2011

Abstract Solar radio emission provides valuable information on the structure and dynamics of the solar atmosphere above the temperature minimum. We review the background and most recent observational and theoretical results on the quiet Sun and active region studies, covering the entire radio range from millimeter to decameter wavelengths. We examine small- and large-scale structures, at short and long time scales, as well as synoptic aspects. Open questions and challenges for the future are also identified.

Keywords Chromosphere, quiet · Corona, quiet · Polarization, radio · Radio emission, quiet · Active regions, radio

1. Introduction

Solar radio emission is traditionally divided into a background component (quiet sun, QS), a slowly varying component (associated mainly with active regions), and a sporadic (burst) component (Kundu, 1965). Both short- and long-lived transient emissions can occur in quiet regions, thus the 'quiet' emission may not be that quiet after all – it can be intense and show considerable time variation.

Energy Storage and Release through the Solar Activity Cycle – Models Meet Radio Observations
Guest Editors: Christophe Marqué and Alexander Nindos

K. Shibasaki
Nobeyama Solar Radio Observatory/NAOJ, Nagano 384-1305, Japan
e-mail: shibasaki@nro.nao.ac.jp

C.E. Alissandrakis
Section of Astro-Geophysics, Department of Physics, University of Ioannina, 45110 Ioannina, Greece
e-mail: calissan@cc.uoi.gr

S. Pohjolainen (✉)
Department of Physics and Astronomy, Tuorla Observatory, University of Turku, 21500 Piikkiö, Finland
e-mail: silpoh@utu.fi

Compared to the other components of the solar radio emission, as well as to quiet emission in other spectral ranges (*e.g.*, the EUV), the radio quiet Sun has the advantage of being fairly well understood as originating from thermal bremsstrahlung in local thermodynamic equilibrium (LTE). It can thus be used as a powerful diagnostic of the physical conditions in a wide range of atmospheric layers. As the opacity increases with the wavelength of observation, the effective height of formation moves from the temperature minimum (at sub-millimeter waves) to the low corona (at meter waves), or even higher into the corona and the heliosphere (longer wavelengths being observed from space).

In the metric range, in addition to opacity effects, we have to consider refraction and total reflection. The latter makes solar radio emission a useful tool in determining source heights; radio waves from a particular atmospheric layer will reach the observer only if their frequency is higher than the local plasma frequency. As the electron density decreases with height, so does the critical plasma frequency. Thus, both opacity and refraction contribute so that we can form a 3D-view of the solar atmosphere from multi-wavelength observations.

In this review we describe how our understanding of the quiet Sun has evolved over the years and give emphasis on recent results on the quiet Sun phenomena, based on radio observations and comparisons to other wavelength data. For reviews on earlier work see, *e.g.*, Alissandrakis (1994), Gary (1996), Alissandrakis and Einaudi (1997), Lantos (1999), Shibasaki (1999), and Keller and Krucker (2004).

In Section 2 we discuss the formation of radiation and the radio spectrum, in Section 3 we give an overview of high-resolution radio imaging observations, and in Section 4 we describe the quiet Sun features that have been studied extensively using radio observations. In Section 5 we discuss briefly the active region emission, in order to present a complete picture of the non-flaring Sun. The last section is devoted to future challenges in radio studies of the quiet Sun, where we discuss the present and future radio instrumentation, questions on data analysis, and the current theoretical issues.

2. Radiative Transfer and Modeling

2.1. The Basics

The quiet Sun emission at radio wavelengths is the simplest case for modeling for a number of reasons. The radiation is of thermal origin, from the ambient plasma; the mechanism is bremsstrahlung (free–free) from electrons interacting with ions in the presence of a relatively weak magnetic field. The magnetic field is weak enough so that gyro-resonance emission can safely be ignored; since the opacity at the 4th and higher harmonics layers is negligible, the condition is that the observing frequency should be higher than three times the local gyrofrequency, which gives B (in G) $< f$ (in MHz)/8.4, *i.e.* 3600 G at 1 cm and 36 G at 1 m.

The electrons are always in LTE, and thus one can assume that the source function is equal to the Planck function. Moreover, thanks to the Rayleigh–Jeans approximation, it is convenient to express the specific intensity of radiation in terms of brightness temperature, T_b, and the Planck function in terms of electron temperature T_e. Thus, the solution of the transfer equation in a magnetoionic medium takes the form

$$T_{\mathrm{b},j} = \int_0^\infty T_\mathrm{e}(\tau_j) e^{-\tau_j}\, \mathrm{d}\tau_j, \tag{1}$$

where $j = 1$ for the extraordinary mode and $j = 2$ for the ordinary mode, $T_e(\tau_j)$ is the distribution of the electron temperature along the line of sight, and τ_j is the optical depth, related to the path ℓ along the line of sight through

$$\mathrm{d}\tau_j = -k_j(T_e, N_e)\,\mathrm{d}\ell, \tag{2}$$

where $T_e(\ell)$ and $N_e(\ell)$ represent the variations of electron temperature and density along the line of sight. In the case of a plane-parallel atmosphere, ℓ and τ_j in Equations (1) and (2) can be replaced by ℓ/μ and τ_j/μ, respectively, where $\mu = \cos\theta$, θ being the angle between the direction of the line of sight and the direction of the center of the solar disk.

The free–free absorption coefficient at frequency f has the well-known form (Kundu, 1965)

$$k_j(T_e, N_e) = \xi \frac{N_e^2}{n_j f^2 T_e^{3/2}} A_j(B, \theta), \tag{3}$$

where ξ is a slowly varying function of T_e and N_e, n_j the index of refraction ($\sim$1 at short wavelengths) and $A_j(B, \theta)$ is a function of the magnitude of the magnetic field B and its angle with the line of sight, θ (Zheleznyakov, 1970).

In the absence of magnetic field, $A_j = 1$. Otherwise the extraordinary mode opacity is slightly above that of the unmagnetized case, whereas the ordinary opacity is below; since in the region of formation of radiation the temperature increases with height, the same thing happens to the brightness temperature. The net effect is weakly polarized emission in the sense of the extraordinary mode, which is a powerful diagnostic of the magnetic field (see, *e.g.*, Grebinskij *et al.*, 2000). We should note that, far from the center of the disk, the observed circular polarization is influenced by propagation effects (Cohen, 1960), which are stronger at longer wavelengths. Thus, the circular polarization is sometimes more useful as a diagnostic of the magnetic field in the propagation region, rather than as a diagnostic of the physical conditions at the source (Alissandrakis, 1999; Ryabov *et al.*, 1999)

It follows from the above that all we need to know in order to compute the total intensity $I = \frac{1}{2}(T_{b,1} + T_{b,2})$, and the circular polarization $V = \frac{1}{2}(T_{b,1} - T_{b,2})$ of the radiation, is $T_e(\ell)$, $N_e(\ell)$, and, if applicable, $B(\ell)$, and $\theta(\ell)$. There is no need to consider non-LTE effects, abundances, ionization and excitation equilibria that are common sources of uncertainty in other parts of the spectrum.

Things are not always that simple though. One inconvenience is the large radial extent of the region of formation of radiation at cm-waves, as the chromosphere–corona transition region (TR) is too thin to provide any sizable opacity. This makes it difficult to define the effective height of formation.

Complications also arise at longer wavelengths, where the index of refraction departs from unity. This does not only affect the ray paths, but also decreases the brightness, as the rays divert away from high density regions and the total optical depth drops below unity (*cf.* Figure 1 of Alissandrakis, 1994). Scattering of the radiation in inhomogeneities can have similar effects (see Thejappa and MacDowall, 2008 and earlier works by Aubier, Leblanc, and Boischot, 1971; Hoang and Steinberg, 1977; Thejappa and Kundu, 1992; 1994; Bastian, 1994). Scattering also smoothes structures of small angular scale and increases the apparent size of the radio Sun.

2.2. One-Dimensional Models of the Radio Spectrum

Anyone who has seen a solar image will agree that the Sun is rich in fine structures. As a matter of fact, inhomogeneities become more important as we move from the photosphere to the corona. Under these circumstances, it is rather surprising that one-dimensional models, which treat the physical parameters as a function of height only and assume hydrostatic equilibrium, have any resemblance to the observations at all. The physical reason behind their success is the strong radial stratification of the solar atmosphere, due to gravity. Besides the average QS, coronal holes are another favorite for one-dimensional models, as we will discuss in Section 4.5.

There is a long tradition of empirical solar models. The Bilderberg Continuum Atmosphere (BCA; Gingerich and de Jager, 1968), which superseded the Utrecht Reference Photosphere (URP; Heintze, Hubenet, and de Jager, 1964), was the first to take into account mm-wave observations together with EUV; a comparison between BCA-predicted brightness temperatures and observations in the range of 0.0086 – 15.8 mm was presented in Noyes, Beckers, and Low (1968), in their Figure 7. The Harvard Smithsonian Reference Atmosphere (HSRA; Gingerich *et al.*, 1971) followed, and the brightness computations and comparison with data were extended to 10 cm (their Figure 3).

However, these models did not extend beyond the low chromosphere; they all stopped around $T_e \simeq 10^4$ K, which is too low for model computations at longer wavelengths. The VAL models that followed (Vernazza, Avrett, and Loeser, 1973, 1976, 1981) extended the temperature range up to 3×10^4 K, into the lower part of the transition region. The subsequent FAL models (Fontenla, Avrett, and Loeser, 1990, 1991, 1993) went up to 10^5 K, while the model of Fontenla, Avrett, and Loeser (2002) reached the low corona at 1.2×10^6 K. These models also developed further the multi-component approach, first introduced by Vernazza, Avrett, and Loeser (1981), to represent different quiet and active regions on the Sun.

In the more recent models, by Fontenla *et al.* (2006), Fontenla, Balasubramaniam, and Harder (2007), and Fontenla *et al.* (2009), the emphasis has shifted again to the lower atmospheric layers (up to 10^4 K). Still, Avrett and Loeser (2008) produced a model of the average quiet Sun chromosphere and TR up to 1.6×10^6 by extending model C (average quiet Sun) of Fontenla *et al.* (2006). The authors computed, among other things, the spectrum in the range of 0.04 – 40 mm. Their results are shown in Figure 1, together with the temperature *vs.* height variation of their model and the contribution functions at several wavelengths, including some in the short microwave range. Note that the microwave spectrum reproduces, as expected, the shape of the $T_e(z)$ curve. According to this figure, most of the emission at 3 cm and shorter wavelengths originates below the TR, with a small contribution from the latter and no contribution from the corona.

Figure 1 also shows the computations of Loukitcheva *et al.* (2004). They tried a number of static FAL models from Fontenla, Avrett, and Loeser (1993), and also made computations on the basis of the dynamic simulations of Carlsson and Stein (2002). Their conclusion was that the dynamic picture of the solar internetwork chromosphere is consistent with the currently available mm- and sub-mm brightness observations.

In order to compute brightness spectra at longer wavelengths, one has to add a coronal contribution. Zirin, Baumert, and Hurford (1991) found that their measurements, which extended up to 21 cm, could simply be reproduced by a two-component model: an optically thick chromosphere and an isothermal corona. We know, however, that there is a TR between these two. In a recent work, Selhorst, Silva, and Costa (2005) used a hybrid model (combination of models for the photosphere, chromosphere, and corona) to reproduce the observed features in Nobeyama Radioheliograph images at 17 GHz. To obtain acceptable brightness temperature values and the observed solar radius, they had to include absorbing chromospheric structures, such as spicules, into the model. Their computed spectrum is shown in Figure 2, together with a model by Bastian, Dulk, and Leblanc (1996) that covers approximately the same frequency range.

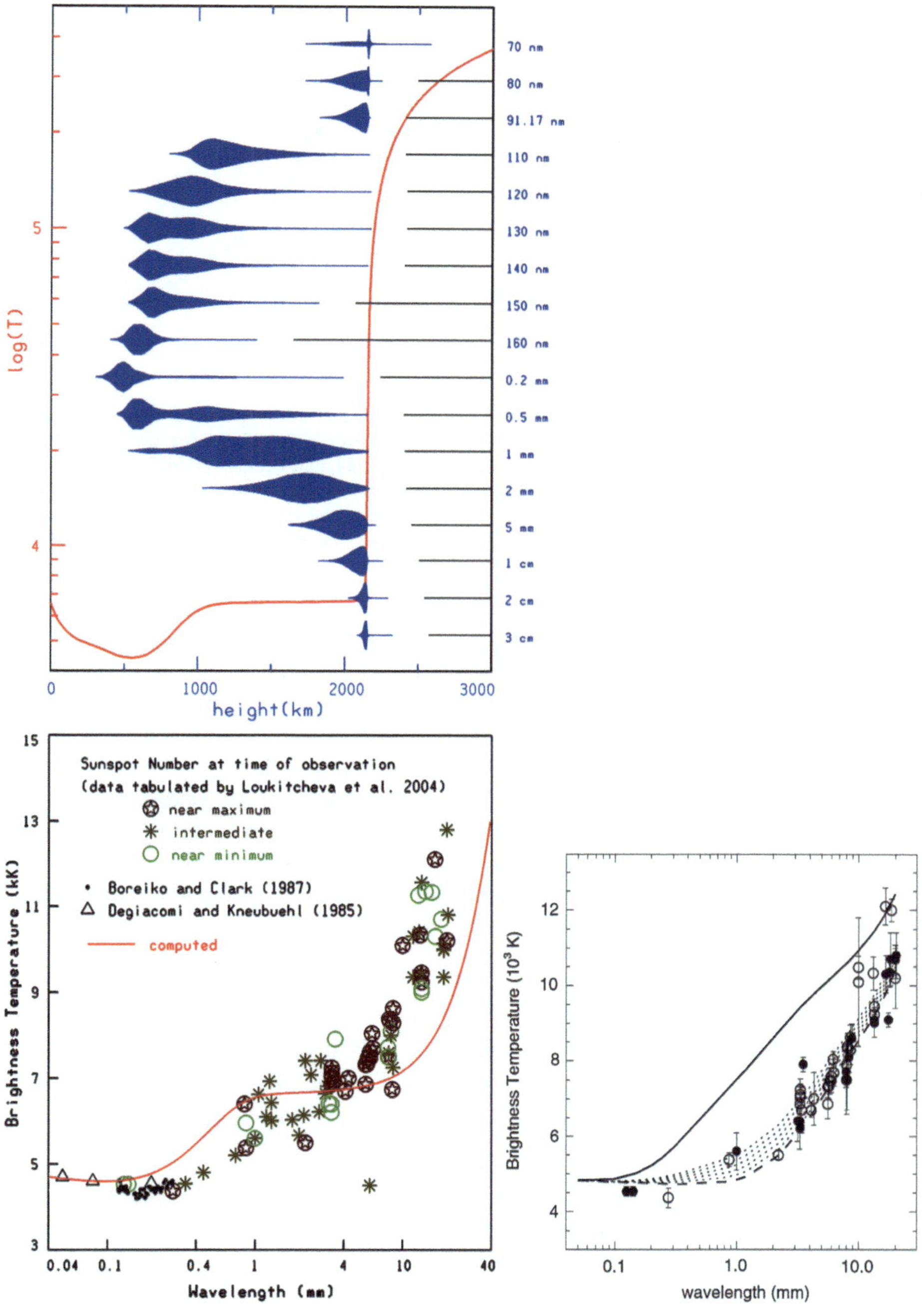

Figure 1 Contribution functions at several wavelengths, including microwave, according to the model of Avrett and Loeser (2008) (top panel); the corresponding spectrum is shown in the bottom left panel. Reproduced by permission of the AAS. The bottom right panel shows spectra computed by Loukitcheva *et al.* (2004); solid line: on the basis of model F (network) of Fontenla, Avrett, and Loeser (1993); dashed line: on the basis of dynamic simulations by Carlsson and Stein (2002); dotted lines: combinations of the two. Reproduced with permission © ESO. Very similar results were obtained using model A (cell interior) of Fontenla, Avrett, and Loeser (1993), instead of the Carlsson and Stein model. Only data obtained near solar maximum and solar minimum are presented in this plot, as open and filled circles, respectively.

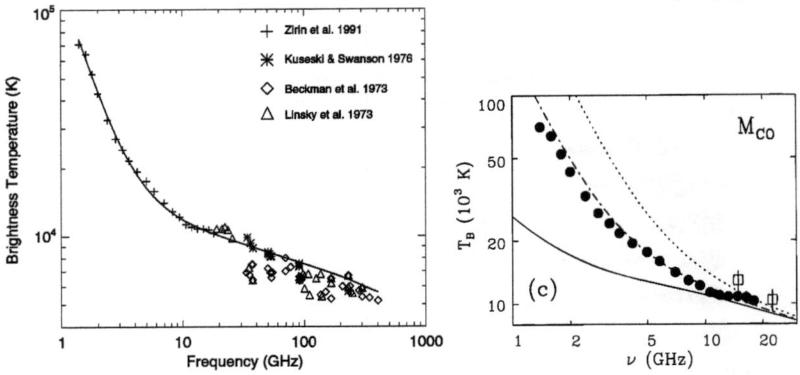

Figure 2 Left: Observed brightness temperatures of the solar disk center (symbols) and the hybrid model fit (solid line) of Selhorst, Silva, and Costa (2005), presented over a wide frequency range. Reproduced with permission © ESO. Right: Spectra computed by Bastian, Dulk, and Leblanc (1996); solid line: on the basis of a model by Avrett (1995); dot-dashed line: with the addition of a minimum corona inferred by Zirin, Baumert, and Hurford (1991); dotted line: with the addition of three times the minimum corona. Filled circles are the measurements of Zirin, Baumert, and Hurford, squares are VLA measurements by Bastian, Dulk, and Leblanc. Reproduced by permission of the AAS.

The results presented in Figures 1 and 2 show that a model can be found that fits most microwave data, although there is a large scatter in the mm to sub-mm range due to inherent difficulties in the measurements and possible solar cycle effects. The main problem with these models is that the radio data are better fitted with models that represent the *cell interior*, rather than with average models, which predict too much radio flux.

All computations presented above start from standard atmospheric models. In a different approach, several authors have tried to use direct information from the EUV part of the spectrum to compute the radio brightness, which is reasonable since the radiation in both wavelength ranges is formed in the same atmospheric layers. In one approach the emission measure, EM, is used:

$$EM = \int_0^L N_e^2 \, d\ell \tag{4}$$

which appears in the expression for the intensity of EUV lines:

$$I = \frac{1}{4\pi} \int_0^L G(T_e, N_e) \, N_e^2 \, d\ell \simeq \frac{1}{4\pi} \langle G(T_e, N_e) \rangle EM, \tag{5}$$

where $G(T_e, N_e)$ is the contribution function for the line. The emission measure also appears in the expression for the radio brightness temperature; for an isothermal layer and negligible refraction we get from Equations (2) and (3):

$$\tau = \xi \frac{1}{f^2 T_e^{3/2}} \int_0^L N_e^2 \, d\ell = \xi \frac{EM}{f^2 T_e^{3/2}}. \tag{6}$$

Furthermore, if the layer is optically thin,

$$T_b \simeq \tau T_e = \xi \frac{EM}{f^2 T_e^{1/2}}. \tag{7}$$

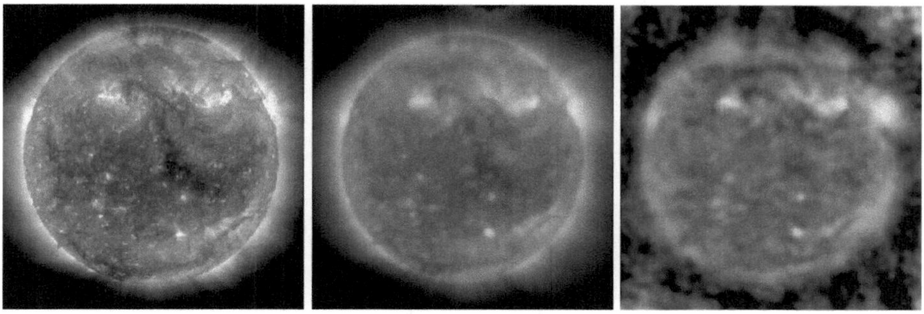

Figure 3 EIT image at 195 Å (left), computed brightness (middle), and observed (VLA) brightness at 20 cm for November 11, 1997. Figure adapted from Zhang *et al.* (2001), reproduced by permission of the AAS.

It is a well-known practice to use two EUV lines for an estimate of the plasma temperature and the emission measure. Zhang *et al.* (2001) used three EIT images, at 171, 195 and 284 Å, to derive the emission measure in a two-temperature model. They used this information to compute the emission at 6 and 20 cm wavelengths, which they compared with their VLA observations. Figure 3 shows their results at 20 cm. Although the model image looks very much like the observed, the computed brightness temperature was twice the observed at both wavelengths. They attributed the discrepancy to errors in the coronal abundances used to infer the radio flux from the EIT data. Their results also raise the question of accuracy in temperature and emission measure computations from broad-band EUV images.

The differential emission measure (DEM) is even better than the emission measure; it is defined as

$$\varphi(T_e) = N_e^2 \frac{d\ell}{dT_e} \tag{8}$$

and represents the distribution of electron density over a temperature range T_1 to T_2. Equations (5) and (6) are then replaced by

$$I = \frac{1}{4\pi} \int_{T_1}^{T_2} G(T_e, N_e)\varphi(T_e)\, dT_e \tag{9}$$

$$\tau = \frac{\xi}{f^2} \int_{T_1}^{T_2} T_e^{-3/2}\varphi(T_e)\, dT_e \tag{10}$$

and the isothermal assumption is dropped. Obviously, one needs many EUV lines formed over the appropriate temperature range in order to make good use of the DEM.

Landi and Chiuderi Drago (2003) used the DEM values derived from UV and EUV spectral line intensities observed by SUMER and CDS, and showed that a TR model *for the cell interior* – excluding any network contribution – could give an agreement with the observed radio brightness temperatures. In a subsequent work Landi and Chiuderi Drago (2008) showed that radio observations provide a much more reliable diagnostic tool for the determination of the DEM than UV and EUV lines at $T < 30\,000$ K, since the latter are optically thick. Moreover, they extended the DEM down to 5600 K using the radio spectrum from 1.5 to 345 GHz, and obtained very good agreement with the radio data (Figure 4).

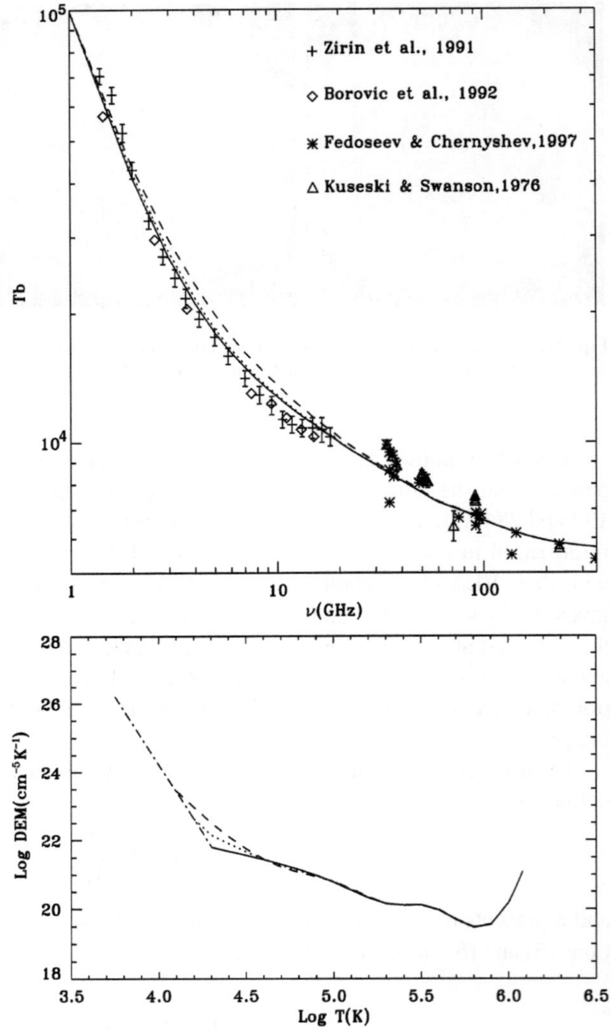

Figure 4 Spectra computed by Landi and Chiuderi Drago (2008) (top) and the corresponding differential emission measure DEM (bottom). Reproduced by permission of the AAS.

In other studies, a profound decrease in the computed temperatures have been obtained by assuming a tail of suprathermal electrons present in the solar atmosphere. Chiuderi and Chiuderi Drago (2004) modeled this with a two-component Maxwellian electron distribution and with a so-called kappa-function, and found both working. More recently, the non-thermal κ-distributions have been found to change EUV filter responses quite significantly (Dudík *et al.*, 2009), further suggesting a need to revise the atmospheric models.

At longer (metric) wavelengths there is a marked departure of observed brightness temperature below the coronal electron temperature; moreover, the brightness temperature shows a maximum of $\leq 10^6$ K near 2 – 3 m (see Figures 4 and 5 of Lantos, 1999). This is attributed to refraction and scattering effects. Thejappa and MacDowall (2008) used Monte Carlo simulations to show that the one-order of magnitude lower-than-predicted brightness temperatures at low frequencies could be due to scattering of the quiet Sun emission by random density fluctuations.

3. High-Resolution Imaging Observations

3.1. From Microwaves to Sub-millimeter Waves

Low spatial resolution is a well-known obstacle for radio observations. It is, therefore, not surprising that solar radio astronomers exploit every possibility to use the most powerful aperture synthesis instruments available. In this context, Kundu *et al.* (1979) were the first to obtain quiet Sun images at 6 cm wavelength with arc-second resolution with the Westerbork Synthesis Radio Telescope (WSRT), after its successful use for active region studies (Kundu and Alissandrakis, 1975a) and following a number of interferometric (see, *e.g.*, Kundu and Velusamy, 1974; Kundu and Alissandrakis, 1975b) and one-dimensional (see, *e.g.*, Bogod and Korolkov, 1975; Gelfreikh *et al.*, 1977) observations of the QS.

The first WSRT quiet Sun images showed a clear association of the microwave emission with the chromospheric network, something that had already been suggested on the basis of interferometric data (Kundu and Alissandrakis, 1975b). This conclusion was subsequently verified with the Very Large Array (VLA) observations at 6 and 20 cm by Gary and Zirin (1988), and by Gary, Zirin, and Wang (1990) at 3.6 cm. In the mid 90's the VLA was used for quiet Sun observations in the short cm-range (1.2, 2.0 and 3.6 cm) by Bastian, Dulk, and Leblanc (1996), Benz *et al.* (1997), and Krucker *et al.* (1997).

At present, the Nobeyama Radioheliograph (NoRH, Nakajima *et al.*, 1994) carries out full-disk routine observations of the Sun at 17 and 34 GHz. NoRH has been operating since 1992, with only a two-month gap in observations in 1995. The available data cover almost 18 years, and observations run eight hours per day with one-second time cadence. The Siberian Solar Radio Telescope (SSRT, Grechnev *et al.*, 2003) is also performing two-dimensional mapping, concurrently with one-dimensional images obtained in the additive mode, since 1996 at 5.7 GHz. The correlation mode with high-sensitivity permits low-contrast features on the Sun to be investigated.

High-resolution images of the quiet Sun in the mm-range were first produced by White, Loukitcheva, and Solanki (2006) and Loukitcheva, Solanki, and White (2006). They used the 10-element Berkeley–Illinois–Maryland Association Array (BIMA) in its most compact D-configuration, to obtain $\sim 10''$ resolution. BIMA was followed by CARMA (Combined Array for Research in Millimeter-wave Astronomy, a fusion of BIMA and OVRO) and White (communication at the CESRA 2010 meeting, La Roche-en-Ardenne, Belgium) used the 15-element array to observe the quiet Sun at 3.3 mm. In both cases extreme care was taken to minimize atmospheric effects and image reconstruction problems.

BIMA and CARMA images at 3.5 and 3.3 mm wavelengths are shown in Figure 5, together with VLA images at 1.3 and 2.0 cm, for comparison. In all cases the chromospheric network, delineated in the TRACE continuum images or the photospheric magnetograms, is the dominant structure in the radio images. Network structures appear broader in radio, but this effect almost disappears if one degrades the TRACE images to the resolution of the radio images.

3.2. Metric Range Imaging

There is a noticeable lack of imaging in the decimetric range, from 30 cm (1 GHz) to 70 cm (450 MHz). At the long wavelength end of the spectrum one has to be content with arc-minute resolution. Moreover, the Sun must be without any burst activity if the quiet Sun component is to be imaged.

The first images in the metric range were obtained with the now-extinct Culgoora Radioheliograph at 80 MHz (3.75 m) and 160 MHz (1.88 m), followed by the also extinct Clark

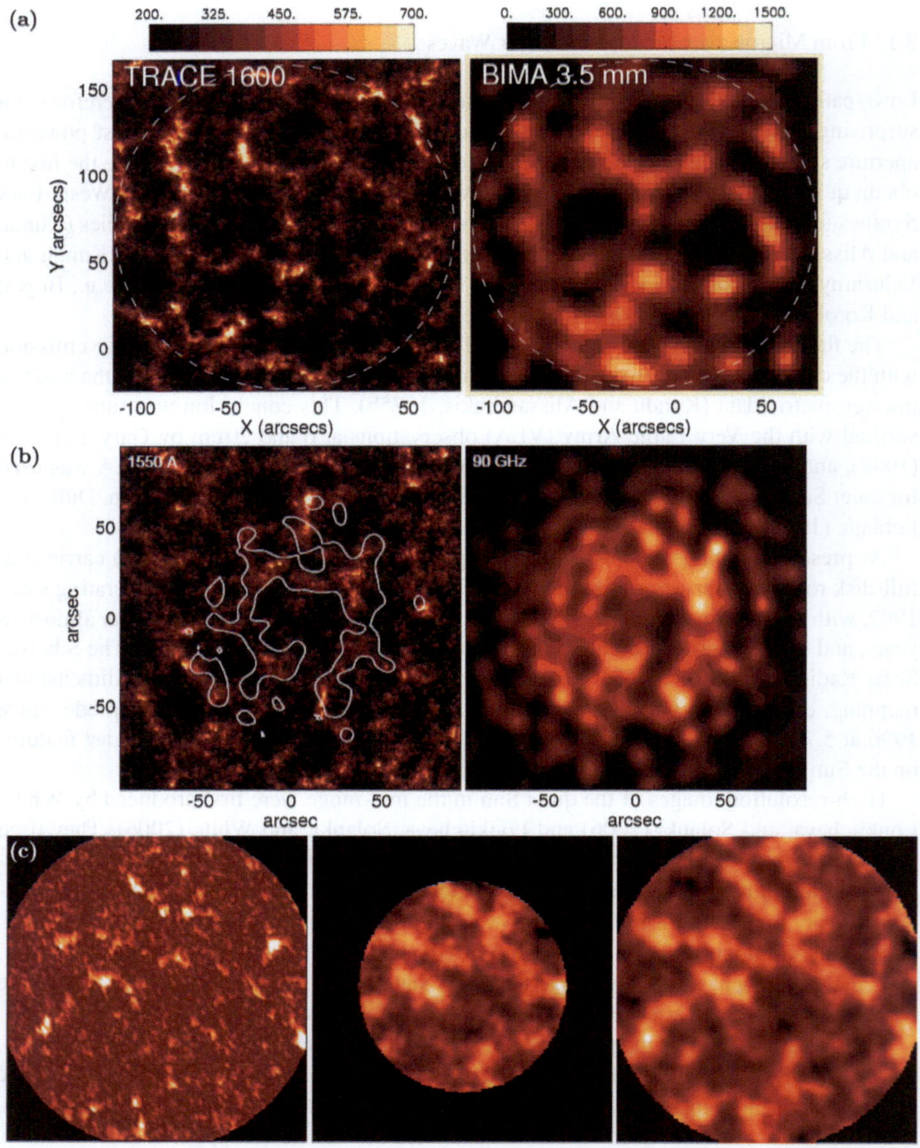

Figure 5 High-resolution radio images of the quiet Sun in the mm-wave to short cm-wave range. (a): TRACE 1600 Å image together with the BIMA radio image at 3.5 mm (85 GHz) on May 18, 2004 (adapted from Loukitcheva, Solanki, and White, 2010, reproduced with permission © SAIt); (b): TRACE 1550 Å image together with the 3.3 mm (90 GHz) image obtained with CARMA on September 26, 2008 by White (communication at the CESRA 2010 meeting, La Roche-en-Ardenne, Belgium) and (c): Image of the absolute value of the photospheric magnetic field from KPNO (left), together with VLA radio images at 1.3 cm (23 GHz, middle) and at 2.0 cm (15 GHz, right) on September 23, 1992 (adapted from Bastian, Dulk, and Leblanc, 1996, reproduced by permission of the AAS). Fields of view are 195″by 195″ in (a), 190 by 190″ in (b) and 165 by 165″ in (c). The spatial resolution is 12″ for the mm-wave images in (a) and (b), and 3.3″ and 5″ at 1.3 and 2 cm, respectively, in (c). The dashed circles in (a) show the 96″ radius of the BIMA field-of-view where the fluxes were estimated to be reliable.

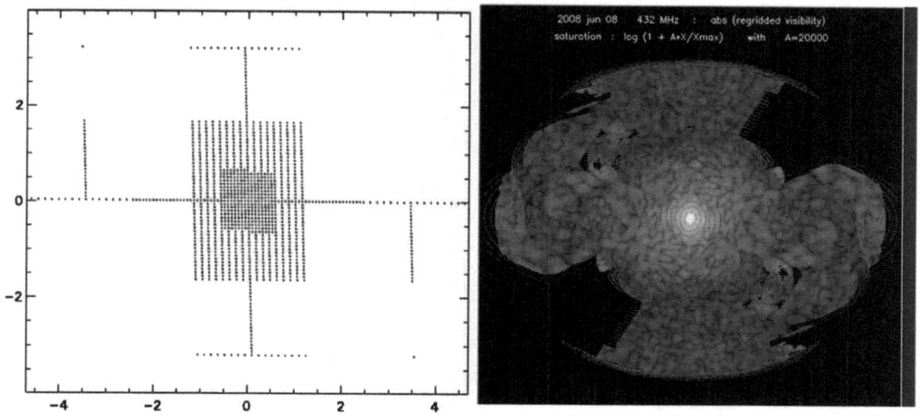

Figure 6 (a) Instantaneous coverage of the u–v plane by the Nançay Radioheliograph. The units are in 10^3 rad^{-1} (adapted from Mercier and Chambe, 2009, reproduced by permission of the AAS). (b) Gridded amplitude of the visibility function after six hours of synthesis at 432 MHz on June 8, 2008. Amplitude values cover a range of five orders of magnitude (Mercier, communication at the CESRA 2010 meeting, La Roche-en-Ardenne, Belgium).

Lake Radioheliograph at 73.8 MHz (4.07 m), 50 MHz (6 m) and 30.9 MHz (9.7 m). The Nançay Radioheliograph (NRH) operated as two independent arrays (E–W and N–S) at 169 MHz (1.78 m) until Alissandrakis, Lantos, and Nicolaidis (1985) employed aperture synthesis to combine 6 hours of observations into a 2D-image with 1.2′ by 4.2′ resolution.

The NRH has evolved gradually to its present state of 2D synthesis instrument (Kerdraon and Delouis, 1997), providing images at 10 frequencies from 450 to 150 MHz (67 cm to 2 m) with a cadence of 0.25 sec. The instantaneous images, however, cannot exploit the full resolution of the system but only use the densely sampled inner part of the u–v plane, see Figure 6. In order to exploit the full resolution one has to resort to full-day synthesis, which improves the resolution by a factor of ~ 2.5. This was done by Marqué (2004) with an emphasis on filament cavities and, more recently, by Mercier and Chambe (2009) in a systematic study of the quiet Sun. They used self-calibration and a modified CLEAN algorithm to correct and process the NRH visibilities. We note that Mercier *et al.* (2006) also computed instantaneous burst images at 327 MHz by combining NRH and Giant Metrewave Radio Telescope (GMRT) data.

Figure 7 shows some examples of synthesis images from Mercier and Chambe (2009), together with earlier work by Marqué *et al.* (1999) and a set of 1992 images from Lantos (1999). The NRH covers a broad range of frequencies, the ratio of the maximum to minimum frequency being ~ 3; it can thus probe an altitude range from the upper TR to the low corona. The altitude range goes lower in coronal holes. They are the most prominent feature at long dm wavelengths and, in agreement with previous observations (see, *e.g.*, Lantos *et al.*, 1987), their contrast decreases at longer wavelengths (see Section 4.5).

4. Observations and Interpretation of Quiet Sun Features

4.1. Center-to-Limb Variation

In the optical range, the decrease of temperature with height in the photosphere leads to the well-known limb darkening effect. In contrast, the temperature rise in the chromosphere and

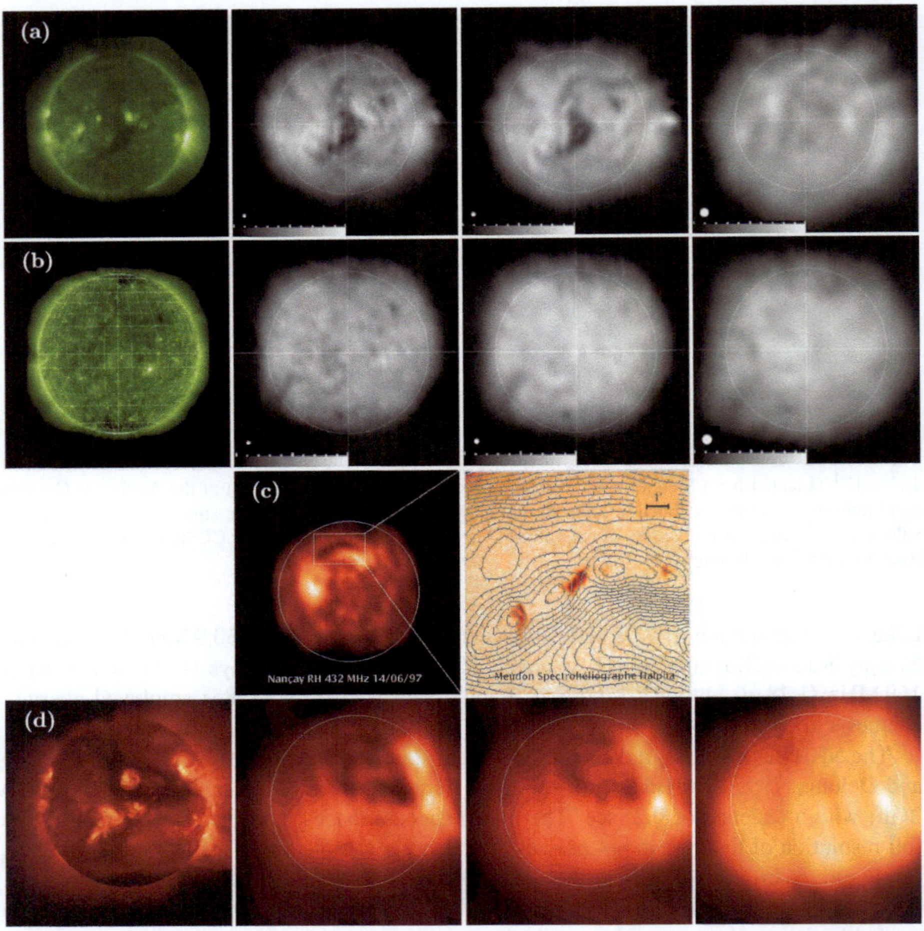

Figure 7 Synthesis images with the Nançay Radioheliograph: (a) Images at 432, 327 and 164 MHz on June 27, 2004, together with a GOES SXI image; (b) Images at 432, 327 and 173 MHz on June 6, 2008, together with an EIT 195 Å image; (c) Image of a filament at 432 MHz and Hα, and (d) Images at 432, 327 and 410 MHz on May 26, 1992, together with a *Yohkoh* SXT image. White dots in (a) and (b) show the resolution. (a) and (b) were adapted from Mercier and Chambe (2009), reproduced by permission of the AAS, (c) from Marqué *et al.* (1999), and (d) from Lantos (1999).

the corona is expected to cause limb brightening at radio wavelengths. Thus observations of the radio brightness variation from the disk center to the limb provide a good test for solar atmospheric models.

These conclusions are valid if the atmosphere is stratified, *i.e.* the temperature increase/decrease depends only on height. However, the actual solar atmosphere has complicated structures such as spicules and loop-like structures, filled with low or high temperature plasma. For example, Belkora *et al.* (1992) measured the solar limb brightness profile at 3 mm during a solar eclipse and found a close correspondence between the 3-mm limb profile and the height of Hα spicules.

It is therefore no surprise that the observed center-to-limb variation shows less brightening than the homogeneous models predict, or even shows darkening, which cannot be attributed to the limited instrumental resolution. This is usually interpreted in terms of absorbing features, such as chromospheric spicules (see, *e.g.* Lantos and Kundu, 1972). More recently, Selhorst, Silva, and Costa (2005) calculated the brightness distribution at 17 GHz with an atmospheric model they developed and obtained a too-sharp brightening around the limb, even after convolving with the instrumental beam. This changed drastically when they introduced spicules. They could then successfully fit the observed east–west brightness profiles with the model.

Enhanced limb brightening causes an increase of the solar disk size observed in radio. Selhorst, Silva, and Costa (2004) measured the solar radius at 17 GHz using images taken by the NoRH and showed a $3''$ variation of mean radius during one solar cycle, which is smaller than previous studies. Recently, Krissinel (2005) used SSRT observations to derive two-dimensional brightness distributions and to calculate the radio radius as a function of position angle for several wavelengths from 4 to 31.6 cm.

Limb brightening shows solar activity dependence due to the change of structure of the upper atmosphere. The radio limb is enhanced when active regions are closer to the limb. Not only active regions, but also the polar atmosphere depends on solar activity, as discussed in the following subsection.

4.2. Polar Brightening

Polar regions are brighter than the low-latitude quiet Sun at short cm-waves to mm-waves. This phenomenon was found by Efanov *et al.* (1980), and it was confirmed by Kosugi, Ishiguro, and Shibasaki (1986) with higher spatial resolution observations. The brightness in the polar regions can reach that of active regions. Many studies, including these two, have been carried out to find counterparts of the radio brightening at other wavelengths, see *e.g.*, Nindos *et al.* (1999), Pohjolainen, Portier-Fozzani, and Ragaigne (2000), and Selhorst *et al.* (2010). Figure 8 shows one example of the observed brightness enhancements. General agreement so far is that the radio brightness in polar regions is anti-correlated with solar activity, represented by sunspot numbers. Polar faculae have similar activity dependence, but no one-to-one correspondence has been found between faculae and radio bright patches (see, *e.g.*, Riehokainen *et al.*, 2001).

The annual variation of averaged polar brightening was studied by Shibasaki (1998), who found that averaged polar brightening consists of two components; one depends on the distance to the limb and the other depends on the distance to the poles. This can be seen by plotting north–south brightness distribution with and without B_0, heliographic latitude of the disk center, correction. Both plots have annual variation with similar amplitude. This result can be interpreted as follows: The averaged polar brightening is a limb brightening and it is modulated by the distance from the poles due to B_0 variation.

Around solar activity minimum, polar regions are dominated by strong unipolar magnetic field and hence the magnetic field lines are expected to be radial. Under such conditions, we can assume that the atmosphere is mainly in stratified structure with a smaller number of closed loops and/or tilted spicules filled with low temperature plasma. As the strong unipolar field is associated with the solar polar regions, limb brightening will be modulated annually by B_0. In contrast, during solar activity maximum the polar magnetic field becomes weak and changes its polarity. This means that unipolar field will be suppressed and complex magnetic field dominates. Under this condition, we expect that limb brightening will be suppressed. This can explain the anti-correlation of polar brightening with solar activity.

For systematic studies of solar cycle dependence of polar brightening, long-term full-disk observations are necessary. Daily images taken by NoRH can be used for such studies, and for example Shibasaki (1998) synthesized a radio butterfly diagram by using images from

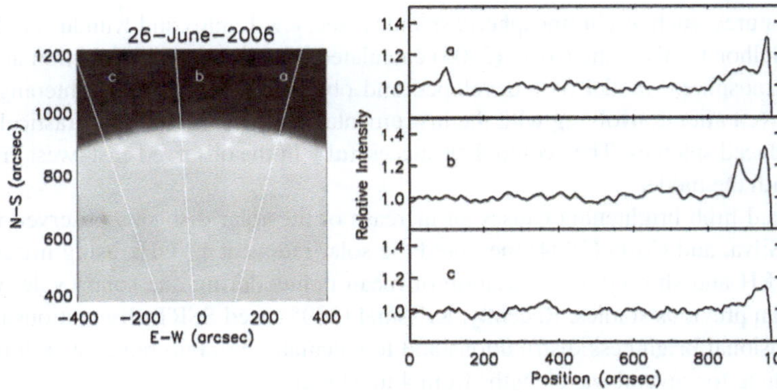

Figure 8 Bright patches are observed near the north pole during solar activity minimum at 17 GHz (NoRH observations). White lines represent three scans in the map, the profiles of which are shown on the right. Scan (a) goes through one bright patch, scan (b) through two patches, and scan (c) shows only the background limb brightening (Selhorst *et al.*, 2010, reproduced with permission © ESO).

Figure 9 Radio butterfly diagram synthesized from daily NoRH images between 1992 and 2010 at 17 GHz.

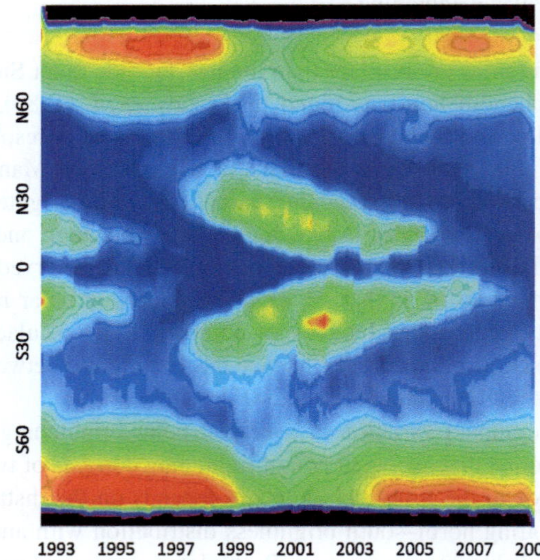

half a solar cycle. He found a systematic increase of the bright polar area as the solar activity declined.

At present, 18 years of NoRH imaging data are available, and Figure 9 shows a butterfly diagram with 17 GHz data taken between 1992 and 2010. To suppress annual variation, one-year running mean values were used in constructing the plot. In the diagram, we can clearly see that polar brightening is anti-correlated with low-latitude activities. In this display, polar regions are much brighter than active regions due to averaging along longitude. From the diagram, we notice easily the north–south asymmetry, as the southern pole is brighter than the northern pole. It can also be noticed that the polar brightness during activity minimum, between solar Cycles 23 and 24, is weaker than that during the previous minimum, at both poles. This suggests that the polar magnetic field during Cycles 23 – 24 minimum is weaker than that of Cycles 22 – 23 minimum. At lower latitudes, we notice the long delay of the appearance of active regions in the 24th Cycle. There must be some relation between the magnetic field in the polar regions and in the active region belts. It will be most useful to follow how the solar Cycle 24 will develop and to study the link between cycles.

4.3. Chromospheric Network, Transient Phenomena and "Heating" Events

The observational aspects of the network were treated in Section 3.1. In addition to the spatial coincidence of bright radio features with the chromospheric network, it is important to measure the intensity and size of the bright features as a function of wavelength. We should note that interferometric/synthesis observations cannot measure the background level, which should be provided by other means. Thus, the most appropriate measure of the intensity fluctuations is their amplitude or, even better, their rms variation.

In all reported measurements both the amplitude and the rms increase with wavelength. According to Benz *et al.* (1997), the rms varies from 290 K at 1.3 cm to 930 K at 3.6 cm; these values are consistent with the ones given by Bastian, Dulk, and Leblanc (1996) at 1.3 and 2 cm and they correspond to 2.8% and 5.8% variations, respectively, if reasonable values for the background are considered. The actual increase should be larger, due to the decrease of spatial resolution with wavelength. At 6.1 cm, Gary and Zirin (1988) give 8300 K and 21400 K above the background as the average value of individual source peak brightness in two regions, while Kundu *et al.* (1979) give absolute values of $\sim 2.5 \times 10^4$ and $\sim 1.5 \times 10^4$ K for typical network elements and cell interiors, respectively, at 6 cm. In the mm range, Loukitcheva, Solanki, and White (2009) give 1100 K for the intensity variation.

These results have not been fully exploited in multi-component models. Older computations by Chiuderi Drago, Kundu, and Schmahl (1983), based on the Vernazza, Avrett, and Loeser (1981) model, do show an increase of brightness difference between network and cell interiors with wavelength, while the values reported in recent works are close to those computed by the above authors for the VAL model D (average network).

It is well known from EUV data that the network becomes diffuse in the upper transition region and disappears in the low corona, thus a similar behavior is expected for the radio network. The main problem here is the variable spatial resolution. Bastian, Dulk, and Leblanc (1996) reported no detectable change between 1.3 and 2 cm, after smoothing the 1.3 cm image to match the 2 cm resolution. There is a spectral gap between 6 and 20 cm; however, the few published QS images at 20 cm (see, *e.g.* Gary and Zirin, 1988, see also the 20 cm image in Figure 3) do not show much of a network.

It is fairly well established (Erskine and Kundu, 1982; Bastian, Dulk, and Leblanc, 1996) that the radio fine structure presents time variability comparable to that of the chromospheric network, on time scales of minutes to hours. Apart from that, it is important to check for oscillatory behavior, which could be a signature of shock waves heating the upper solar atmosphere. This is not an easy task, due to the inherent difficulties in the instantaneous mapping of complex structures.

In the mm-wavelength range, White, Loukitcheva, and Solanki (2006) reported intensity oscillations (actually wave trains lasting typically for 1 – 3 wave periods) in the frequency range 1.5 – 8 mHz (periods of 125 – 700 s), with rms brightness variation of 50 – 150 K; they noted that network regions had a tendency to exhibit longer period oscillations than the internetwork.

A number of authors have also looked for impulsive/transient events that may have a bearing on the heating of the upper atmosphere, as signatures of magnetic reconnection. Note that the radio range is very sensitive to small non-thermal electron populations that are expected as a by-product of the reconnection process. We must also note that impulsive brightenings might be associated with X-ray bright points rather than with the network

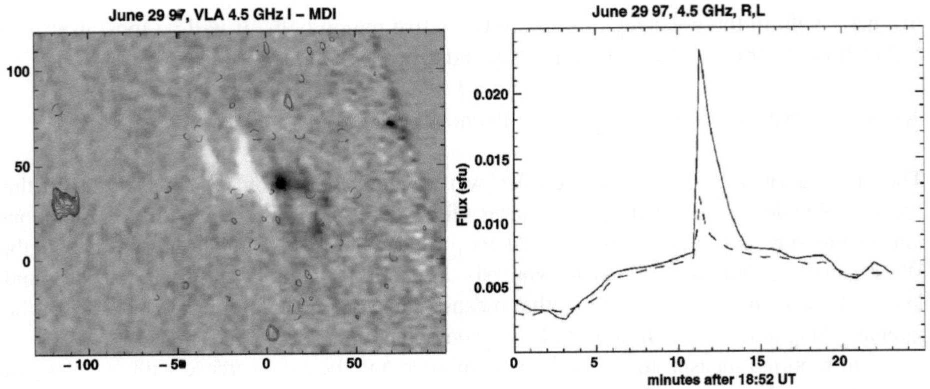

Figure 10 Left: Contours of a transient brightening at 4.53 GHz observed by the VLA, overlaid on an MDI magnetogram (axes are in arc seconds). Right: time profiles of right- and left-hand polarized microwave emission (solid and dashed lines). The event occurred well away from any active region and had no associated soft X-ray emission (adapted from Nindos, Kundu, and White, 1999, reproduced by permission of the AAS).

proper; moreover, the limited spatial and temporal resolution could make it difficult to discriminate between a single source and multiple sources and this will have a bearing on the number of sources detected.

In this context, Krucker *et al.* (1997) detected four Yohkoh soft X-ray events associated with radio emission at 2 cm in 10 s VLA snapshot images; they estimated that roughly one such event occurs on the Sun every 3 seconds, all together providing a total power of $\sim 2 \times 10^{25}$ erg s^{-1}, which is five times less than the total power radiated by the quiet corona in X-rays. They also estimated the number of radio sources without any associated soft X-ray emission to be about 10 000 at any given time on the Sun, which appear at a rate of about 20 per second. In a similar study Gary, Hartl, and Shimizu (1997), using one OVRO antenna and Yohkoh SXT images, found that only 5 out of 35 transient brightenings detected with SXT in an active region had no radio counterpart; they also found a delay of 1 to 2 min of the X-ray peak with respect to the radio. The relative timing of events was also studied by Benz and Krucker (1999), using data at 2, 3.6, and 6 cm from VLA together with EIT, CDS and SUMER data from SOHO. They found that the radio peaks preceded the coronal emission and lagged behind O V emission (see also Krucker and Benz, 2000).

Some of these transient events have flare-like characteristics. Indeed, there appears to be a continuity between bursts, the radio counterparts of X-ray bright points (see Keller and Krucker, 2004, and references therein), and smaller events such as those described in the previous paragraph. Transient brightenings are observed in microwaves both within and well away from any active regions (Figure 10), and both thermal and non-thermal emission mechanisms have been suggested (White *et al.*, 1995; Nindos, Kundu, and White, 1999). Selection methods may have influenced these findings, as the thermal transient brightenings in microwaves were selected on the basis of their simultaneous soft X-ray emission, and the non-thermal brightenings were microwave-selected. In any case, it seems evident that electrons are accelerated to non-thermal energies even in the quiet solar atmosphere.

Short-duration, isolated metric radio bursts that are not associated with any radio continuum (noise storms) and are located outside active regions have been reported (Crosby *et al.*, 1997; Ramesh *et al.*, 2010). The burst locations have been resolved from radioheliograph images, but no counterparts have been found when compared with X-ray, EUV, and

UV structures or events. Further analysis would require simultaneous high time-resolution imaging data at all wavelengths.

Decametric radio observations present numerous faint frequency-drifting emissions, similar to "solar S bursts" reported by McConnell (1982). No specific relationship between the occurrence of these emissions and the solar cycle or presence of flares have been found (Briand *et al.*, 2008). Briand *et al.* (2008) suggest that a moderate, localized, time-dependent heating leads to low velocity electron clouds, which can in turn generate Langmuir waves and electromagnetic signals by nonlinear processes.

4.4. Solar Rotation and Tracers

Structures visible at radio waves can be used to determine solar rotation rates and source heights, especially at high solar latitudes. Since the radio structures are located in the solar chromosphere and corona, projection effects play a role in the position determination. Thus, Alissandrakis, Lantos, and Nicolaidis (1985) measured a height of $0.23\,R_\odot$ for emission sources at 169 MHz. A method of solar rotation stereoscopy was developed by Aschwanden and Bastian (1994), to enable height determination. The method was tested on Owens Valley microwave data (Aschwanden *et al.*, 1995), which showed that the height dependence of the coronal magnetic field and the plasma temperature can be inverted from the stereoscopic altitude spectra and the observed brightness temperature spectra.

Later, Rosa *et al.* (1998) developed a method for the simultaneous determination of the solar synodic rotation velocity and the tracer height. Using this method, Brajša *et al.* (2009) determined that the low brightness-temperature regions observed at 8 mm wavelength were located at heights of about 45 600 km. These heights are in agreement with filament heights, but disagree with our current understanding on coronal hole structures. Brajša *et al.* also concluded that thermal bremsstrahlung can explain both the brightness temperature depressions (prominences and coronal condensations) and enhancements (ordinary active regions). Brajša *et al.* (2000) found earlier that the differential rotation of brightness depressions is more rigid than that obtained tracing magnetic features and measuring Doppler shifts. There were also cycle-related changes and north–south asymmetry in the measured solar rotation velocities.

4.5. Coronal Holes, Streamers, and Filaments

Coronal holes are regions of lower density and therefore they are usually observed as brightness depressions at radio wavelengths (*e.g.*, Lantos and Alissandrakis, 1999; Borovik and Medar, 1999). At mm- and cm-waves, the average brightness temperature of coronal holes is not much different from the quiet Sun. Both diffuse emission and local brightness enhancements inside some of the holes and at coronal hole borders have been observed (Kosugi, Ishiguro, and Shibasaki, 1986; Gopalswamy *et al.*, 1999; Pohjolainen, Portier-Fozzani, and Ragaigne, 2000; Moran *et al.*, 2001; Riehokainen, Valtaoja, and Pohjolainen, 2003). Increased Hα brightness was detected inside some equatorial coronal holes that also showed enhanced radio emission, see Figure 11, but comparisons to UV and EUV line observations did not reveal any correspondence (Moran *et al.*, 2001). However, the inadequate spatial resolution of single-dish radio observations and the difficulty of interferometers to resolve faint diffuse sources have made these comparisons difficult.

Magnetic network elements, similar to those responsible for the enhanced radio emission in bright points (Benz *et al.*, 1997), have been suggested as the origin for coronal hole brightenings. The electron-cyclotron maser mechanism has been excluded as a possibility,

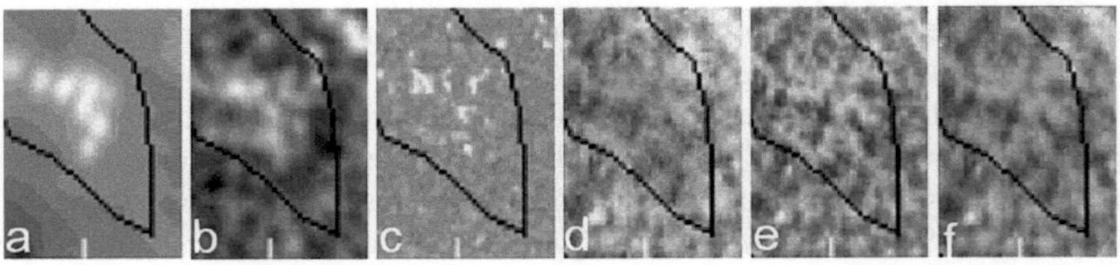

Figure 11 Radio-bright structures observed inside an equatorial coronal hole: (a) NoRH observations at 17 GHz, compared with (b) BBSO Hα image, (c) MDI magnetogram, and (d) – (f) selected UV spectral lines observed by SUMER. (Figure adapted from Moran *et al.*, 2001, reproduced with permission © ESO.)

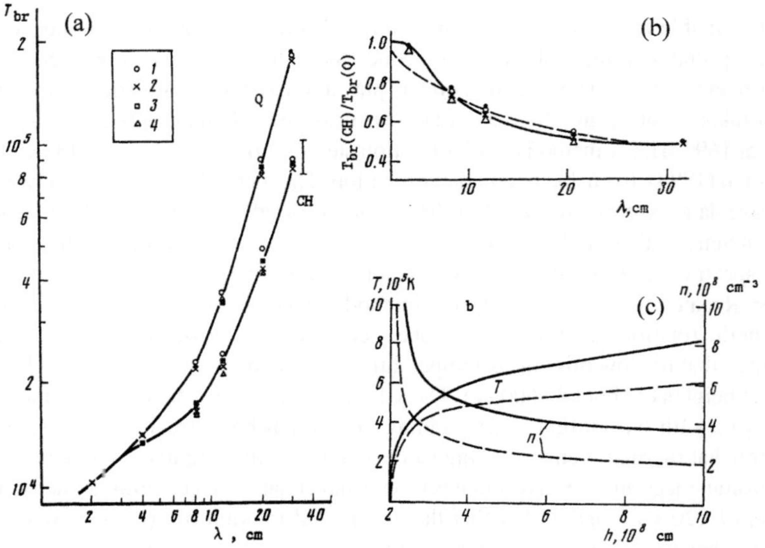

Figure 12 (a): RATAN-600 measurements of the brightness temperature for the quiet Sun (Q) and coronal holes (CH). (b): the contrast of coronal holes with respect to the quiet Sun, the dashed line shows the model computations. (c): the corresponding electron temperature and density models. (Adapted from Borovik *et al.*, 1990.)

as it is not able to provide significant radio wave emission from the coronal funnels at the supergranular network (Vocks and Mann, 2004). Krissinel *et al.* (2000) proposed that a difference in the electron density and temperature in different holes and in different parts of a hole can account for the variable brightness temperatures. They also noted that the influence of these parameters is strongest in the 2 – 5 cm wavelength range. A recent study by Brajša *et al.* (2007) suggests that an increased density in the transition region and low corona could explain both the brightness temperature enhancements and depressions at millimeter waves.

A number of computations of the coronal hole radio emission (see, *e.g.*, Borovik *et al.*, 1990; Chiuderi Drago *et al.*, 1999) and calculations of the expected brightness temperatures (see, *e.g.*, Pohjolainen, 2000) have been published. Figure 12 shows the RATAN-600 spectral observations of four coronal holes in the wavelength range of 2 to 32 cm, together with quiet Sun data; the brightness difference between the two becomes appreciable above ~ 4 cm. In the best-fit models the coronal holes are cooler than the background and less dense by a factor of two. Mercier (communication at the CESRA 2010 meeting, La Roche-en-Ardenne, Belgium) presented measurements obtained with the NRH from 70 cm to 1.8 m,

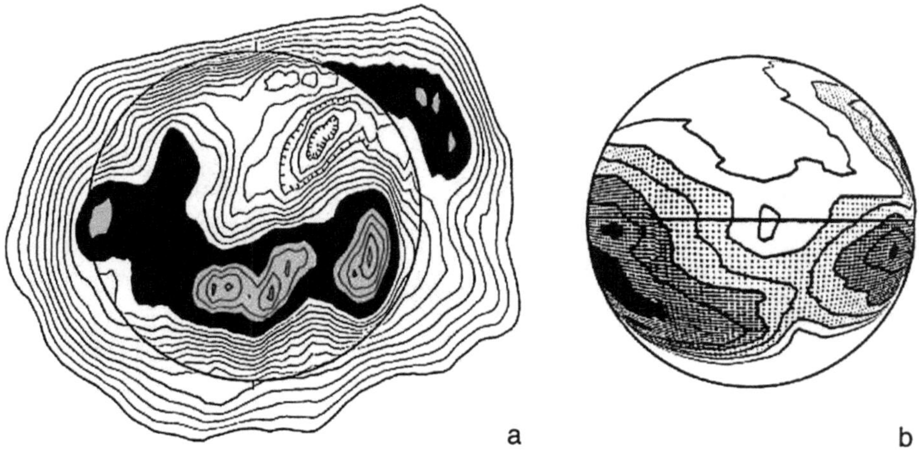

Figure 13 (a): NRH map of the Sun at 169 MHz on July 15, 1984. The black region is the *coronal plateau*. A coronal hole is marked by hatched contours. (b): The corresponding K-corona synoptic map, on spherical projection. From Lantos and Alissandrakis (1999), reproduced with permission © ESO.

together with models that assumed isothermal corona and chromosphere in hydrostatic equilibrium with no transition region. They, too, found a density decrease in the holes by a factor of two with respect to the quiet Sun.

At decameter waves coronal holes are sometimes seen in emission, Dulk and Sheridan (1974) reported such a case. Lantos *et al.* (1987) observed a bright emission source at decametric wavelengths, located above an extended coronal hole, in a region where a depression was seen at meter waves. A possible interpretation is in terms of refraction effects (Alissandrakis, 1994, *cf.* Figure 3 of Lantos, 1999) and/or scattering in inhomogeneities.

In a systematic study of emission sources observed with the NRH at 169 MHz, Lantos and Alissandrakis (1999) came to the conclusion that the large-scale emission is dominated by the *coronal plateau* (Figure 13). This is an intermediate brightness area forming a belt around the Sun and surrounding almost all local emission sources (Lantos, Alissandrakis, and Rigaud, 1992). It is visible both in daily images and in synoptic charts, and has a close association with the coronal plasma sheet observed with K-corona instruments (Figure 13b). The diffuse emission of the *coronal plateau* could be due to a high altitude loop system which overrides the principal neutral line of the general solar magnetic field at the base of the heliosheet, with a possible contribution of loops connecting active regions to surrounding quiet areas. The same authors reported that medium scale local sources include both faint noise storm continua and thermal sources. Most of the latter are located between faculae and neutral lines, possibly in the upper leg of large-scale loops. A small number of thermal sources is closer to neutral lines, which might be loops at the base of isolated coronal streamers.

Coronal streamers are best visible at decametric wavelengths (see Lantos, 1999). They are less prominent in the meter and decimeter ranges, where one sees loops at the base of streamers rather than proper streamers, as pointed out in the previous paragraph. From the circularly polarized thermal emission of streamers observed with the Gauribidanur radioheliograph at 77 and 109 MHz, Ramesh, Kathiravan, and Sastry (2010) were able to estimate magnetic field strengths in the range of 5 – 6 G at 1.5 – 1.7 solar radii.

Large filaments are observed as depressed regions within the solar disk at mm- and cm-waves (Chiuderi Drago, 1990) and sometimes at decimeter-meter waves (Marqué *et al.*,

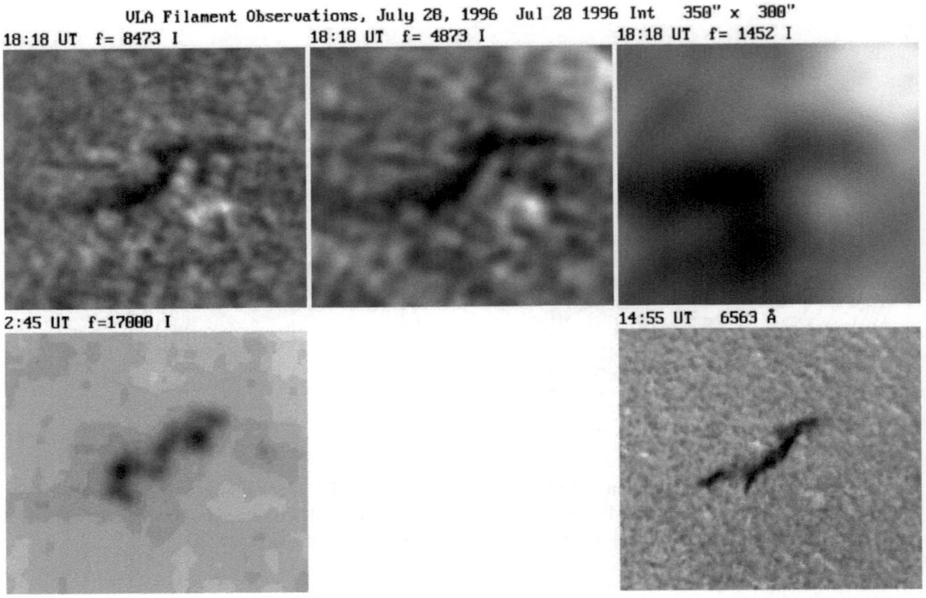

Figure 14 A filament observed at radio wavelengths (VLA and NoRH) and in Hα. From Chiuderi Drago *et al.* (2001).

1999; *cf.* Figure 7c, Section 3.2). Above the limb, projected against the sky, filaments are seen in emission. It is possible to calculate electron temperatures and densities within the filament using multi-frequency radio observations (Irimajiri *et al.*, 1995), especially when the filaments erupt. A systematic study of the filament environment in the metric radio range was presented by Marqué (2004), where it was concluded that the most likely source of the radio depression is the cavity that surrounds the filament. The on-the-disk filament observations at microwaves, analyzed by Chiuderi Drago *et al.* (2001), see Figure 14, support a prominence model of cool threads embedded in the hot coronal plasma, with a sheath-like TR around them.

5. Active Regions

Active region associated emission dominates in the cm range. The main emission mechanisms of non-flaring active regions are the gyro-resonance radiation above sunspots and the free–free radiation from hot plasma trapped in active region loops. Both emissions are due to thermal electrons. We note, however, that quasi-steady non-thermal emission has been invoked in the interpretation of *peculiar sources* on top of neutral lines of the magnetic field (Alissandrakis *et al.*, 1993) and the *decimetric halo component* (Gelfreikh, 1998). At longer decimetric and metric wavelengths no sunspot-associated emission is visible, presumably due to the high opacity of the overlaying plasma and refraction effects.

The gyro-resonance emission mechanism is unique to radio and it is a useful tool to study the atmospheric layers above sunspots and measure the magnetic field (Gelfreikh, 1998). The emission is generated in thin layers around iso-Gauss surfaces with magnetic field strength such that the observing frequency is equal to the second or third harmonic of the local gyro-frequency. Due to the high opacity of these layers, strong emission is observed if they are located in the transition region or the corona.

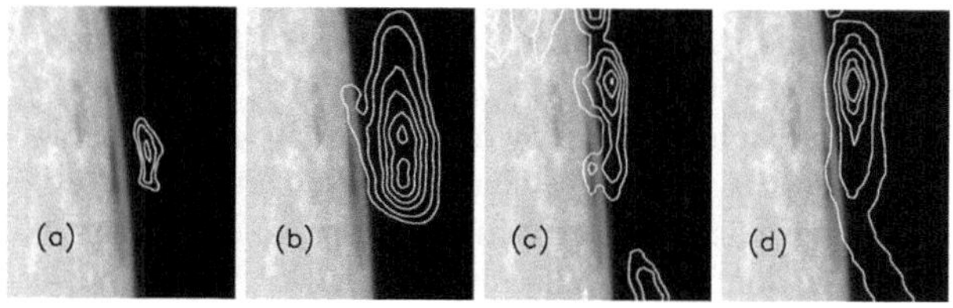

Figure 15 Co-aligned white-light (TRACE), EUV (from CDS), and radio (VLA) observations of an active region near the solar limb: (a) 15 GHz, (b) 8 GHz, (c) O V, and (d) Ne VI intensity contours plotted over a white-light image (adapted from Brosius and White, 2006, reproduced by permission of the AAS).

The opacity (Kakinuma and Swarup, 1962; Zheleznyakov, 1962) is a complicated function of the temperature, the density, the intensity and the direction of the magnetic field and the wave mode. It is much higher in the extraordinary mode than in the ordinary one, it is also much higher at the second harmonic than at the third; thus, the third harmonic is opaque in the extraordinary and transparent in the ordinary mode, while the second harmonic is opaque in both modes.

Consequently, if the photospheric field strength is low enough (or the frequency is high enough) both harmonic layers are below the TR and no sunspot-associated emission is observed. For higher field or lower frequency, the third harmonic enters into the TR and strong, highly polarized emission is observed (Shibasaki *et al.*, 1994). For still higher field strengths the second harmonic also enters the TR and the polarization is reduced.

Due to the thin resonance layers and the strong dependence of the opacity on the angle between the line-of-sight and magnetic field, gyro-resonance sources show center-to-limb variation; the dependence of the source flux on the angular distance from the disk center can be described by a cosine curve with the top "cut off" (Borovik, Gelfreikh, and Lubyshev, 1975; Lubyshev, 1977). The level of the cut depends on the wavelength and increases at shorter wavelengths (Gelfreikh and Lubyshev, 1979). For example, at cm-waves the flux has a dip around the central meridian, making the source appear weak around the disk center, brighter at mid-distance, and weak again near the limb.

The strong radio emission over sunspots and sunspot groups has stimulated a large amount of theoretical and observational work over a long period of time, particularly after the first high-resolution observations by Kundu and Alissandrakis (1975a) and the first detailed modeling by Alissandrakis, Kundu, and Lantos (1980). Recent works on gyro-resonance emission are reviewed by White (2004) and Lee (2007).

A lot of recent work is based on NoRH imaging observations, and a consensus exists that at 17 GHz the gyro-resonance mechanism plays the most important role (Nindos *et al.*, 2000; Vourlidas, Gary, and Shibasaki, 2006; Selhorst, Silva-Válio, and Costa, 2008). At higher frequencies, imaging observations are more limited and the results are still inconclusive. The 34 GHz emission, for example, was successfully modeled as completely free–free radiation by Selhorst, Silva-Válio, and Costa (2008). At sub-mm waves, Silva *et al.* (2005) report a brightness excess of 3 – 20% over active regions and that the flux density spectra are increasing toward the THz range.

High-resolution multi-wavelength observations of sunspots can be used to test in detail the magnetic field model extrapolated from measurements at the photosphere (Lee *et al.*, 1998). For example, Brosius and White (2006) calculated coronal magnetic field strengths of 1750 G at a surprising large height (8000 km) above a large sunspot at the west solar limb, using radio observations from the VLA together with TRACE, CDS, EIT, and MDI observations (Figure 15). The observations can also provide valuable diagnostics of the active region atmosphere and magnetic field, in particular if high spatial resolution spectral data are available (see, *e.g.* Tun, Gary, and Georgoulis, 2011).

Radio observations are valuable in providing information on temporal variations of the emission from active regions. Generally, signals of oscillations and waves are weak compared to the background quiet Sun and active regions, and intensity fluctuations caused by terrestrial atmosphere dominate. Radio interferometric observations can suppress atmospheric effects and 3-minute umbral oscillations of gyro-resonance sources are clearly observed by the NoRH (Gelfreikh *et al.*, 1999). Radio pulsations show a large variety of periods, bandwidths, amplitudes, temporal and spatial signatures, and on this extensive field of research we refer to the review by Nindos and Aurass (2007). Recently, Sych *et al.* (2010) found three-minute propagating waves in a sunspot observed with the NoRH and TRACE, near the footpoints of locally open magnetic field lines.

On long time scales, the total radio flux integrated over the whole disk is known to be a good index of solar activity, especially around 10 cm wavelength (2.8 GHz), as it shows good correlation with relative sunspot numbers, due to contributions from both emission components. The spectral peak of radio emission from non-flaring active regions is located around the 10 cm wavelength. Weather conditions such as clouds and rain do not influence the measurements at wavelengths longer than 10 cm, and ionospheric disturbances do not influence at wavelengths shorter than 10 cm. Therefore the 10 cm wavelength is favorable for accurate ground-based measurements because it is free from both influences. Also, a very robust absolute calibration method can be established (Tanaka *et al.*, 1973), to minimize the instrumental effects. Thus, it is possible to use the total radio flux as a long-term activity index of the Sun, similar to or even better than the relative sunspot numbers.

For studies of anomalous solar activity during and after the minimum period of the solar Cycles 23 and 24, total flux data will play an important role. More than 60 years of records with uniform calibration exist nowadays (Tapping *et al.*, 2007). Figure 16 presents averaged total flux measurements at four microwave frequencies between 1951 and 2010, observed by radio observatories in Toyokawa and Nobeyama in Japan.

6. Future Challenges

6.1. Atmospheric Structure and Dynamics

The radio data provide valuable information on the structure of the solar atmosphere above the temperature minimum. To this end reliable absolute measurements of the brightness temperature are required. The existing data show very much scatter, most prominent in the mm and sub-mm wavelengths which are formed below the TR. The scatter could be due to inherent difficulties in the measurements and/or solar cycle variations, although the latter are not expected to be very important at the relevant atmospheric layers. In any case, there is an obvious need for more consistent observations and these can only be obtained with single-dish instruments. Observations from space would be the ideal, since they are not hindered by the terrestrial atmosphere.

As pointed out in Section 2.2, a model can be found that fits most microwave data, the main problem being that the radio data are better fitted with models of the *cell interior*, rather than the average Sun. In addition to the fact that radio data are less affected by non-LTE effects and other uncertainties of the optical-UV range, this is probably due to different contributions of network and cell regions to the average intensity in the two domains. Indeed, Gary and Zirin (1988) estimated that the network contributes only 5 – 10% of the total brightness at 6 cm.

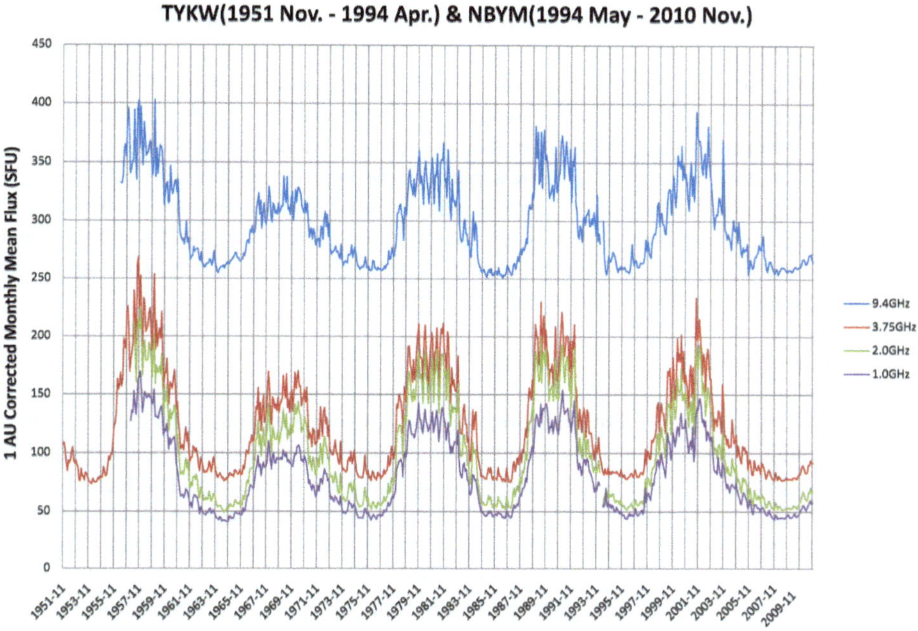

Figure 16 Corrected monthly mean total solar flux at 1, 2, 3.75, and 9.4 GHz during 1951 – 2010 (Toyokawa and Nobeyama recordings).

There is a noticeable lack of multi-component atmospheric models that are based on observations of the network. The existing models are based on center-to-limb variations and employ absorbing structures, such as spicules, to explain the lack of brightening. Spicules, however, probably have *higher* temperature than the ambient chromosphere (*cf.* Beckers, 1972), so they could not be absorbing features. The limb darkening might be due to the ragged structure of the chromosphere-corona TR. In any case, the network/cell contrast and its variation with the wavelength could make the basis for a two-component model. Here we should point out again the lack of high-resolution observations in the range of 6 to 20 cm wavelength, which would reveal the gradual fading of the network in the upper TR and low corona.

The low value of the circular polarization in QS structures, combined with instrumental limitations, makes the measurement of the magnetic field in the network quite difficult. In coronal holes, Borovik, Medar, and Korzhavin (1999) reported values of 0.2 – 3% for the degree of polarization, and 2.5 – 10 G for the magnetic field. Surprisingly, they found an increase of the magnetic field with wavelength. Similar values, but at larger heights, have been reported for streamers (Ramesh, Kathiravan, and Sastry, 2010, see Section 4.5). Thus, there is a lot of important information to be obtained from improved measurements of circular polarization.

Radio observations can provide important input to the problem of the heating of the chromosphere and the corona, both in the wave and nano-flare heating scenarios (Section 4.3). Instruments with dense coverage of the u–v plane as well as improved reconstruction techniques are necessary, in order to provide accurate instantaneous images of regions with complex structure.

The study of large-scale structures, such as coronal holes, loop systems, streamers and the *coronal plateau*, will benefit from systematic observations such as those with the NRH

(Section 3.2). Improved modeling in terms of refraction and scattering effects is also required.

Important diagnostics of active regions in the TR and the low corona, too many to be discussed here in detail, have been provided by radio observations. Still, the nature of apparently non-thermal, quasi-stationary emission sources remains unclear.

Finally, long time scales and synoptic aspects can be approached from the radio perspective, too. The total solar flux is an excellent index of solar activity (Figure 16), while radio butterfly diagrams show very well not only the migration of active regions toward the equator, but also the long-term behavior of polar brightening (Figure 9). It is thus important that such observations continue for the decades to come.

6.2. Instrumentation

The present-day solar radio instrumentation consists of single antennas and interferometer-aperture synthesis instruments. The antenna and receiver characteristics depend strongly on the wavelengths used. At millimeter waves and beyond, very good antenna surface accuracy and receiver stability is required. The Solar Submillimeterwave Telescope SST (Kaufmann *et al.*, 2008) uses a multiple beam technique to overcome the mapping limitations of a single-dish and single-receiver instrument. At long wavelengths like meter waves, much simpler and much more cost-effective techniques can be used. For example, the e-Callisto spectrometers (Benz *et al.*, 2009) have been installed at dozens of locations, with simple antenna structures, since the first installation in 2002.

The spatial resolution of a single antenna increases with the antenna size (*i.e.*, bigger antennas produce more narrow beams), which sets upper limits to fully steerable constructions. For example, the Nobeyama 45-meter diameter antenna has been used to image an eruptive prominence (Irimajiri *et al.*, 1995), but steering the antenna and scanning the solar disk with a 15″ beam (at 110 GHz) proved to be very challenging. An example of a non-traditional design is the 600-meter diameter RATAN-600 reflector-type telescope, which can observe the Sun at high spatial resolution near the local noon. One-dimensional scans can be used to build two-dimensional maps (Nindos *et al.*, 1996) and from the observed circular polarization magnetic field strengths can be calculated (Akhmedov *et al.*, 1982).

Radio interferometers are providing high-resolution imaging at selected wavelengths. At present, there are only three solar-dedicated interferometers that observe the Sun on daily basis: NoRH, SSRT, and NRH (see Section 3). Also non-solar-dedicated interferometers, such as the WSRT and the VLA, have been used for solar observations regularly. We note, however, that general-purpose large radio interferometers have large elements and as a result their field-of-view is limited to a small part of the Sun. Dedicated solar radio interferometers consist of many small dishes to cover the full solar disk. Non-imaging total flux measurements also require small aperture dishes to cover the full solar disk uniformly.

To obtain imaging spectroscopy over a wide frequency range, a Frequency Agile Solar Radiotelescope (FASR) was designed in the early 2000. The Fourier synthesis telescope would provide such frequency, temporal, and angular resolution that both QS and flare observations are possible. Since then, the project has suffered from budget cuts, but the ongoing upgrade of OVSA and its use as a testbed will hopefully help to build FASR.

Of the new radio interferometers designed for astronomical use, the Low Frequency Array (LOFAR), the Murchison Widefield Array (MWA), the Long Wavelength Array (LWA), and the Atacama Large Millimeter/submillimeter Array (ALMA) can all be used for solar observations. The LWA, LOFAR, and MWA frequency ranges (10 – 90, 30 – 240, and 80 – 300 MHz, respectively) make them ideal for observing both the quiet Sun and solar transients such as coronal mass ejections, but with the available spectroscopy (Oberoi *et al.*, 2011) also general plasma processes in the solar corona can be studied. We should note, however, that imaging at very low frequencies presents a significant challenge due to ionospheric effects. The ALMA frequency range, from 84 to 950 GHz, will open new opportunities in high-resolution imaging and spectroscopy at millimeter and submillimeter waves (Karlický *et al.*, 2011). Since the quiet Sun emission at these wavelengths originates from the chromosphere, ALMA observations will help refine atmospheric models (Sections 2.2 and 4.3).

6.3. Data Analysis

Many of the solar-dedicated radio instruments provide archives and data bases of their data. These databases can be used for various studies, such as flares, active regions, oscillations, prominence eruptions, quiet Sun and global solar cycle variations. The continued operation of instruments like the NoRH the SSRT and the NRH is essential both for case studies and long-term analysis and, not least, because the calibrated data are comparable independently of the year of observation. Especially for studies of anomalous solar activity during the rising phase of Cycle 24, well-calibrated uniform data covering more than one solar cycle are essential. After the long minimum of activity between Cycles 23 and 24, activity increases extremely slowly compared to previous cycles. Continuous full-disk observation by NoRH is very important.

The search for time variability in the quiet Sun radio emission has turned out to be a challenging task. With single-dish observations the flux is summed from the beam area where the antenna is pointing, and often simultaneous source tracking and source region imaging are not possible. Interferometers can overcome this problem but as they are less sensitive to the disk emission, new data analysis techniques have to be used for the deconvolution and building of images.

Many authors use a maximum entropy deconvolution technique, but this requires that there is significant flux present in the solar disk for a relatively long period of time (background positivity constraint). To completely resolve between cell interiors and network, we still need better deconvolution techniques and more closely-spaced antenna arrays.

Finally, it is important to develop tools for data analysis. For example, a web-based system for automated interactive detection of wave and oscillatory phenomena in imaging data sets was developed and tested with NoRH microwave maps and TRACE EUV images (Sych *et al.*, 2010; see also Section 5).

6.4. Concluding Remarks

In this review we have tried to present a picture, as complete as possible, of what the radio domain can tell us about the quiet Sun and active regions. In principle, the radio range can provide us with as much information as the rest of the spectrum for the solar atmosphere, from the temperature minimum up to the low corona; moreover, it can do that from the ground. A definite advantage of the radio range is that, for thermal processes as well for those non-thermal processes that are due to accelerated electrons, we understand the physics fairly well and we do not have to worry about such things as non-LTE effects, excitation and ionization equilibria, abundances, *etc.*, which are well-known burdens in the rest of the spectrum.

Acknowledgements The authors wish to thank the organizers and the participants of the CESRA2010 Workshop, which took place in La Roche-en-Ardenne in June 2010. This review took inspiration from the many presentations and working group discussions, as well as the more informal exchanges of ideas during the meeting. The authors also thank A. Nindos for his comments on the manuscript and the anonymous referee for careful review. Solar radio data were obtained from the archives of Nobeyama Solar Radio Observatory, at http://solar.nro.nao.ac.jp, and from the collaborative radio monitoring project at Paris Observatory-Meudon, at http://secchirh.obspm.fr/, and we thank their personnel for creating and developing these databases.

References

Akhmedov, S.B., Gelfreikh, G.B., Bogod, V.M., Korzhavin, A.N.: 1982, *Solar Phys.* **79**, 41.
Alissandrakis, C.E.: 1994, *Adv. Space Res.* **14**, 81.
Alissandrakis, C.E.: 1999, In: Bastian, T., Gopalswamy, N., Shibasaki, K. (eds.) *Solar Physics with Radio Observations, NRO Report* **479**, 53.
Alissandrakis, C.E., Einaudi, G.: 1997, In: Trottet, G. (ed.) *Coronal Physics from Radio and Space Observations, Lecture Notes in Physics* **483**, 53.
Alissandrakis, C.E., Kundu, M.R., Lantos, P.: 1980, *Astron. Astrophys.* **82**, 30.
Alissandrakis, C.E., Lantos, P., Nicolaidis, E.: 1985, *Solar Phys.* **97**, 267.
Alissandrakis, C.E., Gel'Frejkh, G.B., Borovik, V.N., Korzhavin, A.N., Bogod, V.M., Nindos, A., Kundu, M.R.: 1993, *Astron. Astrophys.* **270**, 509.
Aschwanden, M.J., Bastian, T.S.: 1994, *Astrophys. J.* **426**, 434.
Aschwanden, M.J., Lim, J., Gary, D.E., Klimchuk, J.A.: 1995, *Astrophys. J.* **454**, 512.
Aubier, M., Leblanc, Y., Boischot, A.: 1971, *Astron. Astrophys.* **12**, 435.
Avrett, E.H.: 1995, In: Kuhn, J., Penn, M. (eds.) *Infrared Tools for Solar Astrophysics: What's Next? Proc. 15th NSO Sac Peak Workshop*, 303.
Avrett, E.H., Loeser, R.: 2008, *Astrophys. J. Suppl.* **175**, 229.
Bastian, T.S.: 1994, *Astrophys. J.* **426**, 774.
Bastian, T.S., Dulk, G.A., Leblanc, Y.: 1996, *Astrophys. J.* **473**, 539.
Beckers, J.M.: 1972, *Ann. Rev. Astron. Astrophys.* **10**, 73.
Belkora, L., Hurford, G.J., Gary, D.E., Woody, D.P.: 1992, *Astrophys. J.* **400**, 692.
Benz, A.O., Krucker, S., Acton, L.W., Bastian, T.S.: 1997, *Astron. Astrophys.* **320**, 993.
Benz, A.O., Krucker, S.: 1999, *Astron. Astrophys.* **341**, 286.
Benz, A.O., Monstein, C., Meyer, H., Manoharan, P.K., Ramesh, R., Altyntsev, A., Lara, A., Paez, J., Cho, K.-S.: 2009, *Earth Moon Planets* **104**, 277.
Bogod, V.M., Korolkov, D.V.: 1975, *Pisma Astron. Zh.* **1**, 25 (*Soviet Astron. Lett.* **1**, 205).
Borovik, V.N., Gelfreikh, G.B., Lubyshev, B.I.: 1975, *Astron. Zh.* **52**, 97 (*Soviet Astron.* **19**, 57).
Borovik, V.N., Kurbanov, M.S., Livshits, M.A., Ryabov, B.I.: 1990, *Astron. Zh.* **67**, 1038 (*Soviet Astron.* **34**, 522).
Borovik, V.N., Medar, V.G., Korzhavin, A.N.: 1999, *Astron. Lett.* **25**, 250.
Borovik, V.N., Medar, V.G.: 1999, In: Vial, J.-C., Kaldeich-Schümann, B. (eds.) *Plasma Dynamics and Diagnostics in the Solar Transition Region and Corona, 8th SOHO Workshop, ESA* **SP-446**, 185.
Brajša, R., Benz, A.O., Temmer, M., Jurdana-Šepić, R., Šaina, B., Wöhl, H.: 2007, *Solar Phys.* **245**, 167.
Brajša, R., Romštajn, I., Wöhl, H., Benz, A.O., Temmer, M., Roša, D.: 2009, *Astron. Astrophys.* **493**, 613.
Brajša, R., Ruždjak, V., Vršnak, B., Wöhl, H., Pohjolainen, S., Upro, S.: 2000, *Solar Phys.* **196**, 279.
Briand, C., Zaslavsky, A., Maksimovic, M., Zarka, P., Lecacheux, A., Rucker, H.O., Konovalenko, A.A., Abranin, E.P., Dorovsky, V.V., Stanislavsky, A.A., Melnik, V.N.: 2008, *Astron. Astrophys.* **490**, 339.
Brosius, J.W., White, S.M.: 2006, *Astrophys. J.* **641**, L69.
Carlsson, M., Stein, R.F.: 2002, *Astrophys. J.* **572**, 626.
Chiuderi, C., Chiuderi Drago, F.: 2004, *Astron. Astrophys.* **422**, 331.
Chiuderi Drago, F.: 1990, In: Ruzdjak, V., Tandberg-Hanssen, E. (eds.) *Dynamics of Quiescent Prominences, Proc. of the No. 117 Colloquium of the IAU, Lecture Notes in Physics* **363**, 70.
Chiuderi Drago, F., Kundu, M.R., Schmahl, E.J.: 1983, *Solar Phys.* **85**, 237.
Chiuderi Drago, F., Alissandrakis, C.E., Bastian, T., Bocchialini, K., Harrison, R.A.: 2001, *Solar Phys.* **199**, 115.
Chiuderi Drago, F., Landi, E., Fludra, A., Kerdraon, A.: 1999, *Astron. Astrophys.* **348**, 261.
Cohen, M.H.: 1960, *Astrophys. J.* **131**, 664.
Crosby, N., Vilmer, N., Chiuderi Drago, F., Pick, M., Kerdraon, A., Khan, J., Schwenn, R.: 1997, In: Wilson, A. (ed.) *Fifth SOHO Workshop: The Corona and Solar Wind Near Minimum Activity, ESA SP-404*, 299.
Dudík, J., Kulinová, A., Dzifčáková, E., Karlický, M.: 2009, *Astron. Astrophys.* **505**, 1255.

Dulk, G.A., Sheridan, K.V.: 1974, *Solar Phys.* **36**, 191.
Efanov, V.A., Moiseev, I.G., Nesterov, N.S., Stewart, R.T.: 1980, In: Kundu, M.R., Gergely, T.E. (eds.) *Radio Physics of the Sun*, Reidel, Dordrecht, 141.
Erskine, F.T., Kundu, M.R.: 1982, *Solar Phys.* **76**, 221.
Fontenla, J.M., Avrett, E.H., Loeser, R.: 1990, *Astrophys. J.* **355**, 700.
Fontenla, J.M., Avrett, E.H., Loeser, R.: 1991, *Astrophys. J.* **377**, 712.
Fontenla, J.M., Avrett, E.H., Loeser, R.: 1993, *Astrophys. J.* **406**, 319.
Fontenla, J.M., Avrett, E.H., Loeser, R.: 2002, *Astrophys. J.* **572**, 636.
Fontenla, J.M., Avrett, E., Thuillier, G., Harder, J.: 2006, *Astrophys. J.* **639**, 441.
Fontenla, J.M., Balasubramaniam, K.S., Harder, J.: 2007, *Astrophys. J.* **667**, 1243.
Fontenla, J.M., Curdt, W., Haberreiter, M., Harder, J., Tian, H.: 2009, *Astrophys. J.* **707**, 482.
Gary, D.E.: 1996, In: Taylor, A.R., Paredes, J.M. (eds.) *Radio Emission from the Stars and the Sun, ASP Conf. Series* **93**, 387.
Gary, D.E., Hartl, M.D., Shimizu, T.: 1997, *Astrophys. J.* **477**, 958.
Gary, D.E., Zirin, H.: 1988, *Astrophys. J.* **329**, 991.
Gary, D.E., Zirin, H., Wang, H.: 1990, *Astrophys. J.* **355**, 321.
Gelfreikh, G.B.: 1998, In: Alissandrakis, C.E., Schmieder, B. (eds.) *Three-Dimensional Structure of Solar Active Regions, 2nd Advances in Solar Physics Euroconference, ASP Conf. Series* **155**, 110.
Gelfreikh, G.B., Kaidanovskii, M.N., Kononovich, E.V., Smirnova, O.B., Stotskii, A.A., Trushkin, S.A.: 1977, *Pisma Astron. Zh.* **3**, 550 (*Soviet Astron. Lett.* **3**, 300).
Gelfreikh, G.B., Grechnev, V., Kosugi, T., Shibasaki, K.: 1999, *Solar Phys.* **185**, 177.
Gelfreikh, G.B., Lubyshev, B.I.: 1979, *Astron. Zh.* **56**, 562 (*Soviet Astron.* **23**, 316).
Gingerich, O., de Jager, C.: 1968, *Solar Phys.* **3**, 5.
Gingerich, O., Noyes, R.W., Kalkofen, W., Cuny, Y.: 1971, *Solar Phys.* **18**, 347.
Gopalswamy, N., Shibasaki, K., Thompson, B.J., Gurman, J., DeForest, C.: 1999, *J. Geophys. Res.* **104**, 9767.
Grebinskij, A., Bogod, V., Gelfreikh, G., Urpo, S., Pohjolainen, S., Shibasaki, K.: 2000, *Astron. Astrophys. Suppl. Ser.* **144**, 169.
Grechnev, V.V., Lesovoi, S.V., Smolkov, G.Y., Krissinel, B.B., Zandanov, V.G., Altyntsev, A.T., Kardapolova, N.N., Sergeev, R.Y., Uralov, A.M., Maksimov, V.P., Lubyshev, B.I.: 2003, *Solar Phys.* **216**, 239.
Heintze, J.R.W., Hubenet, H., de Jager, C.: 1964, *SAO Spec. Rep.* **167**, 240.
Hoang, S., Steinberg, J.L.: 1977, *Astron. Astrophys.* **58**, 287.
Irimajiri, Y., Takano, T., Nakajima, H., Shibasaki, K., Hanaoka, Y., Ichimoto, K.: 1995, *Solar Phys.* **156**, 363.
Kakinuma, T., Swarup, G.: 1962, *Astrophys. J.* **136**, 975.
Karlický, M., Bárta, M., Dąbrowski, B.P., Heinzel, P.: 2011, *Solar Phys.* **268**, 165.
Kaufmann, P., Levato, H., Cassiano, M.M., Correia, E., Costa, J.E.R., Giménez de Castro, C.G., Godoy, R., Kingsley, R.K., Kingsley, J.S., Kudaka, A.S., Marcon, R., Martin, R., Marun, A., Melo, A.M., Pereyra, P., Raulin, J.-P., Rose, T., Valio, A. Silva, Walber, A., Wallace, P., Yakubovich, A., Zakia, M.B.: 2008, In: *Society of Photo-Optical Instrumentation Engineers (SPIE) Conference Series* **7012**, 70120L.
Keller, C.U., Krucker, S.: 2004, In: Gary, D.E., Keller, C.U. (eds.) *Solar and Space Weather Radiophysics*, Kluwer Academic, Dordrecht, 287.
Kerdraon, A., Delouis, J.-M.: 1997, In: Trottet, G. (ed.) *Coronal Physics from Radio and Space Observations, Lecture Notes in Physics* **483**, 192.
Kosugi, T., Ishiguro, M., Shibasaki, K.: 1986, *Publ. Astron. Soc. Japan* **38**, 1.
Krissinel, B.B.: 2005, *Astron. Rep.* **49**, 939.
Krissinel, B.B., Kuznetsova, S.M., Maksimov, V.P., Prosovetsky, D.V., Grechnev, V.V., Stepanov, A.P., Shishko, L.F.: 2000, *Publ. Astron. Soc. Japan* **52**, 909.
Krucker, S., Benz, A.O.: 2000, *Solar Phys.* **191**, 341.
Krucker, S., Benz, A.O., Bastian, T.S., Acton, L.W.: 1997, *Astrophys. J.* **488**, 499.
Kundu, M.R.: 1965, *Solar Radio Astronomy*, Interscience, New York.
Kundu, M.R., Alissandrakis, C.E.: 1975a, *Nature* **257**, 465.
Kundu, M.R., Alissandrakis, C.E.: 1975b, *Mon. Not. Roy. Astron. Soc.* **173**, 65.
Kundu, M.R., Rao, A.P., Erskine, F.T., Bregman, J.D.: 1979, *Astrophys. J.* **234**, 1122.
Kundu, M.R., Velusamy, T.: 1974, *Solar Phys.* **34**, 125.
Landi, E., Chiuderi Drago, F.: 2003, *Astrophys. J.* **589**, 1054.
Landi, E., Chiuderi Drago, F.: 2008, *Astrophys. J.* **675**, 1629.
Lantos, P., Kundu, M.R.: 1972, *Astron. Astrophys.* **21**, 119.
Lantos, P.: 1999, In: Bastian, T., Gopalswamy, N., Shibasaki, K. (eds.) *Solar Physics with Radio Observations, NRO Report* **479**, 11.
Lantos, P., Alissandrakis, C.E.: 1999, *Astron. Astrophys.* **351**, 373.
Lantos, P., Alissandrakis, C.E., Gergely, T., Kundu, M.R.: 1987, *Solar Phys.* **112**, 325.

Lantos, P., Alissandrakis, C.E., Rigaud, D.: 1992, *Solar Phys.* **137**, 225.

Lee, J.: 2007, *Space Sci. Rev.* **133**, 73.

Lee, J., McClymont, A.N., Mikic, Z., White, S.M., Kundu, M.R.: 1998, *Astrophys. J.* **501**, 853.

Loukitcheva, M., Solanki, S.K., Carlsson, M., Stein, R.F.: 2004, *Astron. Astrophys.* **419**, 747.

Loukitcheva, M., Solanki, S.K., White, S.: 2006, *Astron. Astrophys.* **456**, 713.

Loukitcheva, M., Solanki, S.K., White, S.M.: 2009, *Astron. Astrophys.* **497**, 273.

Loukitcheva, M., Solanki, S.K., White, S.M.: 2010, *Mem. Soc. Astron. Ital.* **81**, 592.

Lubyshev, B.I.: 1977, *Astron. Zh.* **54**, 130 (*Soviet Astron.* **21**, 74).

Marqué, C.: 2004, *Astrophys. J.* **602**, 1037.

Marqué, C., Lantos, P., Delouis, J.M., Alissandrakis, C.E.: 1999, In: Vial, J.-C., Kaldeich-Schümann, B. (eds.) *Plasma Dynamics and Diagnostics in the Solar Transition Region and Corona, 8th SOHO Workshop, ESA* **SP-446**, 483.

McConnell, D.: 1982, *Solar Phys.* **78**, 253.

Mercier, C., Chambe, G.: 2009, *Astrophys. J. Lett.* **700**, L137.

Mercier, C., Subramanian, P., Kerdraon, A., Pick, M., Ananthakrishnan, S., Janardhan, P.: 2006, *Astron. Astrophys.* **447**, 1189.

Moran, T., Gopalswamy, N., Dammasch, I.E., Wilhelm, K.: 2001, *Astron. Astrophys.* **378**, 1037.

Nakajima, H., Nishio, M., Enome, S., Shibasaki, K., Takano, T., Hanaoka, Y., Torii, C., Sekiguchi, H., Bushimata, T., Kawashima, S., Shinohara, N., Irimajiri, Y., Koshiishi, H., Kosugi, T., Shiomi, Y., Sawa, M., Kai, K.: 1994, *Proc. IEEE* **82**, 705.

Nindos, A., Alissandrakis, C.E., Gelfreikh, G.B., Borovik, V.N., Korzhavin, A.N., Bogod, V.M.: 1996, *Solar Phys.* **165**, 41.

Nindos, A., Aurass, H.: 2007, In: Klein, K.-L., MacKinnon, A.L. (eds.) *The High Energy Solar Corona: Waves, Eruptions, Particles, Lecture Notes in Physics* **725**, 251.

Nindos, A., Kundu, M.R., White, S.M.: 1999, *Astrophys. J.* **513**, 983.

Nindos, A., Kundu, M.R., White, S.M., Gary, D.E., Shibasaki, K., Dere, K.P.: 1999, *Astrophys. J.* **527**, 415.

Nindos, A., Kundu, M.R., White, S.M., Shibasaki, K., Gopalswamy, N.: 2000, *Astrophys. J. Suppl.* **130**, 485.

Noyes, R.W., Beckers, J.M., Low, F.J.: 1968, *Solar Phys.* **3**, 36.

Oberoi, D., Matthews, L.D., Cairns, I.H., Emrich, D., Lobzin, V., Lonsdale, C.J., Morgan, E.H., Prabu, T., Vedantham, H., Wayth, R.B., Williams, A., Williams, C., White, S.M., *et al.*: 2011, *Astrophys. J. Lett.* **728**, L27.

Pohjolainen, S.: 2000, *Astron. Astrophys.* **361**, 349.

Pohjolainen, S., Portier-Fozzani, F., Ragaigne, D.: 2000, *Astron. Astrophys. Suppl. Ser.* **143**, 227.

Ramesh, R., Kathiravan, C., Barve, I.V., Beeharry, G.K., Rajasekara, G.N.: 2010, *Astrophys. J. Lett.* **719**, L41.

Ramesh, R., Kathiravan, C., Sastry, C.V.: 2010, *Astrophys. J.* **711**, 1029.

Riehokainen, A., Urpo, S., Valtaoja, E., Makarov, V.I., Makarova, L.V., Tlatov, A.G.: 2001, *Astron. Astrophys.* **366**, 676.

Riehokainen, A., Valtaoja, E., Pohjolainen, S.: 2003, *Astron. Astrophys.* **402**, 1103.

Rosa, D., Vrsnak, B., Bozic, H., Brajsa, R., Ruzdjak, V., Schroll, A., Wohl, H.: 1998, *Solar Phys.* **179**, 237.

Ryabov, B.I., Pilyeva, N.A., Alissandrakis, C.E., Shibasaki, K., Bogod, V.M., Garaimov, V.I., Gelfreikh, G.B.: 1999, *Solar Phys.* **185**, 157.

Selhorst, C.L., Silva, A.V.R., Costa, J.E.R.: 2004, *Astron. Astrophys.* **420**, 1117.

Selhorst, C.L., Silva, A.V.R., Costa, J.E.R.: 2005, *Astron. Astrophys.* **433**, 365.

Selhorst, C.L., Silva-Válio, A., Costa, J.E.R.: 2008, *Astron. Astrophys.* **488**, 1079.

Selhorst, C.L., Giménez de Castro, C.G., Varela Saraiva, A.C., Costa, J.E.R.: 2010, *Astron. Astrophys.* **509**, A51.

Shibasaki, K.: 1998, In: Balasubramaniam, K.S., Harvey, J., Rabin, D. (eds.) *Synoptic Solar Physics, ASP Conference Series* **140**, 373.

Shibasaki, K.: 1999, In: Bastian, T., Gopalswamy, N., Shibasaki, K. (eds.) *Solar Physics with Radio Observations, NRO Report* **479**, 1.

Shibasaki, K., Enome, S., Nakajima, H., Nishio, M., Takano, T., Hanaoka, Y., Torii, C., Sekiguchi, H., Kawashima, S., Bushimata, T., Shinohara, N., Koshiishi, H., Shiomi, Y., Irimajiri, Y., Leka, K.D., Canfield, R.C.: 1994, *Publ. Astron. Soc. Japan* **46**, L17.

Silva, A.V.R., Laganá, T.F., Castro, C.G. Gimenez, Kaufmann, P., Costa, J.E.R., Levato, H., Rovira, M.: 2005, *Solar Phys.* **227**, 265.

Sych, R.A., Nakariakov, V.M., Anfinogentov, S.A., Ofman, L.: 2010, *Solar Phys.* **266**, 349.

Tanaka, H., Castelli, J.P., Covington, A.E., Krüger, A., Landecker, T.L., Tlamicha, A.: 1973, *Solar Phys.* **29**, 243.

Tapping, K.F., Boteler, D., Charbonneau, P., Crouch, A., Manson, A., Paquette, H.: 2007, *Solar Phys.* **246**, 309.

Thejappa, G., Kundu, M.R.: 1992, *Solar Phys.* **140**, 19.
Thejappa, G., Kundu, M.R.: 1994, *Solar Phys.* **149**, 31.
Thejappa, G., MacDowall, R.J.: 2008, *Astrophys. J.* **676**, 1338.
Tun, S.D., Gary, D.E., Georgoulis, M.K.: 2011, *Astrophys. J.* **728**, 1.
Vernazza, J.E., Avrett, E.H., Loeser, R.: 1973, *Astrophys. J.* **184**, 605.
Vernazza, J.E., Avrett, E.H., Loeser, R.: 1976, *Astrophys. J. Suppl.* **30**, 1.
Vernazza, J.E., Avrett, E.H., Loeser, R.: 1981, *Astrophys. J. Suppl.* **45**, 635.
Vocks, C., Mann, G.: 2004, *Astron. Astrophys.* **419**, 763.
Vourlidas, A., Gary, D.E., Shibasaki, K.: 2006, *Publ. Astron. Soc. Japan* **58**, 11.
White, S.M.: 2004, In: Gary, D.E., Keller, C.U. (eds.) *Solar and Space Weather Radiophysics*, Kluwer Academic, Dordrecht, 89.
White, S.M., Kundu, M.R., Shimizu, T., Shibasaki, K., Enome, S.: 1995, *Astrophys. J.* **450**, 435.
White, S.M., Loukitcheva, M., Solanki, S.K.: 2006, *Astron. Astrophys.* **456**, 697.
Zhang, J., Kundu, M.R., White, S.M., Dere, K.P., Newmark, J.S.: 2001, *Astrophys. J.* **561**, 396.
Zheleznyakov, V.V.: 1962, *Astron. Zh.* **39**, 5 (*Soviet Astron.* **6**, 3).
Zheleznyakov, V.V.: 1970, *Radio Emission of the Sun and Planets*, Pergamon Press, Oxford.
Zirin, H., Baumert, B.M., Hurford, G.J.: 1991, *Astrophys. J.* **370**, 779.

Solar Phys (2011) 273:339–361
DOI 10.1007/s11207-011-9875-6

Origin of the Submillimeter Radio Emission During the Time-Extended Phase of a Solar Flare

G. Trottet · J.-P. Raulin · G. Giménez de Castro · T. Lüthi · A. Caspi · C.H. Mandrini · M.L. Luoni · P. Kaufmann

Received: 19 April 2011 / Accepted: 25 September 2011 / Published online: 8 November 2011

Abstract Solar flares observed in the 200 – 400 GHz radio domain may exhibit a slowly varying and time-extended component which follows a short (few minutes) impulsive phase and can last for a few tens of minutes to more than one hour. The few examples discussed in the literature indicate that such long-lasting submillimeter emission is most likely thermal bremsstrahlung. We present a detailed analysis of the time-extended phase of the 27 October 2003 (M6.7) flare, combining 1 – 345 GHz total-flux radio measurements with X-ray, EUV, and Hα observations. We find that the time-extended radio emission is, as expected,

Energy Storage and Relase through the Solar Activity Cycle – Models Meet Radio Observations
Guest Editors: Christophe Marqué and Alexander Nindos

A. Caspi now at: Laboratory for Atmospheric and Space Physics, University of Colorado, Boulder, CO, 80303, USA.

G. Trottet (✉)
Observatoire de Paris, LESIA-CNRS UMR 8109, Univ. P & M Curie and Paris-Diderot, Observatoire de Meudon, 92195 Meudon, France
e-mail: gerard.trottet@obspm.fr

G. Trottet · J.-P. Raulin · G. Giménez de Castro · P. Kaufmann
CRAAM Universidade Presbiteriana Mackenzie, São Paulo, Brazil

J.-P. Raulin
e-mail: raulin@craam.mackenzie.br

G. Giménez de Castro
e-mail: guigue@craam.mackenzie.br

P. Kaufmann
e-mail: kaufmann@craam.mackenzie.br

T. Lüthi
Hexagon Metrology, Leica Geosystem AG, Moenchmattweg 5, 5035 Unterenfelden, Switzerland
e-mail: thomas.Luethi@leica-geosystems.com

A. Caspi
Space Sciences Laboratory, University of California, Berkeley, CA 94720-7450, USA
e-mail: amir.caspi@lasp.colorado.edu

radiated by thermal bremsstrahlung. Up to 230 GHz, it is entirely produced in the corona by hot and cool materials at 7 – 16 MK and 1 – 3 MK, respectively. At 345 GHz, there is an additional contribution from chromospheric material at a few 10^4 K. These results, which may also apply to other millimeter–submillimeter radio events, are not consistent with the expectations from standard semiempirical models of the chromosphere and transition region during flares, which predict observable radio emission from the chromosphere at all frequencies where the corona is transparent.

Keywords Radio bursts, association with flares · Radio bursts, microwave · X-ray bursts, association with flares · Flares, relation to magnetic field · Chromosphere, active

1. Introduction

Since 2000, new instrumentation has allowed us to observe solar flares at submillimeter wavelengths. Such observations are routinely carried out by the *Solar Submillimeter Telescope* (SST; Kaufmann *et al.*, 2008) at 212 and 405 GHz, and studies of ten major bursts detected by this instrument have been reported in the literature. Measurements at 230 and 345 GHz have been obtained for two more flares by the *Köln Observatory for Submillimeter and Millimeter Astronomy* (KOSMA; Lüthi, Magun, and Miller, 2004; Lüthi, Lüdi, and Magun, 2004) telescope; one of them has also been observed at 210 GHz by the *Bernese Multibeam Radiometer for KOSMA* (BEMRAK).

Above 200 GHz, these events exhibit an impulsive phase lasting for a few minutes, which is sometimes followed by a slowly varying and time-extended emission (the "gradual" phase). While, for some events, the >200 GHz emission appears as the extension toward high frequencies of the gyrosynchrotron emission seen in the microwave domain (Trottet *et al.*, 2002; Lüthi, Magun, and Miller, 2004; Raulin *et al.*, 2004; Giménez de Castro *et al.*, 2009), other events exhibit an unexpected upturn toward the terahertz (THz) domain (*e.g.*, Kaufmann *et al.*, 2004). Such spectra with positive slopes in the millimeter–submillimeter domain have been measured during both the impulsive phase (Kaufmann *et al.*, 2004, 2009; Silva *et al.*, 2007) and the gradual phase (Trottet *et al.*, 2002; Lüthi, Magun, and Miller, 2004; Lüthi, Lüdi, and Magun, 2004).

Although various theoretical ideas have appeared in the literature (*e.g.*, Kaufmann and Raulin, 2006; Fleishman and Kontar, 2010), the emission mechanism responsible for a positive slope in the sub-THz domain remains uncertain, at least during the impulsive phase. During the gradual phase, thermal bremsstrahlung from the chromosphere and the corona

C.H. Mandrini · M.L. Luoni
Instituto de Astronomía y Física del Espacio, CONICET-UBA, CC. 67 Suc. 28, 1428, Buenos Aires, Argentina

C.H. Mandrini
e-mail: mandrini@iafe.uba.ar

M.L. Luoni
e-mail: mluoni@iafe.uba.ar

C.H. Mandrini
Facultad de Ciencias Exactas y Naturales, FCEN-UBA, Buenos Aires, Argentina

P. Kaufmann
Centro de Componentes Semicondutores, Universidade Estadual de Campinas, Campinas, Brazil

may account for the observed radio spectrum, particularly during the late decay of the flare. This is the case for the 22 March 2000 (X1.1) flare (Trottet *et al.*, 2002) and for the 28 October 2003 (X17.2) flare (Trottet *et al.*, 2008), where the radio spectrum, late in the gradual phase, is roughly flat below 200 GHz and increases at higher frequencies. Such spectra were interpreted as optically thin thermal bremsstrahlung from the corona below $\sim$200 GHz, and it was suggested that the excess emission measured at higher frequencies arose from thermal bremsstrahlung from the lower atmosphere.

The best example of a thermal phase at 230 and 345 GHz reported so far is the time-extended component of the 12 April 2001 (X2.1) flare, which lasts for more than two hours and during which the radio spectrum is flat, first up to 230 GHz and later up to 89 GHz, with a positive slope above these frequencies (Lüthi, Magun, and Miller, 2004). However, although the radio emission during the gradual phase is conjectured to arise from separate contributions of coronal and chromospheric sources, the few available observations have not allowed a characterization of the plasmas involved.

This paper presents an analysis of radio observations of the 27 October 2003 (M6.7) flare at $\sim$12:30 UT obtained in the 1 – 405 GHz range by the United States Air Force (USAF) Radio Solar Telescope Network[1] (RSTN), the Bumishus patrol telescopes (Institute of Applied Physics of Bern University), BEMRAK, the KOSMA telescope, and the SST. Above 200 GHz, the radio event is similar to the 12 April 2001 radio burst. Indeed, *i*) it exhibits a short impulsive phase followed by a time-extended component, *ii*) the impulsive radio emission is gyrosynchrotron radiation from relativistic electrons, and *iii*) the gradual phase is optically thin (8 – 230 GHz) and optically thick (345 GHz) thermal bremsstrahlung. The main goal of this study is to combine the radio observations with soft X-ray, UV/EUV, and optical data in order to discuss the origins of the thermal radio emission during the gradual phase. The paper is organized as follows. Section 2 describes the radio observations and flux calibration of the SST, KOSMA, and BEMRAK measurements. The results are discussed in Section 3, and conclusions are drawn in the final section.

2. Observations and Data Analysis

The 27 October 2003 radio event at $\sim$12:30 UT was associated with a *Geostationary Operational Environmental Satellite* (GOES) M6.7-class soft X-ray (SXR) flare and an Hα subflare which occurred in active region AR 10486 at S17 E25. The flare was well observed in the UV and EUV, from $\sim$12:28 to $\sim$13:00 UT, by the *Transition Region and Coronal Explorer*[2] (TRACE; Handy *et al.*, 1999). Unfortunately, only the late part of the flare (after $\sim$12:46 UT) was observed in the hard X-ray domain (> 20 keV), by the *Reuven Ramaty High Energy Solar Spectroscopic Imager*[3] (RHESSI; Lin *et al.*, 2002). Radio observations were available over a wide spectral domain from 10 kHz up to 405 GHz. Figure 1 shows the time evolution of the GOES 1 – 8 Å SXR flux and of the radio flux densities[4] at selected frequencies in the 1 – 400 GHz domain. Total flux densities were obtained by RSTN at eight frequencies in the 0.245 – 15 GHz range and by the Bumishus patrol telescopes at 8.4, 11.8, 19.6, 35, and 50 GHz (until 12:57 UT). Measurements at 212 and 405 GHz were made

[1] http://www.ngdc.noaa.gov/stp/solar/solarradio.html.

[2] http://trace.lmsal.com/.

[3] http://hesperia.gsfc.nasa.gov/hessi.

[4] 1 sfu = 10^{-22} W m^{-2} Hz^{-1}.

Figure 1 Time histories of X-ray and radio emissions during the 27 October 2003 flare at ∼ 12:30 UT. From top to bottom: 1 – 8 Å soft X-ray flux (background subtracted) from the GOES-12 X-ray Sensor; radio flux densities at 1.415 GHz from RSTN; at 19.6 GHz from the Bumishus telescopes (the blue curve shows the time extended phase with an enlarged flux density scale); at 210 GHz from BEMRAK (red), 230 GHz from KOSMA (blue), and at 212 GHz from SST (black); and at 345 GHz from KOSMA.

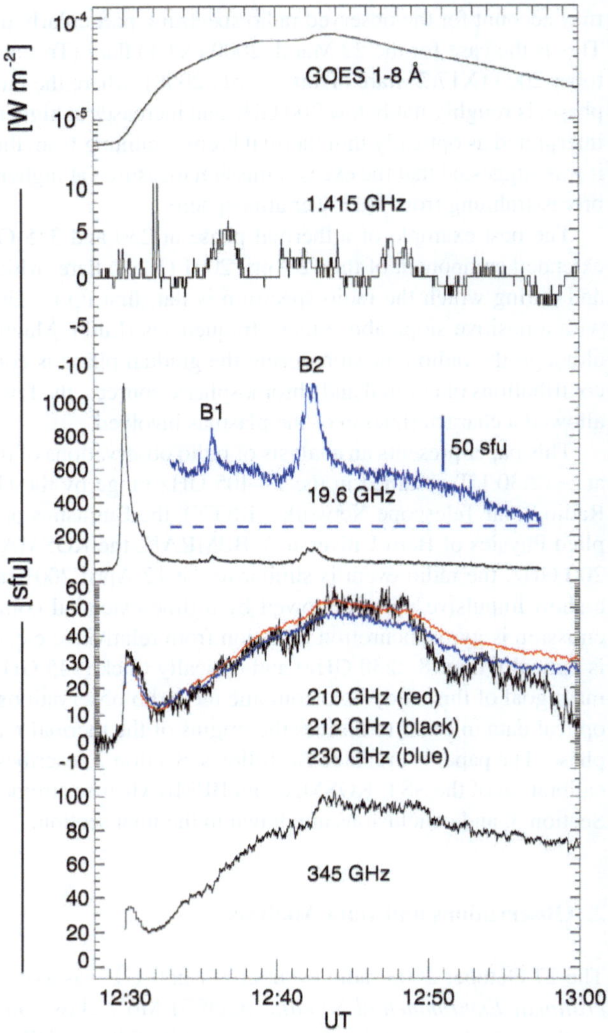

by the SST. KOSMA and BEMRAK provided measurements at 230 and 345 GHz, and at 210 GHz, respectively, until 13:00 UT.

The radio event exhibits two phases: an impulsive burst which lasts for about three minutes (∼ 12:29 – 12:32 UT) and a slowly varying, time-extended emission which lasts up to about 13:15 UT at 212 GHz. Note that: *i*) although this time-extended emission is more pronounced and larger than the impulsive burst in the 200 – 400 GHz range, it is also observed at all other frequencies above 1.415 GHz, except at 50 GHz due to strong atmospheric absorption; and *ii*) at 212 GHz, the slowly varying emission starts slightly before the impulsive rise, simultaneously with the SXR emission. This time sequence is similar to that observed in other events with > 200 GHz emission (Lüthi, Magun, and Miller, 2004; Trottet *et al.*, 2008; Giménez de Castro *et al.*, 2009). At microwave frequencies, two impulsive bursts, B1 and B2, with maxima at ∼ 12:35:49 UT and ∼ 12:42:20 UT, respectively, are superimposed on the slowly varying component.

Figure 1 shows that no significant emission is measured at 1.415 GHz. Although radio activity was observed in the decimeter–hectometer range (see Radio Monitoring[5]), these radio emissions are not associated with AR 10486, so the flare under study is radio-silent at frequencies below ~ 2 GHz. Indeed:

- There is a permanent noise storm which exhibits an enhancement around 12:30 UT, *i.e.*, close to the maximum of the impulsive phase of the event under study. Radio imaging by the *Nançay Radioheliograph* (NRH) in the 150 – 450 MHz range shows that this noise storm activity is spatially associated with AR 10488 (at N09 E12), not with AR 10486. A decameter–hectometer type III storm is also associated with this noise storm, as is sometimes the case (see Elgarøy, 1977).
- Between 12:30 and 14:00 UT, there are some decimeter–hectometer type III bursts, indicating that some electron beams accelerated in the low corona reach interplanetary space. Here, again, NRH images show that these type IIIs are not associated with the flare under study, but rather with activity from AR 10484 (at N07 W46).

The 27 October 2003 flare thus belongs to the class of "confined microwave events" (Klein *et al.*, 2011), where microwave-emitting electrons remain confined in the low corona. This is consistent with the analysis of the magnetic field topology of this flare by Luoni *et al.* (2007), which indicates that energy release occurred in compact loops at a magnetic null very low in the corona (~ 3 Mm above the photosphere).

2.1. Flux Densities at 212 and 405 GHz

The SST observed the 27 October 2003 event with five independent total-power receivers: channels 1 – 4 at 212 GHz and channel 5 at 405 GHz. Figure 2 shows the 50%-level contours for beams 2 – 5 while SST was tracking AR 10486, overlaid on a TRACE 1600 Å image obtained at $\sim$12:46 UT, close to the maximum of the time-extended phase. Beam 1 is outside the field of view of the figure. At 212 GHz, beams 1 – 4 may be approximated by circular Gaussians with a half-power beamwidth (HPBW) of $\sim 4'$. The 405 GHz channel was oblate for this period, and is better represented by an elliptical Gaussian with an HPBW of $\sim 2' \times 4'$.

Due to the low solar elevation angle $\theta_{\mathrm{el}} \approx 35°$ at the time of the flare, along with the rather large zenith opacity, the antenna temperature time profiles at 212 and 405 GHz are quite similar, dominated by atmospheric fluctuations, as illustrated in Figure 3. However, we note that the signal from beam 3 shows additional temporal features compared to those in beams 2, 4, and 5. In particular, we observe a rapid variation at 12:30 UT followed by a gradual increase between 12:32 and 13:00 UT, which correspond, respectively, to the impulsive and extended phases of the flare, shown in Figure 1. Therefore, the flare was detected mainly within beam 3, with negligible contributions in beams 2, 4, and 5. This is consistent with the beam positions relative to the UV flaring structures, as shown in Figure 2.

Since the signal from beam 4 is very noisy, including spiky interferences, we instead consider that the signal from beam 2 is the one which is characteristic of the emission of the whole active region. Figure 3 shows that, before the flare, the signals from beams 2 and 3 are identical. This indicates that the pre-flare (background) signal does not depend critically on the position of the whole active region within the beam. The signal of beam 2 has thus

[5]http://secchirh.obspm.fr/survey.php?dayofyear=20031027&composite=1 and http://secchirh.obspm.fr/survey.php?dayofyear=20031027&composite=2.

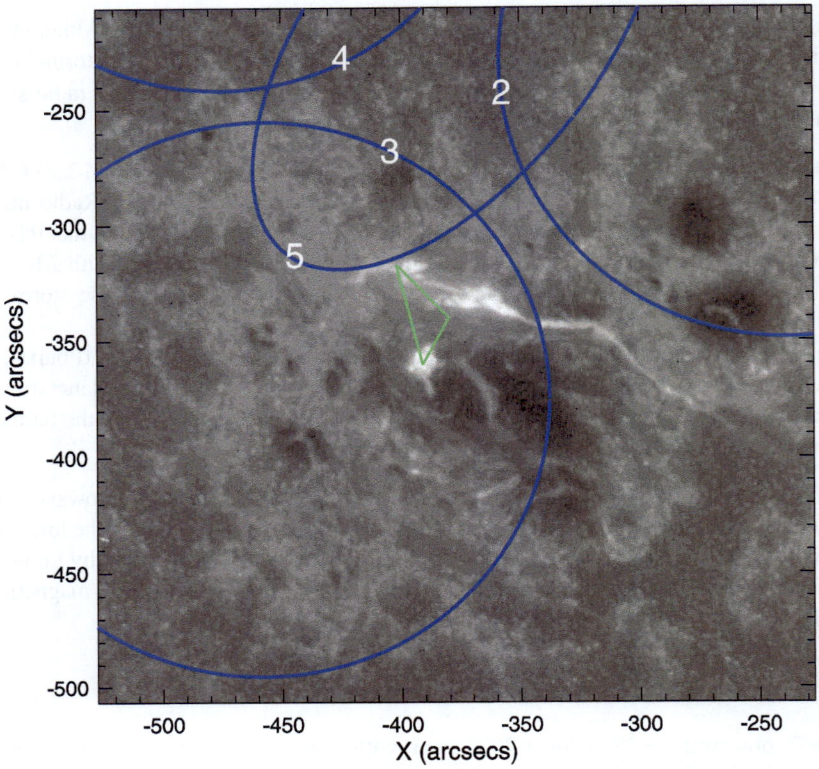

Figure 2 Configuration of the SST beams during the flare. The circles and the ellipse represent the 50% contour (HPBW) of beams 2–5, overlaid on a TRACE 1600 Å image taken at 12:46:21 UT. The numbers indicate the beam. The green triangle represents the region where the 212 and 405 GHz flare sources may be located without being detected by beam 2 at 212 GHz or by beam 5 at 405 GHz (see text).

been subtracted from that of beam 3 in order to suppress variations due to atmospheric fluctuations and to obtain the excess antenna temperature T_a due to the flare.

The flux density at 212 GHz, S_{212}, is then given by

$$S_{212} = \frac{2k_B T_a}{A_a \eta_a} \exp\bigl(\tau_z / \sin(\theta_{el})\bigr), \tag{1}$$

where k_B is the Boltzmann constant, A_a is the antenna surface area, η_a the aperture efficiency, and τ_z the zenith opacity of the Earth's atmosphere. We estimated $\tau_{z(212)} \approx 0.46$ and $\tau_{z(405)} \approx 1.9$ at 212 and 405 GHz, respectively, using the method described in Melo *et al.* (2005). For the present observation, the antenna aperture efficiency η_a is 20% at 212 GHz. The main sources of uncertainty are temporal fluctuations of τ_z and the aperture efficiency, which result in a total uncertainty of 20%. Figure 1 shows the 212 GHz flux density as computed using Equation (1).

In Equation (1), we have assumed that the source is located over the center of beam 3. In order to account for the actual source location, we need to multiply S_{212} by the factor $\exp(\theta_{off}^2/2\sigma^2)$. Here, θ_{off} is the angular distance between the center positions of the source and of beam 3, and σ represents the standard deviation of the convolution of a circular Gaussian source by the beam. We estimated that the maximum flux density at 405 GHz

Figure 3 Antenna temperatures observed by the different SST beams at 212 GHz (beam 2: red; beam 3: thick black; beam 4: dashed black) and at 405 GHz (beam 5: blue). Arrows indicate the maxima of the flare impulsive and extended phases at 12:30 UT and 12:45 UT, respectively.

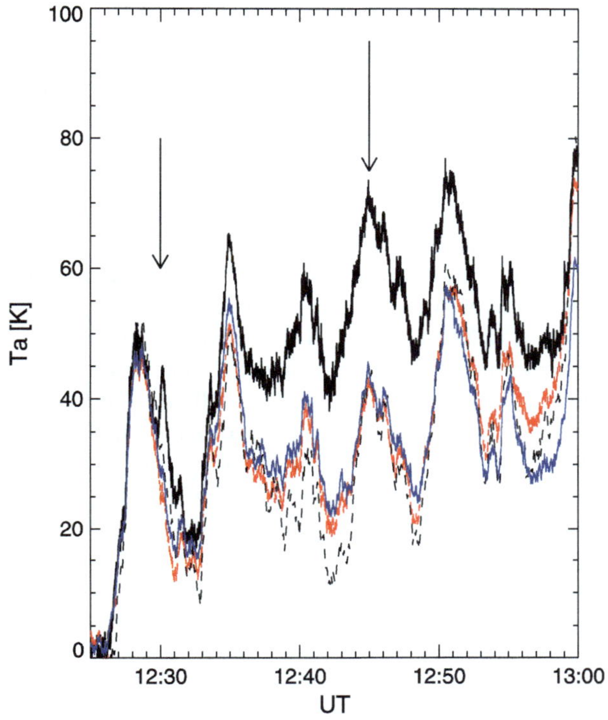

should have been ~ 125 sfu by extrapolating the increasing 212 – 345 GHz flux spectrum up to 405 GHz (see Figure 1). However, the event was detected only in beam 3 at 212 GHz, and was not detected in beam 5 at 405 GHz within the root mean square (rms) noise of 30 – 40 sfu. We used these two facts to constrain the source size and location, and to estimate θ_{off}. To do so, we further supposed that the center of the 405 GHz-emitting source should be near the maximum of the SXR emission observed by RHESSI (see Section 2.3) and between the UV ribbons observed by TRACE (see Figure 2). We then computed the emission that should be detected at 405 GHz by convolving a circular Gaussian source of 125 sfu with beam 5. We found that sources with sizes ranging from $10''$ to $70''$, with centers located within the green triangle in Figure 2, fulfill the preceding two requirements, *i.e.*, no burst detection in beam 5 at 405 GHz or in beam 2 at 212 GHz. The correction factor for S_{212} in Equation (1) is thus between 25% and 40%, including a $20''$ indetermination in antenna pointing. At 212 GHz, the maximum uncertainties in the flux density are thus taken as -30% and $+40\%$.

2.2. Flux Densities at 210, 230, and 345 GHz

On 27 October 2003, the KOSMA target was AR 10486, where the flare under study was located. KOSMA uses a tiltable subreflector which periodically deflects the antenna beams by $6'$ in elevation, from the flaring active region to a quiet reference region on the Sun. The "1 second on/0.1 second off" observation cycle allows for compensation of atmospheric attenuation changes with time scales greater than 1 s (Lüthi, Magun, and Miller, 2004). The absolute flux density calibration was obtained from hourly observations of the sky. Unfortunately, before 12:31 UT the subreflector was wobbling the wrong way, staying 10 s on the quiet reference region and 0.1 s on AR 10486. With such a short sampling of the active region, we could determine neither the excess flux of the pre-flare active region with respect to the quiet reference flux, nor the onset of the flare itself. Significant flux density excesses at 210, 230, and 345 GHz relative to the quiet reference region were thus measured only after 12:29:56 UT, *i.e.*, slightly before the maximum of the impulsive phase at 212 GHz (see Figure 1).

In order to estimate the net flux density excess due to the flare alone, we proceeded as follows. We first assumed that the flare emission started at the same time at 210, 230, and 345 GHz as at 212 GHz. This is generally the case for events detected in the 200 – 400 GHz range (see, *e.g.*, Kaufmann *et al.*, 2004, 2009; Lüthi, Magun, and Miller, 2004; Raulin *et al.*, 2004). The growth of the time-extended emission at 212 GHz is well fit by a straight line between 12:32 and 12:35 UT. By extrapolating this straight line back to the time t_0 where it crosses the pre-event flux density level, we get an estimate of the starting time of the burst: $t_0 =$ 12:38:37 UT ± 5 s. At 210 and 230 GHz, straight lines have been fit to the same time interval as at 212 GHz and extrapolated back to t_0, providing the actual pre-flare flux densities at 210 and 230 GHz. These are 36 ± 6 and 47 ± 7 sfu at 210 and 230 GHz, respectively. These uncertainties are estimated by taking into account the uncertainty on t_0, the uncertainties on the fit parameters, and the amplitude of the fluctuations of the measured flux density. Figure 1 shows that the derived flare excesses at 210 and 230 GHz agree well with that measured at 212 GHz, justifying the procedure described above.

At 345 GHz, the early increase of the time-extended emission cannot be fit by a single straight line. We obtained upper and lower limits for the contribution of the active region before the flare by fitting straight lines to the 345 GHz flux density excess measured between 12:32 and 12:32:34 UT, and between 12:32:34 and 12:33:50 UT, respectively, and extrapolating back to t_0. The contribution of the active region is thus found to be between 74 ± 8 and 98 ± 10 sfu. For the following sections, we adopt the mean value, 86 ± 14 sfu.

Following Lüthi, Magun, and Miller (2004), we adopt maximum uncertainties of -30% and $+40\%$ at 210 and 230 GHz. An absolute error of ± 14 sfu is added at 345 GHz (see previous paragraph). Figure 1 shows that, during the time-extended phase, the time evolution of the 210 – 230 GHz flux and of the 345 GHz fluxes are different, with the flux density at 345 GHz being systematically larger (about $2\times$ after 12:40 UT) than that at 210 – 230 GHz, even when using the upper limit of the pre-flare active region contribution. As the above uncertainties constitute conservative upper limits of the flux accuracy, we consider the difference between the 210 – 230 GHz and 345 GHz flux densities to be significant, although it is only slightly larger than the uncertainties.

2.3. RHESSI Spectral and Imaging Observations

We used RHESSI X-ray spectra and images to quantitatively characterize the hot ($\gtrsim 10$ MK) thermal plasma in the 27 October 2003 flare. While the GOES SXR data provide some information about the flare plasma temperature and emission measure (see Section 3.2.1), the broadband response over only two channels (1 – 8 and 0.5 – 4 Å, corresponding to ~ 1.6 – 12 and ~ 3.1 – 25 keV, respectively) yields only a single, average measurement with no spatial information. On the other hand, in the X-ray range (~ 3 – 100 keV), RHESSI's ~ 1 keV full width at half-maximum (FWHM) spectral resolution enables precise measurements of the hottest plasma, which can then be combined with the GOES observations to also measure cooler components and thus more accurately determine the temperature distribution in the flare. RHESSI also provides spatial information for the hot plasma, with an angular resolution as good as $\sim 2''$.

RHESSI observations of the 27 October 2003 flare are available only in the late decay, from $\sim$12:46 to $\sim$13:04 UT. Because of the limited observing period and the rapidly decaying hard X-ray flux (the incident $\sim$6–25 keV flux decreases by about two orders of magnitude over these $\sim$18 minutes), we analyzed only a single time interval, from 12:47 to 12:49 UT. The 3 – 100 keV spatially integrated spectrum was accumulated over this two-minute period using the single best detector, G4, with 1/3-keV energy bins (the instrument channel width; Smith *et al.*, 2002), and the non-solar background was then subtracted (*cf.* Caspi, 2010); above $\sim$3 keV, the solar pre-flare background is negligible compared to the flare emission. The spectrum decreases steeply with energy (*e*-folding of $\sim$2 keV), with no appreciable flare emission above $\sim$33 keV in the analyzed period.

For precise spectroscopy, we employed the low-energy instrument response calibration improvements described by Caspi (2010). Following the method of Caspi and Lin (2010), the $\sim$4.67–33 keV spectrum was forward-fit with a photon model consisting of two isothermal continua (a single isothermal did not yield an acceptable fit) and two Gaussian features (representing unresolved excitation lines of highly ionized Fe and Ni). The spectrum was well fit by this model (reduced $\chi^2 \approx 0.71$) and yielded, for each of the two thermal components, a best-fit temperature and emission measure of $T_1 \approx 24$ MK, $EM_1 \approx 0.24 \times 10^{49}$ cm^{-3}, $T_2 \approx 16$ MK, and $EM_2 \approx 1.7 \times 10^{49}$ cm^{-3}, respectively.

An image at 6 – 30 keV was accumulated for the same time interval, using the CLEAN image reconstruction algorithm with uniform weighting (Hurford *et al.*, 2002) and grids 3, 4, 5, 6, 8, and 9 (grid 7 was excluded due to its $\sim$20 keV low-energy threshold) to achieve a $\sim 10''$ spatial resolution (see Section 3.2.1). Although it is not possible to image the two thermal components separately, we can nevertheless derive their centroid positions relative to the combined emission (*cf.* Caspi and Lin, 2010). From the spatially integrated spectral model, the $\sim$24 MK source was found to contribute $\sim$45%, $\sim$63%, and $\sim$90% of the total count flux at 6.3 – 7.3, 9 – 12, and 17 – 18 keV, respectively, and imaging shows that the centroid position of the emission in each energy band is correlated with this fractional contribution. From this, we extrapolated the centroid positions of the $\sim$24 and $\sim$16 MK thermal plasmas; they are separated by $\sim 11'' \pm 3.2''$. The 6 – 30 keV source is dominated ($\gtrsim$80%) by counts from the $\sim$24 MK component, and encompasses a projected area (*cf.* Caspi, 2010) of $\sim$280 – 310 square arcsec within the 50% contour, and $\sim$560 – 660 square arcsec within the 30% contour (both values have been corrected for broadening by the instrument point-spread function). Although the $\sim$16 MK component contributes only weakly to the imaged emission, the morphology and centroid separation suggest that it is of a similar size.

3. Results and Discussion

3.1. Impulsive Phase

Figure 4 displays the radio spectrum observed in the 1 – 400 GHz range around the maximum of the impulsive phase at 212 GHz. This spectrum is reminiscent of the usual gyrosynchrotron emission from nonthermal electrons, commonly observed in the microwave domain. It peaks around 10 GHz, has a sharp low frequency cutoff between 1.415 and 2.695 GHz, and has a positive spectral index of $\sim$4 below 5 GHz. This is steeper than the index of 2.5 – 2.9 predicted by gyrosynchrotron self-absorption. Such steep spectra are generally ascribed to the suppression of microwave emission by the ambient medium (the Razin effect). In magnetic fields of a few hundred gauss, the Razin effect becomes significant at a few gigahertz for ambient densities of at least a few 10^{10} cm^{-3}. For stronger magnetic

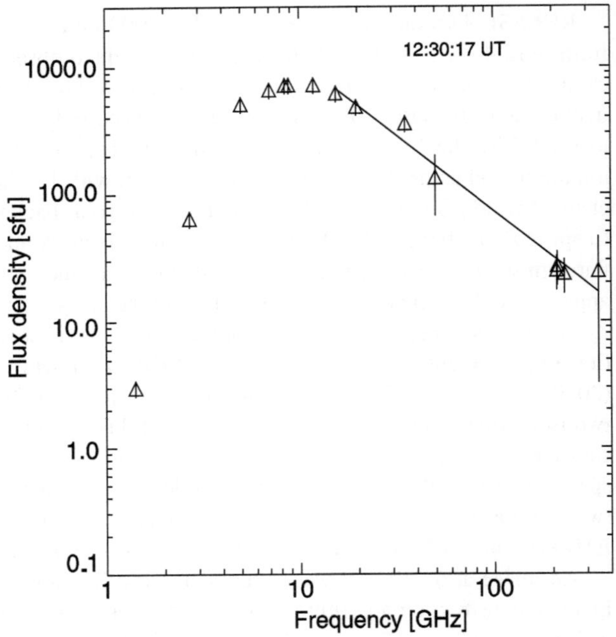

Figure 4 The radio spectrum near the maximum of the 212 GHz emission during the impulsive phase of the 27 October 2003 event. The solid line is a power law fit to the decreasing part of the radio spectrum, with spectral index $\alpha = -1.2$.

fields, higher ambient densities are required for the same level of suppression. Such high densities inferred for the microwave-emitting region further support the statement that the 27 October 2003 flare is a "confined microwave event" (see Section 2).

The optically thin, decreasing part of the spectrum above 15 GHz is well represented by a power law, as shown by the solid line in Figure 4. The power law spectral index α is about -1.2 around the maximum of the impulsive phase at 212 GHz. This optically thin radiation is primarily emitted by MeV electrons. Considering the ultra-relativistic case as a gross approximation, the spectral index δ of the instantaneous distribution of radio-emitting electrons is $\delta = 2\alpha - 1$ (Ginzburg and Syrovatskii, 1965),[6] thus $\delta \approx -3.4$ at the maximum of the impulsive phase. Such a hard electron spectrum is characteristic of gyrosynchrotron events observed up to millimeter–submillimeter wavelengths (see, *e.g.*, Trottet *et al.*, 2002; Lüthi, Magun, and Miller, 2004; Raulin *et al.*, 2004; Giménez de Castro *et al.*, 2009 and references therein).

Figure 4 indicates that, within the large uncertainties, the emission at 210–230 and 345 GHz appears to be the high-frequency part of the gyrosynchrotron spectrum observed in the microwave domain. Such a continuation of the microwave spectrum to the 300–400 GHz domain has been previously observed, although for larger, GOES X-class flares.

3.2. Gradual Phase

As stated in Section 2, the slowly varying and time-extended ("gradual") radio emission starts to rise close to the onset of the SXR emission measured by GOES (see Figure 1). Superimposed on this slowly varying component, two impulsive bursts, labeled B1 and B2 on Figure 1, are observed at frequencies up to 35 GHz but without obvious counterparts in

[6] Ginzburg and Syrovatskii (1965) define $\alpha, \delta > 0$; we have modified their equation to eliminate this assumption, taking $\alpha \to -\alpha$ and $\delta \to -\delta$.

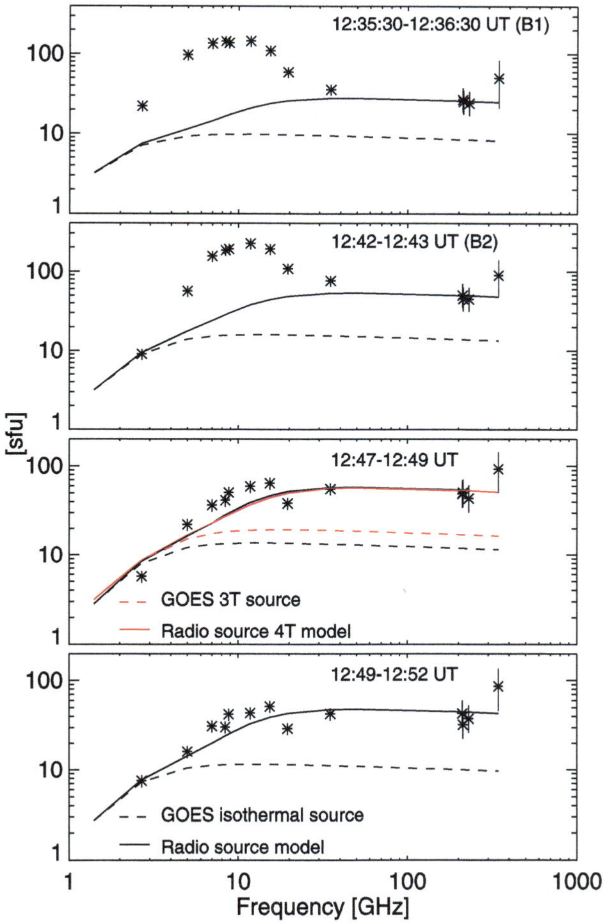

Figure 5 The mean radio spectrum observed at different times during the time-extended (gradual) phase of the 27 October 2003 flare (asterisks with errors bars). From top to bottom, the spectra correspond to bursts B1 and B2 (see Figure 1), and to the plateau and decay of the 210 – 230 GHz emission, respectively. The dashed lines indicate the radio spectrum expected purely from an isothermal plasma with temperature and emission measure derived from the GOES SXR measurements, assuming a circular source of 40″ diameter, while the solid lines indicate the spectrum expected from a two-temperature model (see text).

the 200 – 400 GHz domain. Figure 5 shows the mean radio spectrum (asterisks) observed during B1 and B2 and during two later time intervals. For each of the four time intervals shown, the dashed curve indicates the mean bremsstrahlung spectrum expected from a single isothermal source of assumed 40″ diameter with temperature T_g and emission measure EM_g as derived from the GOES SXR fluxes (see Section 3.2.1). The main points to be drawn from Figure 1 and Figure 5 are as follows.

- During the two last intervals, when no impulsive bursts are present, the observed radio spectrum is almost flat between ~ 8 and 230 GHz, with flux densities approximately three to four times greater than those expected from the single thermal source derived from GOES SXRs.
- During bursts B1 and B2, a gyrosynchrotron component is superimposed on a thermal bremsstrahlung spectrum. B1 has no counterpart in the millimeter–submillimeter range, but the > 200 GHz radiation exhibits an additional increase during B2.
- Throughout the gradual phase, the 345 GHz flux density is significantly higher than that measured in the 210 – 230 GHz range (see Section 2.2).

We discuss each of these points in the following sections.

3.2.1. The Flat Radio Spectrum

The flat spectrum observed between ~8 and 230 GHz throughout the gradual phase is reminiscent of optically thin thermal bremsstrahlung emission. Below ~8 GHz, the flux density decreases with decreasing frequency, indicating that the radio source gradually becomes optically thick at low frequencies. To characterize the thermal plasma seen by GOES, its temperature T_g and emission measure EM_g were derived from the GOES SXR fluxes, assuming an isothermal source, using the CHIANTI database (Dere *et al.*, 1997; Young, Landi, and Thomas, 1998) and coronal elemental abundances. T_g peaks at ~20 MK just after the maximum of the impulsive phase, then decreases to ~12 MK. EM_g initially increases, passes through a maximum of $\sim 3 \times 10^{49}\ \mathrm{cm}^{-3}$ approximately at the end of B2, and then falls to $\sim 10^{49}\ \mathrm{cm}^{-3}$.

We computed the radio spectrum expected from the GOES thermal source as follows. At time t and frequency ν, the brightness temperature $T_{\mathrm{br}}(t,\nu)$ of the radio emission from an isothermal, homogeneous source at temperature $T(t)$, emission measure $EM(t)$, and projected surface area A (normal to the line of sight) is given by

$$T_{\mathrm{br}}(t,\nu) = T(t)\left(1 - \mathrm{e}^{-\tau(t,\nu)}\right), \tag{2}$$

where, for $T > 0.2$ MK, the optical thickness $\tau(t,\nu)$ is given by (Dulk, 1985):

$$\begin{aligned} \tau(t,\nu) &= \frac{K(t,\nu)}{\nu^2 A}\frac{EM(t)}{T(t)^{3/2}}, \\ K(t,\nu) &\equiv 9.78 \times 10^{-3} \times (24.5 + \ln T(t) - \ln \nu) \quad \text{(cgs units)}. \end{aligned} \tag{3}$$

The expected radio flux density $S_{\mathrm{exp}}(t,\nu)$ from the source is then given by

$$S_{\mathrm{exp}}(t,\nu) = \frac{2k_{\mathrm{B}} T_{\mathrm{br}}(t,\nu)\nu^2 A}{c^2 R^2}, \tag{4}$$

where c is the speed of light and R is the Sun–Earth distance. We note that for optically thin ($\tau \ll 1$) regimes, Equations (2) and (4) reduce to

$$T_{\mathrm{br}}(t,\nu) \approx T(t)\tau(t,\nu) \implies S_{\mathrm{exp}}(t,\nu) = \frac{2k_{\mathrm{B}} K(t,\nu)}{c^2 R^2}\frac{EM(t)}{T(t)^{1/2}}, \tag{5}$$

whereby the expected radio flux spectrum is independent of the source area A and only weakly dependent on frequency ν.

Using $T_g(t)$ and $EM_g(t)$ derived for the GOES thermal plasma, and assuming a projected area equivalent to a circle of 40″ diameter (the mean acceptable source area, per Section 2.1; see also the discussion below), we calculated the radio flux density $S_{\mathrm{exp}}(t,\nu)$ expected for this source, shown by the dashed black curves in Figure 5. Like Lüthi, Magun, and Miller (2004), we find a qualitative similarity between the time evolutions of the observed and predicted flux densities in the 200 – 400 GHz range (where the emission is optically thin and where no impulsive bursts are detected), but the observed values are approximately three to four times higher at 210 – 230 GHz and approximately seven times higher at 345 GHz. Even using photospheric rather than coronal abundances to derive T_g and EM_g, the computed radio fluxes are still approximately two and four times lower than the observed values at 8 – 230 and 345 GHz, respectively.

In contrast, Pohjolainen, Valtaoja, and Urpo (1996) reported gradual radio bursts at 37 GHz for which the observed and computed flux excesses were comparable. A similar agreement was obtained from case studies at 8.8 and 15.4 GHz by, *e.g.*, Kundu *et al.* (1994), up to 15.4 GHz by Trottet *et al.* (2002), and up to 86 GHz by Raulin *et al.* (1999). However, we note that in these earlier studies, T_g and EM_g were derived using the Thomas, Crannell, and Starr (1985) polynomial approximation, which closely resembles the GOES response to *photospheric*-abundance CHIANTI models for temperatures up to $\sim$15 MK (White, Thomas, and Schwartz, 2005). The GOES plasma likely resides in coronal loops, not chromospheric footpoints, and as shown above, the use of photospheric abundances results in substantially larger expected radio flux densities than those obtained using the more appropriate coronal abundances. Thus, this suggests that, for most of the gradual bursts reported in the cited works, the measured flux densities in the centimeter–millimeter range are actually higher than those expected from the SXR-emitting plasma observed by GOES, similar to our results.

White and Kundu (1992) emphasized that the above discrepancy between observed radio emission and that computed from SXR observations mainly reflects that: *i*) the assumption of an isothermal coronal plasma is not correct during a flare, and *ii*) radio observations are more sensitive to cooler material than is GOES because, in contrast to SXR intensity, the intensity of optically thin radio emission decreases with increasing temperature. They concluded that a $0.1-1$ MK source would produce a stronger response in the centimeter–millimeter domain than would (hotter) plasma observed by GOES. Chertok *et al.* (1995) reached similar conclusions in a joint analysis of GOES observations and radio bursts measured at $3-80$ GHz.

Following these authors, we consider that, in addition to the coronal SXR source SG observed by GOES, there is an additional, cooler, radio-emitting coronal source SR at plasma temperature $T_r < T_g$ with emission measure EM_r. For simplicity, we assume that SR has the same surface area A, normal to the line of sight, as SG and that it is located directly below it. Then, by modifying Equation (2) for this geometry, we obtain the brightness temperature of the combined source:

$$T_{br}(t,\nu) = T_g(t)\bigl(1-e^{-\tau_g(t,\nu)}\bigr) + T_r(t)\bigl(e^{-\tau_g(t,\nu)} - e^{-(\tau_g(t,\nu)+\tau_r(t,\nu))}\bigr), \tag{6}$$

which, together with Equations (3) and (4), yields the expected coronal flux density $S^{cor}_{exp}(t,\nu)$ for the combined source. If both SG and SR are optically thin ($\tau_g \ll 1$ and $\tau_r \ll 1$), Equations (6) and (4) reduce to

$$\begin{aligned} &T_{br}(t,\nu) \approx T_g(t)\tau_g(t,\nu) + T_r(t)\tau_r(t,\nu), \\ &\Longrightarrow\ S^{cor}_{exp}(t,\nu) = \frac{2k_B}{c^2R^2}\left(\frac{K_g(t,\nu)EM_g(t)}{T_g(t)^{1/2}} + \frac{K_r(t,\nu)EM_r(t)}{T_r(t)^{1/2}}\right). \end{aligned} \tag{7}$$

T_g and EM_g are fixed by the GOES observations; the free parameters T_r, EM_r, and A can then be determined by fitting the model $S^{cor}_{exp}(t,\nu)$ (Equation (4)) to the whole range of observed flux densities $S_{obs}(t,\nu)$.

As a first approach, we varied the free parameters of the cool plasma SR until a reasonable fit to the data was achieved by eye. We constrained the fit parameters by requiring that the cool plasma does not contribute significantly to the GOES SXR flux in either energy channel, since that would conflict with the previously derived T_g and EM_g which were held fixed in the model. In this way, we determined the best-fit parameters of the cool plasma for one-minute time intervals from 12:33 to 12:57 UT; during B1 and B2, when synchrotron radiation adds to the thermal bremsstrahlung up to at least 35 GHz, T_r and EM_r were adjusted to account for the observed $210-230$ GHz flux densities.

Figure 5 illustrates the reasonable agreement of the two-temperature coronal source model (solid black line) with the observations (asterisks) for $A = 6.7 \times 10^{18}$ cm^2, which corresponds to a diameter of 40″ for a uniform circular source. T_r and EM_r, respectively, begin at 1 MK and 7×10^{48} cm^{-3} in the first time interval (12:33 – 12:34 UT), reach a maximum of 2.6 MK and 3.5×10^{49} cm^{-3} in the 12:43 – 12:45 UT time period (decay of burst B2), and decrease to 1.3 MK and 1.2×10^{49} cm^{-3} at the end of the analyzed period (12:57 UT). This 40″, 1 – 3 MK plasma, with $EM_r \approx EM_g$, contributes < 1% of the GOES SXR flux in either channel, as we required. Larger sources, up to 60″, yield similar values of T_r and EM_r, but smaller sources yield higher temperatures and emission measures, resulting in a non-negligible predicted contribution of the SR source to the GOES flux in at least the low-energy channel. A 30″ source, for example, with best-fit $T_r \approx 4.6$ MK and $EM_r \approx 5 \times 10^{49}$ cm$^{-3} \approx 2 \times EM_g$, would contribute only ~0.5% of the SXR flux at 0.5 – 4 Å but a significant ~16% at 1 – 8 Å, which violates our initial constraints. Thus, we adopt a coronal radio source diameter $D_r = 40'' \pm 10''$.

While this simple two-temperature model prediction of the radio spectrum yields reasonable agreement with the observations, the RHESSI data, when available, provide distinct evidence that the true temperature distribution of the thermal flare plasma is more complex. For the specific time interval of 12:47 – 12:49 UT (see Section 2.3), the RHESSI ~3 – 33 keV spectrum is well characterized by two hot, isothermal components with $T_1 \approx 24$ MK, $EM_1 \approx 0.24 \times 10^{49}$ cm^{-3}, $T_2 \approx 16$ MK, and $EM_2 \approx 1.7 \times 10^{49}$ cm^{-3}. Because of the broad-band GOES response, the ~24 MK source will contribute non-negligibly to the measured SXR flux, which will therefore affect the radio spectrum predicted from the GOES measurements. In particular, the two thermal components observed by RHESSI contribute a combined 85% and 98% of the GOES SXR flux observed in the low- and high-energy channels, respectively. The remaining GOES fluxes are consistent with an additional "warm" plasma, with $T_3 \approx 6.7$ MK and $EM_3 \approx 1.0 \times 10^{49}$ cm^{-3}. However, the red dashed curve in Figure 5 shows that the radio emission predicted from this three-temperature (T_1, T_2, T_3) plasma is still below the observed spectrum. Thus, a fourth, cool plasma, with $T_4 \approx 2.0$ MK and $EM_4 \approx 3.0 \times 10^{49}$ cm^{-3}, is still needed to account for the observed radio flux densities (with a diameter $D_r = 40'' \pm 10''$, as before); this cool plasma contributes negligibly (< 1%) to the GOES fluxes, as required.

Figure 5 shows that the four-temperature model prediction (solid red line in the 12:47 – 12:49 UT panel) is nearly identical to that of our previous, simple, two-temperature model (solid black line). Thus, in the absence of RHESSI data, a simple two-temperature scenario nevertheless constitutes a viable working model, representative of the radio spectrum from a multitemperature coronal plasma. However, note that the derived T and EM, and hence the predicted radio contribution, of the cool source do vary between the two models, so additional measurements within the temperature distribution are desirable for optimal accuracy.

TRACE observations at 195 Å and 1600 Å were made between 12:28 and 13:00 UT, with a few observations at 284 Å also made after 12:41 UT. Images at 195 Å (and at 284 Å) show the geometry of the flaring coronal plasma, while the 1600 Å images display the chromospheric footpoints/ribbons, as illustrated in Figure 6 at four instants during the event. The bright 1600 Å kernels are marked r4, r5, r7, and r8, as in Figure 4 of Luoni *et al.* (2007); similar kernels are also seen in Hα (see Figure 3, right panel, of Luoni *et al.*, 2007). The image at 12:30:45 UT, characteristic of the impulsive phase, shows only faint 195 Å emission directly atop the 1600 Å kernels. During the gradual phase, Figure 6 shows two bright 195 Å loop systems, marked L1 and L2, which connect the 1600 Å kernels. The 195 Å intensity, integrated over the area covered by these loops, increases from the beginning of the gradual phase up to at least 12:40 UT; after that time, the images are contaminated by numerous spikes that prevented a reliable measure of the light curve.

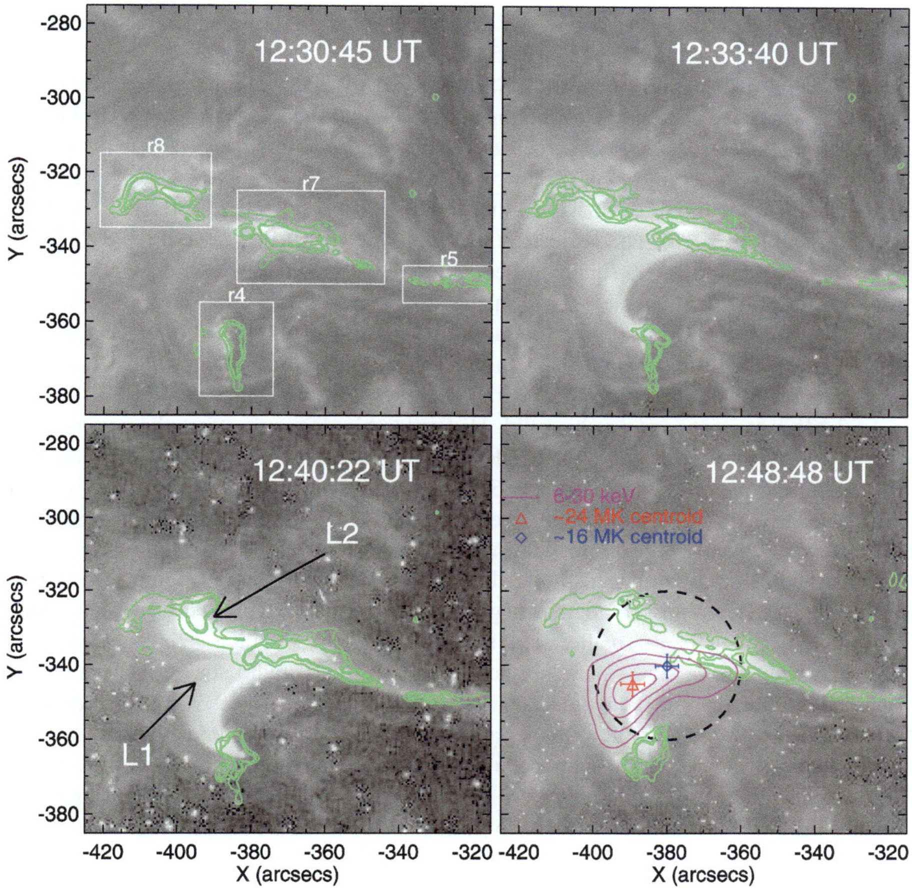

Figure 6 TRACE images at 195 Å (grayscale) and 1600 Å (green contours) at four instants during the 27 October 2003 flare; the TRACE alignment was corrected by $[+6'', -7.5'']$. The 12:30:45 UT image (top left) occurs during the impulsive phase; r4, r5, r7, and r8 refer to the 1600 Å flare ribbons/kernels as identified in Figure 4 of Luoni *et al.* (2007), and the gray boxes indicate the areas over which the kernel intensities have been computed (see Section 3.2.3). The three subsequent images, which exhibit two loop systems marked L1 (south) and L2 (north), occur during the gradual phase. RHESSI 6 – 30 keV image contours (30%, 50%, 70%, 90%) during 12:47 – 12:49 UT are overlaid on the final image (bottom right), along with the derived centroid positions of the two thermal plasmas inferred from the simultaneous, spatially integrated RHESSI spectrum (see Section 2.3). For comparison, the dashed black circle of $40''$ diameter represents the model coronal radio source, arbitrarily centered on the centroid of the ~ 16 MK plasma.

The 195 Å passband is sensitive to spectral line emission from both Fe XII (peak contribution at $\sim 1.1-1.7$ MK, *cf.* Mazzotta *et al.*, 1998) and Fe XXIV ($\sim 13-35$ MK). Thus, the 195 Å images will contain some contribution from the hot ~ 24 and ~ 16 MK plasmas seen by RHESSI. However, per unit emission measure, the sensitivity is $\gtrsim 10$ times stronger at the low (Fe XII) temperatures (*cf.* Handy *et al.*, 1999), and thus the total 195 Å emission in L1 and L2 should be vastly dominated by cool plasma. This is confirmed by the similar loop structures observed at 284 Å (Fe XV, $\sim 1.8-2.6$ MK), and suggests that L1 and L2 are indeed largely representative of the derived $1-3$ MK source considered in our working model, and therefore of the coronal thermal radio-emitting region. The increase of the

195 Å intensity up to at least 12:40 UT is thus consistent with our model, which predicts an increase of T_r and EM_r until 12:43 – 12:45 UT.

The spatial distribution of the RHESSI-observed 6 – 30 keV emission during 12:47 – 12:49 UT, where we applied the four-temperature thermal model, is also shown in Figure 6 (bottom right panel). The X-ray source has a similar shape as and projects onto the L1 loop system. The projected centroid positions of the ∼ 24 and ∼ 16 MK plasmas (see Section 2.3) are located near the top of the brightest loops within L1, as expected for the hottest plasma (*cf.* Caspi and Lin, 2010). The projected (and deconvolved) surface area of the RHESSI X-ray source, ∼ 300 and ∼ 600 square arcsec within the 50% and 30% contours, respectively, is significantly smaller than the ∼ 1250 square arcsec area of the radio source obtained from the model. Since the X-ray source produces only $\lesssim 20\%$ of the radio flux (see Figure 5), its size is not likely representative of the primary radio source, suggesting that the 4-T model assumption of identical sizes for the hot (>16 MK), warm (∼ 6 – 7 MK), and cool (∼ 1 – 3 MK) sources may not be valid. However, this does not affect the results in the optically thin regime, which are independent of source size (*cf.* Equation (7)). Although the 195 Å-emitting region (bounded by a ∼ 60″ × 50″ rectangle) is very inhomogeneous, with a diffuse edge that is difficult to define precisely, we estimate that the projected surface area is between ∼ 800 and ∼ 1150 square arcsec. While the areas of the X-ray and EUV sources are estimated using dissimilar methods, the difference in their areas is significant, and it is noteworthy that the X-ray source has negligible emission overlying the L2 system, which can largely account for its smaller size. However, the 195 Å source area is in reasonable agreement with the ∼ 1250 square arcsec area of the model radio source, consistent with the model finding that the radio emission is primarily from the ∼ 1 – 3 MK plasma observed by TRACE. Unfortunately, the present data do not provide enough constraints to consider a more sophisticated model which accounts for inhomogeneities within the EUV- and X-ray-emitting regions. Observations from UV/EUV and X-ray imagers and spectrometers onboard the *Solar Dynamics Observatory* and *Hinode* satellites may provide the necessary information to develop such complex and realistic models in the future.

In summary, the 1 – 230 GHz radio spectrum observed during the gradual phase is consistent with thermal bremsstrahlung radiated by a coronal source of ∼ 40″ equivalent diameter, filled with hot, ∼ 7 – 24 MK plasma observed by GOES and RHESSI and with cooler, ∼ 1 – 3 MK plasma observed by TRACE. The similarity between the time evolution of the SXR and radio emissions, particularly in the 200 – 400 GHz range (see Figure 1), implies that these emissions likely share a common origin. The increase of EM_g and EM_r from the beginning to the maximum of the gradual phase suggests that this common origin could be chromospheric evaporation.

Since the radio corona is optically thin above ∼ 8 GHz, the observed flux density $S_{\mathrm{obs}}(\nu)$ at frequency $\nu > 8$ GHz can be written as the sum of the emission from the corona, $S_{\mathrm{obs}}^{\mathrm{cor}}(\nu)$, and from the chromosphere, $S_{\mathrm{obs}}^{\mathrm{chr}}(\nu)$. Similarly, the flux density S_0 before the flare is thus $S_0^{\mathrm{cor}} + S_0^{\mathrm{chr}}$. Then, the excess flux density due to the flare, S_{F}, is given by

$$S_{\mathrm{F}}(\nu) = S_{\mathrm{obs}}^{\mathrm{cor}}(\nu) - S_0^{\mathrm{cor}}(\nu) + S_{\mathrm{obs}}^{\mathrm{chr}}(\nu) - S_0^{\mathrm{chr}}(\nu). \tag{8}$$

The observation of optically thin bremstrahlung during the gradual phase leads to the following statements. *i*) The excess flux seen at 345 GHz and the enhanced 1600 Å and Hα emissions (see Section 3.2.3) provide evidence that the chromosphere responded markedly to flare energy deposition. *ii*) However, Equation (8) implies that, since the observed flux densities are well modeled entirely by coronal sources (per above), $S_{\mathrm{obs}}^{\mathrm{chr}}(\nu) \approx S_0^{\mathrm{chr}}(\nu)$; that is, the chromosphere did not contribute substantially to the time-extended radio emission over a large frequency range $\Delta\nu \approx 10-230$ GHz. This range may vary from flare to flare, and even during a single flare, as was the case during the 12 April 2001 event (Lüthi, Magun, and Miller, 2004) for which $\Delta\nu$ varied from $\sim 10-230$ GHz at the maximum of the thermal phase to $\sim 10-100$ GHz in the late decay. Variations of $\Delta\nu$ during the gradual phase likely reflect the dynamics of the chromospheric response to flare energy deposition.

3.2.2. The Bursts B1 and B2

Although a detailed study of bursts B1 and B2 is not of interest here, one must identify the origin and the nature of these bursts to understand the relationship between the radio, UV, and Hα emission discussed in Section 3.2.3.

B1 and the smaller burst just prior to it (see Figure 1) are both observed between 5 and 35 GHz in time coincidence with type III electron beams recorded by the *Nançay Radioheliograph* (NRH). NRH imaging in the metric domain shows that these type IIIs arise from AR 10484 at N07 W46. Furthermore, high-cadence Hα images, obtained with the *Hα Solar Telescope for Argentina* (HASTA; Bagalá *et al.*, 1999; Fernandez Borda *et al.*, 2002), also show simultaneous brightenings in the same active region. We thus conclude that B1 is not part of radio emission arising from the 27 October 2003 flare under study.

In contrast, B2, which is also observed in the $5-35$ GHz range, does arise from AR 10486. Figure 7 shows that B2 is coincident with an impulsive increase of the four 1600 Å kernels marked r4, r5, r7, and r8 in Figure 6. The time derivative of the GOES $1-8$ Å SXR flux, which may be considered as a proxy for nonthermal hard X-ray emission (*cf.* the "Neupert effect;" Neupert, 1968; Dennis and Zarro, 1993), also exhibits a bump coincident with B2 (see Figure 7). Both nonthermal X-rays and the 1600 Å UV continuum are widely believed to result from direct particle injection into the chromosphere (Fletcher and Hudson, 2001; Coyner and Alexander, 2009); thus, B2 is the signature of a new episode of particle acceleration in the AR 10486 magnetic field. After subtraction of the thermal radio spectrum expected from the model described in Section 3.2.1, the remaining B2 spectrum is a typical gyrosynchrotron spectrum with a maximum at $\sim$10 GHz and a power law spectral index of approximately -2 above that frequency. For such an index, the expected flux density above 200 GHz is < 1 sfu at the maximum of B2. This is consistent with the lack of detection of synchrotron emission from B2 at these high frequencies. However, the > 200 GHz emission shows a significant increase from the beginning of B2 until slightly after its maximum. This indicates that the new energy release associated with B2 is not only a signature of a new episode of particle acceleration, but that it also produced a significant increase of the thermal radio emission measured during the gradual phase.

3.2.3. The Flux Density Increase Between 210–230 GHz and 345 GHz

Luoni *et al.* (2007) noted that the 195 Å-emitting loops shown in Figure 6, with footpoint regions r4, r5, r7, and r8, extended above the inversion lines separating an elongated region of negative polarity from positive polarities northward and southward. Similar kernels, which we label H4, H5, H7, and H8, are also seen in the Hα line center. Luoni *et al.* suggested that the flare energy release resulted from magnetic reconnection at a coronal null located at a height of 3.1 Mm, slightly to the east of r7 (see their Figure 6). In order to investigate the chromospheric response to this energy release, we analyzed the time evolution of the 1600 Å and Hα kernels. For this, we computed the mean intensities per pixel over the boxes shown in Figure 6. Figure 7 shows the results for both the 1600 Å kernels (black histograms) and the Hα kernels (red histograms). The bottom panel shows the excess flux at 345 GHz with respect to the 210 GHz flux ($\delta S_{345} \equiv S_{345} - S_{210}$). Figure 7 suggests the following.

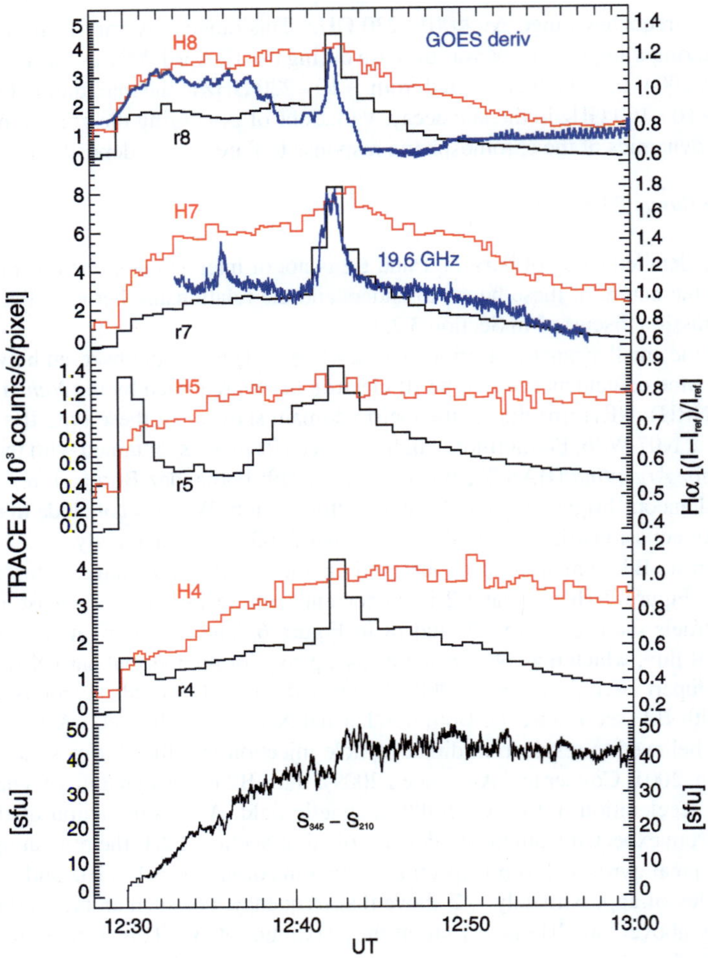

Figure 7 From bottom to top: time evolution of the excess flux density at 345 GHz with respect to that of the flux density at 210 GHz; and time evolution of the four 1600 Å kernels marked r4, r5, r7, and r8 on Figure 6 (black curves) and of the corresponding Hα kernels H4, H5, H7, and H8 (red curves). At 1600 Å, the pre-flare intensity of the kernel has been subtracted. The intensity excess of a given Hα kernel $I - I_{\mathrm{ref}}$ is given with respect to that of a quiet area I_{ref}. The blue curves overplotted on the time histories of kernels r7 (H7) and r8 (H8) are, respectively, the time evolutions of the 19.6 GHz radio emission and of the time derivative of the GOES 1–8 Å SXR flux.

- Kernels r4 and r5 show similar time behaviors; they both exhibit a short peak corresponding to the radio impulsive phase, another peak corresponding to the radio burst B2, and similar time decays. H4 and H5 also evolve similarly, but unlike for r4 and r5, the impulsive phase and B2 are not seen in their time profiles. We emphasize that the H4 and H5 intensities remain roughly constant from ∼ 12:42 UT until the end of the analyzed period.
- There is a close similarity between the r7 and r8 light curves. The marked difference compared to r4 and r5 is that the impulsive phase is not clearly observed, and the decay times are slightly shorter. H7 and H8 evolve together, but unlike H4 and H5, they decay after ∼ 12:43 UT. Note that the region in which kernel r7 (H7) is located is highly dynamic. The elongated shape of this kernel is indicative of the photospheric trace of the fan as-

sociated with the null point found in Luoni *et al.* (2007). Magnetic reconnection at the null point would inject energy into the chromosphere at the footpoints of loops anchored in r7, but at the same time, would induce a disturbance at the section of a long, curved filament lying along the inversion line located in its close vicinity (see, *e.g.*, Figure 2 in Mandrini *et al.*, 2006). The temporal evolution of the pixels included in the box shown in Figure 6 may thus reflect not only the chromospheric response at the loop footpoints, but also the nearby filament activity. Nevertheless, we computed the r7 intensity for different sub-boxes within the box shown in Figure 6 and found that the shapes of the time profiles are not substantially different from the one shown in Figure 7.

These observational findings therefore support the following statements.

- After reconnection and energy release at the magnetic null point, the two sets of 195 Å loops L1 and L2 have footpoints at r4 (H4) and r5 (H5) and at r7 (H7) and r8 (H8), respectively (see Figure 6). Taking into account the magnetic field evolution (*e.g.*, emergence of a new bipole into the pre-existing field configuration), Luoni *et al.* (2007) proposed that magnetic reconnection at the null point would proceed in such a way that field lines connecting r4 to r7 and r5 to r8 would correspond to L1 and L2, contrary to what our analysis of the temporal evolution of the flare kernels show. Note that it is very difficult to clearly separate these two loops in the horizontal "Y-shape" of the TRACE 195 Å brightening. However, computation of the magnetic field topology is only an indication of the magnetic field connectivity and does not tell in which direction the reconnection process proceeds; this is inferred by analyzing the temporal evolution of the associated phenomena. Furthermore, magnetic reconnection may proceed in one direction and, afterwards, in the reverse direction in consecutive events, as was shown by Goff *et al.* (2007). Therefore, we conclude that the magnetic field topology computed by Luoni *et al.* (2007) agrees with the connectivities inferred from our analysis of TRACE and Hα data, although the reconnection process may have proceeded in the reverse sense compared to that proposed by these authors.
- As stated in Section 3.2.2, it is well documented that the 1600 Å peaks, associated with the impulsive phase and B2, constitute signatures of the interaction of flare-accelerated particles with the chromospheric footpoints of loops connected to the acceleration/injection region. Our results thus indicate that, during the impulsive phase, particles are preferentially injected into L1, while during B2, they are injected into both L1 and L2.
- The time profile of δS_{345} closely mimics the H4 and H5 light curves. This strongly suggests that δS_{345} is optically thick thermal bremsstrahlung from the footpoints of L1. Assuming that an equal amount of 345 GHz emission arises from two sources of diameter d (expressed in arcseconds), the projections of which correspond to H4 and H5, we find that δS_{345} leads to a temperature increase of up to $8 \times 10^3 \times (20''/d)^2$ K. For $d = 10''$ to $20''$, *i.e.*, comparable to H4 and H5 (see Figure 6), this yields maximum temperature increases of 3.2×10^4 and 8×10^3 K, respectively. These temperatures are consistent with a chromospheric origin for δS_{345} and with its similarity to the Hα time profiles, the latter being the signature, when seen in emission as during our event, of $\sim 10^4$ K plasma.

The latter point above and the discussion in Section 3.2.1 suggest that, during the gradual phase of the flare, the 230 GHz emission arises from a $1-3$ MK source of $40''$ diameter centered near the top of L1, while the 345 GHz emission arises from both the same coronal source and from two chromospheric sources of $\sim 20''$ diameter that project onto r4 (H4) and r5 (H5). The convolution of this configuration of sources with the KOSMA beams (HPBWs of $117'' \pm 7''$ at 230 GHz and $88'' \pm 5''$ at 345 GHz, *cf.* Lüthi, Magun, and Miller, 2004) yields source sizes of $124''$ at 230 GHz and $107'' \times 96''$ at 345 GHz. If we then assume that the emission comes from unresolved single sources, after deconvolution we find source sizes of 40″ at 230 GHz, as expected, and 37″ × 61″ at 345 GHz, substantially larger than the coronal 230 GHz source. This increase of source size with frequency was previously reported by Lüthi, Magun, and Miller (2004), who, during the late decay of the gradual phase of the 12 April 2001 flare, measured sizes of 42″ ± 20″ and 70″ ± 6″ at 230 and 345 GHz, respectively. A similar explanation as the one proposed here may qualitatively apply to the 12 April 2001 event. Indeed, Figure 4 of Lüthi, Magun, and Miller (2004) indicates that the 345 GHz-emitting source encompasses both of the two chromospheric footpoints seen in hard X-rays, while the 230 GHz source encompasses only the northern one.

4. Conclusions

The 27 October 2003 M6.7 flare at ∼12:30 UT is one of the few events for which a time-extended phase (tens of minutes) following a short (few minutes) impulsive phase has been well observed at millimeter–submillimeter wavelengths. The radio data measured from a few tens of kHz up to 405 GHz shows that the event was confined in the chromosphere and low corona, confirming earlier conclusions from the study of the active region magnetic field topology. Consistent with the confined nature of this event, the 1 – 345 GHz emission detected during the impulsive phase is produced by gyrosynchrotron radiation from high-energy electrons in a dense medium.

The combined analysis of radio total-flux observations, X-ray imaging and spectral data, and UV/EUV and Hα spatially resolved observations has allowed us to determine the nature and origin of the radio emission during the gradual phase of the flare. The main findings can be summarized as follows.

- Except for the small impulsive burst B2, where gyrosynchrotron emission contributes up to 35 GHz, the radio emission is entirely produced by thermal bremsstrahlung.
- Below 230 GHz, the radio emission, which is optically thin above ∼8 GHz, is entirely produced in the corona by hot and cool materials at ∼7 – 16 MK and at ∼1 – 3 MK, respectively. The combination of X-ray data and EUV images at 195 Å and 284 Å indicates that the ∼1 – 3 MK plasma fills two large coronal loop structures, L1 and L2, while the hot X-ray source lies primarily within L1, with the centroid of the ∼16 MK plasma near the top of the brightest portion of the L1 loop system.
- At 345 GHz, in addition to the coronal emission, there is an optically thick component arising from the lower atmosphere. The similarity between the time evolution of this submillimeter excess and that of the Hα footpoints of L1 suggests that the 345 GHz flux density excess is emitted by chromospheric material at a few 10^4 K located in the L1 footpoints.

Importantly, these results show that the chromospheric response to the energy release during the gradual phase produces negligible emission over a substantially extended frequency range, from ∼10 to ∼230 GHz for the studied flare; similar behavior has been observed in other flares as well. The thermal radio emission expected from semiempirical flare models (*cf.* Machado *et al.*, 1980; Mauas, Machado, and Avrett, 1990) is inconsistent with the non-detection of chromospheric radio signatures over such a wide frequency range. Indeed, since the density–temperature structure considered in these models varies quasi-monotonically with height, one would expect the radio flux density to increase with frequency in the spectral range where the corona is transparent. Thus, observations of the radio continuum at centimeter–submillimeter wavelengths and in the far infrared domain can serve as powerful diagnostic tools to investigate both the structure and the dynamics of the low solar atmosphere during flares.

Acknowledgements G.T. acknowledges the FAPESP agency for support during his stay in Brazil (Proc. 2009/15880-0). J.P.R. thanks CNPq (Proc. 305655/2010-8), and G.G.C. is grateful to FAPESP (Proc. 2009/18386-7). C.H.M. and M.L.L. acknowledge financial support from the Argentinean grants UBACyT X127, PIP 2009-100766 (CONICET), and PICT 2007-1790 (ANPCyT). C.H.M. is a member of the Carrera del Investigador Científico (CONICET). M.L.L. is a member of the Carrera del Personal de Apoyo (CONICET). A.C. was supported by NASA grant NNX08AJ18G and NASA contract NAS5-98033. Thanks are due to E. Correia for providing 7 GHz polarimeter data. The authors wish to thank G. Hurford and K.-L. Klein for helpful discussions and comments. We are grateful to the referee, Säm Krucker, for his constructive recommendations.

References

Bagalá, L.G., Bauer, O.H., Fernández Borda, R., Francile, C., Haerendel, G., Rieger, R., Rovira, M.G.: 1999, The new Hα solar telescope at the German–Argentinian solar observatory. In: Wilson, A., *et al.* (eds.) *Magnetic Fields and Solar Processes* **SP-448**, ESA, Noordwijk, 469 – 474.

Caspi, A.: 2010, Super-hot ($T > 30$ MK) thermal plasma in solar flares. Ph.D. thesis, Department of Physics, University of California, Berkeley, CA 94720-7450, USA.

Caspi, A., Lin, R.P.: 2010, RHESSI line and continuum observations of super-hot flare plasma. *Astrophys. J.* **725**, L161 – L166. doi:10.1088/2041-8205/725/2/L161.

Chertok, I.M., Fomichev, V.V., Gorgutsa, R.V., Hildebrandt, J., Krüger, A., Magun, A., Zaitsev, V.V.: 1995, Solar radio bursts with a spectral flattening at millimeter wavelengths. *Solar Phys.* **160**, 181 – 198. doi:10.1007/BF00679104.

Coyner, A.J., Alexander, D.: 2009, Implications of temporal development of localized ultraviolet and hard X-ray emission for large solar flares. *Astrophys. J.* **705**, 554 – 567. doi:10.1088/0004-637X/705/1/554.

Dennis, B.R., Zarro, D.M.: 1993, The Neupert effect – What can it tell us about the impulsive and gradual phases of solar flares? *Solar Phys.* **146**, 177 – 190. doi:10.1007/BF00662178.

Dere, K.P., Landi, E., Mason, H.E., Monsignori Fossi, B.C., Young, P.R.: 1997, CHIANTI – an atomic database for emission lines. *Astron. Astrophys. Suppl.* **125**, 149 – 173. doi:10.1051/aas:1997368.

Dulk, G.A.: 1985, The solar atmosphere, solar magnetism and solar activity. In: McLean, D.J., Labrum, N.R. (eds.) *Solar Radiophysics: Studies of Emission from the Sun at Metre Wavelengths (A87-13851 03-92)*, Cambridge University Press, Cambridge, 19 – 35.

Elgarøy, E.Ø.: 1977, *Solar Noise Storms*, Pergamon, Oxford.

Fernandez Borda, R.A., Mininni, P.D., Mandrini, C.H., Gómez, D.O., Bauer, O.H., Rovira, M.G.: 2002, Automatic solar flare detection using neural network techniques. *Solar Phys.* **206**, 347 – 357.

Fleishman, G.D., Kontar, E.P.: 2010, Sub-THz radiation mechanisms in solar flares. *Astrophys. J.* **709**, L127 – L132. doi:10.1088/2041-8205/709/2/L127.

Fletcher, L., Hudson, H.: 2001, The magnetic structure and generation of EUV flare ribbons. *Solar Phys.* **204**, 69 – 89. doi:10.1023/A:1014275821318.

Giménez de Castro, C.G., Trottet, G., Silva-Valio, A., Krucker, S., Costa, J.E.R., Kaufmann, P., Correia, E., Levato, H.: 2009, Submillimeter and X-ray observations of an X class flare. *Astron. Astrophys.* **507**, 433 – 439. doi:10.1051/0004-6361/200912028.

Ginzburg, V.L., Syrovatskii, S.I.: 1965, Cosmic magnetobremsstrahlung (synchrotron radiation). *Annu. Rev. Astron. Astrophys.* **3**, 297 – 346. doi:10.1146/annurev.aa.03.090165.001501.

Goff, C.P., van Driel-Gesztelyi, L., Démoulin, P., Culhane, J.L., Matthews, S.A., Harra, L.K., Mandrini, C.H., Klein, K.L., Kurokawa, H.: 2007, A multiple flare scenario where the classic long-duration flare was not the source of a CME. *Solar Phys.* **240**, 283 – 299. doi:10.1007/s11207-007-0260-4.

Handy, B.N., Acton, L.W., Kankelborg, C.C., Wolfson, C.J., Akin, D.J., Bruner, M.E., Caravalho, R., Catura, R.C., Chevalier, R., Duncan, D.W., Edwards, C.G., Feinstein, C.N., Freeland, S.L., Friedlaender, F.M., Hoffmann, C.H., Hurlburt, N.E., Jurcevich, B.K., Katz, N.L., Kelly, G.A., Lemen, J.R., Levay, M., Lindgren, R.W., Mathur, D.P., Meyer, S.B., Morrison, S.J., Morrison, M.D., Nightingale, R.W., Pope, T.P., Rehse, R.A., Schrijver, C.J., Shine, R.A., Shing, L., Strong, K.T., Tarbell, T.D., Title, A.M., Torgerson, D.D., Golub, L., Bookbinder, J.A., Caldwell, D., Cheimets, P.N., Davis, W.N., Deluca, E.E., McMullen, R.A., Warren, H.P., Amato, D., Fisher, R., Maldonado, H., Parkinson, C.: 1999, The transition region and coronal explorer. *Solar Phys.* **187**, 229 – 260. doi:10.1023/A:1005166902804.

Hurford, G.J., Schmahl, E.J., Schwartz, R.A., Conway, A.J., Aschwanden, M.J., Csillaghy, A., Dennis, B.R., Johns-Krull, C., Krucker, S., Lin, R.P., McTiernan, J., Metcalf, T.R., Sato, J., Smith, D.M.: 2002, The RHESSI imaging concept. *Solar Phys.* **210**, 61 – 86. doi:10.1023/A:1022436213688.

Kaufmann, P., Raulin, J.P.: 2006, Can microbunch instability on solar flare accelerated electron beams account for bright broadband coherent synchrotron microwaves? *Phys. Plasmas* **13**(7), 070701. doi:10.1063/1.2244526.

Kaufmann, P., Raulin, J., de Castro, C.G.G., Levato, H., Gary, D.E., Costa, J.E.R., Marun, A., Pereyra, P., Silva, A.V.R., Correia, E.: 2004, A new solar burst spectral component emitting only in the terahertz range. *Astrophys. J.* **603**, L121 – L124. doi:10.1086/383186.

Kaufmann, P., Levato, H., Cassiano, M.M., Correia, E., Costa, J.E.R., Giménez de Castro, C.G., Godoy, R., Kingsley, R.K., Kingsley, J.S., Kudaka, A.S., Marcon, R., Martin, R., Marun, A., Melo, A.M., Pereyra, P., Raulin, J., Rose, T., Silva Valio, A., Walber, A., Wallace, P., Yakubovich, A., Zakia, M.B.: 2008, New telescopes for ground-based solar observations at submillimeter and mid-infrared. *Proc. SPIE* **7012**, 70120L. doi:10.1117/12.788889.

Kaufmann, P., Trottet, G., Giménez de Castro, C.G., Raulin, J., Krucker, S., Shih, A.Y., Levato, H.: 2009, Sub-terahertz, microwaves and high energy emissions during the 6 December 2006 flare, at 18:40 UT. *Solar Phys.* **255**, 131 – 142. doi:10.1007/s11207-008-9312-7.

Klein, K.L., Trottet, G., Samwel, S., Malandraki, O.: 2011, Particle acceleration and propagation in strong flares without major solar energetic particle events. *Solar Phys.* **269**, 309 – 333. doi:10.1007/s11207-011-9710-0.

Kundu, M.R., White, S.M., Gopalswamy, N., Lim, J.: 1994, Millimeter, microwave, hard X-ray, and soft X-ray observations of energetic electron populations in solar flares. *Astrophys. J. Suppl.* **90**, 599 – 610. doi:10.1086/191881.

Lin, R.P., Dennis, B.R., Hurford, G.J., Smith, D.M., Zehnder, A., Harvey, P.R., Curtis, D.W., Pankow, D., Turin, P., Bester, M., Csillaghy, A., Lewis, M., Madden, N., van Beek, H.F., Appleby, M., Raudorf, T., McTiernan, J., Ramaty, R., Schmahl, E., Schwartz, R., Krucker, S., Abiad, R., Quinn, T., Berg, P., Hashii, M., Sterling, R., Jackson, R., Pratt, R., Campbell, R.D., Malone, D., Landis, D., Barrington-Leigh, C.P., Slassi-Sennou, S., Cork, C., Clark, D., Amato, D., Orwig, L., Boyle, R., Banks, I.S., Shirey, K., Tolbert, A.K., Zarro, D., Snow, F., Thomsen, K., Henneck, R., McHedlishvili, A., Ming, P., Fivian, M., Jordan, J., Wanner, R., Crubb, J., Preble, J., Matranga, M., Benz, A., Hudson, H., Canfield, R.C., Holman, G.D., Crannell, C., Kosugi, T., Emslie, A.G., Vilmer, N., Brown, J.C., Johns-Krull, C., Aschwanden, M., Metcalf, T., Conway, A.: 2002, The Reuven Ramaty high-energy solar spectroscopic imager (RHESSI). *Solar Phys.* **210**, 3 – 32. doi:10.1023/A:1022428818870.

Luoni, M.L., Mandrini, C.H., Cristiani, G.D., Démoulin, P.: 2007, The magnetic field topology associated with two M flares. *Adv. Space Res.* **39**, 1382 – 1388. doi:10.1016/j.asr.2007.02.005.

Lüthi, T., Lüdi, A., Magun, A.: 2004, Determination of the location and effective angular size of solar flares with a 210 GHz multibeam radiometer. *Astron. Astrophys.* **420**, 361 – 370. doi:10.1051/0004-6361:20035899.

Lüthi, T., Magun, A., Miller, M.: 2004, First observation of a solar X-class flare in the submillimeter range with KOSMA. *Astron. Astrophys.* **415**, 1123 – 1132. doi:10.1051/0004-6361:20034624.

Machado, M.E., Avrett, E.H., Vernazza, J.E., Noyes, R.W.: 1980, Semiempirical models of chromospheric flare regions. *Astrophys. J.* **242**, 336 – 351. doi:10.1086/158467.

Mandrini, C.H., Démoulin, P., Schmieder, B., Deluca, E.E., Pariat, E., Uddin, W.: 2006, Companion event and precursor of the X17 flare on 28 October 2003. *Solar Phys.* **238**, 293 – 312. doi:10.1007/s11207-006-0205-3.

Mauas, P.J.D., Machado, M.E., Avrett, E.H.: 1990, The white-light flare of 1982 June 15 – Models. *Astrophys. J.* **360**, 715 – 726. doi:10.1086/169157.

Mazzotta, P., Mazzitelli, G., Colafrancesco, S., Vittorio, N.: 1998, Ionization balance for optically thin plasmas: Rate coefficients for all atoms and ions of the elements H to NI. *Astron. Astrophys. Suppl.* **133**, 403 – 409. doi:10.1051/aas:1998330.

Melo, A.M., Kaufmann, P., Giménez de Castro, C.G., Raulin, J.P., Levato, H., Marun, A., Giuliani, J.L., Pereyra, P.: 2005, Submillimeter-wave atmospheric transmission at El Leoncito, Argentina Andes. *IEEE Trans. Antennas Propag.* **53**, 1528 – 1534. doi:10.1109/TAP.2005.844435.

Neupert, W.M.: 1968, Comparison of solar X-ray line emission with microwave emission during flares. *Astrophys. J.* **153**, L59 – L64. doi:10.1086/180220.

Pohjolainen, S., Valtaoja, E., Urpo, S.: 1996, Solar microwave burst emission relation to X-ray radiation. *Astron. Astrophys.* **314**, 947 – 956.

Raulin, J.P., Makhmutov, V.S., Kaufmann, P., Pacini, A.A., Lüthi, T., Hudson, H.S., Gary, D.E.: 2004, Analysis of the impulsive phase of a solar flare at submillimeter wavelengths. *Solar Phys.* **223**, 181 – 199. doi:10.1007/s11207-004-1300-y.

Raulin, J., White, S.M., Kundu, M.R., Silva, A.V.R., Shibasaki, K.: 1999, Multiple components in the millimeter emission of a solar flare. *Astrophys. J.* **522**, 547 – 558. doi:10.1086/322974.

Silva, A.V.R., Share, G.H., Murphy, R.J., Costa, J.E.R., de Castro, C.G.G., Raulin, J., Kaufmann, P.: 2007, Evidence that synchrotron emission from nonthermal electrons produces the increasing submillimeter spectral component in solar flares. *Solar Phys.* **245**, 311 – 326. doi:10.1007/s11207-007-9044-0.

Smith, D.M., Lin, R.P., Turin, P., Curtis, D.W., Primbsch, J.H., Campbell, R.D., Abiad, R., Schroeder, P., Cork, C.P., Hull, E.L., Landis, D.A., Madden, N.W., Malone, D., Pehl, R.H., Raudorf, T., Sangsingkeow, P., Boyle, R., Banks, I.S., Shirey, K., Schwartz, R.: 2002, The RHESSI spectrometer. *Solar Phys.* **210**, 33 – 60. doi:10.1023/A:1022400716414.

Thomas, R.J., Crannell, C.J., Starr, R.: 1985, Expressions to determine temperatures and emission measures for solar X-ray events from GOES measurements. *Solar Phys.* **95**, 323 – 329. doi:10.1007/BF00152409.

Trottet, G., Krucker, S., Lüthi, T., Magun, A.: 2008, Radio submillimeter and γ-ray observations of the 2003 October 28 solar flare. *Astrophys. J.* **678**, 509 – 514. doi:10.1086/528787.

Trottet, G., Raulin, J., Kaufmann, P., Siarkowski, M., Klein, K., Gary, D.E.: 2002, First detection of the impulsive and extended phases of a solar radio burst above 200 GHz. *Astron. Astrophys.* **381**, 694 – 702. doi:10.1051/0004-6361:20011556.

White, S.M., Kundu, M.R.: 1992, Solar observations with a millimeter-wavelength array. *Solar Phys.* **141**, 347 – 369. doi:10.1007/BF00155185.

White, S.M., Thomas, R.J., Schwartz, R.A.: 2005, Updated expressions for determining temperatures and emission measures from GOES soft X-ray measurements. *Solar Phys.* **227**, 231 – 248. doi:10.1007/s11207-005-2445-z.

Young, P.R., Landi, E., Thomas, R.J.: 1998, CHIANTI: an atomic database for emission lines. II. Comparison with the SERTS-89 active region spectrum. *Astron. Astrophys.* **329**, 291 – 314.

Solar Phys (2011) 273:363–375
DOI 10.1007/s11207-011-9760-3

Location of Decimetric Pulsations in Solar Flares

Arnold O. Benz · Marina Battaglia · Nicole Vilmer

Received: 30 November 2010 / Accepted: 28 March 2011 / Published online: 11 May 2011

Abstract This work investigates the spatial relation between coronal X-ray sources and coherent radio emissions, both generally thought to be signatures of particle acceleration. Two limb events were selected during which the radio emission was well correlated in time with hard X-rays. The radio emissions were of the type of decimetric pulsations as determined from the spectrogram observed by Phoenix-2 of ETH Zurich. The radio positions were measured from observations with the Nançay Radioheliograph between 236 and 432 MHz and compared to the position of the coronal X-ray source imaged with RHESSI. The radio pulsations originated at least 30 – 240 Mm above the coronal hard X-ray source. The altitude of the radio emission increases generally with lower frequency. The average positions at different frequencies are on a line pointing approximately to the coronal hard X-ray source. Thus, the pulsations cannot be caused by electrons trapped in the flare loops, but are consistent with emission from a current sheet above the coronal source.

1. Introduction

Does the very process of particle acceleration in solar flares directly produce coherent radio emission? Waves such as proposed for current instabilities and stochastic acceler-

Energy Storage and Release through the Solar Activity Cycle – Models Meet Radio Observations
Guest Editors: Christophe Marqué and Alexander Nindos.

A.O. Benz (✉) · M. Battaglia
Institute of Astronomy, ETH Zurich, 8093 Zürich, Switzerland
e-mail: benz@phys.ethz.ch

A.O. Benz
Institute for 4D Technologies, FHNW, 5210 Windisch, Switzerland

M. Battaglia
School of Physics and Astronomy, University of Glasgow, Glasgow G12 8QQ, UK

N. Vilmer
LESIA, Observatoire de Paris, CNRS, UPMC, Université Paris-Diderot, 5 place Jules Janssen, 92195 Meudon Cedex, France

ation may couple into radio waves (Benz and Wentzel, 1981; Karlický and Bárta, 2005; Li and Fleishman, 2009). Alternatively, instabilities produced by non-stochastically accelerated electrons having a non-Maxwellian velocity distribution may be observed by their coherent radio emission within the acceleration site.

In the impulsive phase of flares, very intense radio emissions are often observed. At decimeter wavelengths the peak flux density can reach 10^6 solar flux units (10^{10} Jy, Benz, 2009). Coherent emissions are most intense and to be distinguished from incoherent radio emissions, such as synchrotron or thermal radiation, not studied here (*e.g.* Nindos *et al.*, 2008). There are several types of coherent emission processes. Most intense and very frequent are broadband radiations pulsating at irregular intervals on the time scale of one second. As the emission appears to be a non-thermal process, such decimetric pulsations are generally assumed to be tracers of non-thermal electrons and their acceleration.

Radio 'outbursts' at decimeter wavelength during flares have been detected very early with single frequency instruments (Lehany and Yabsley, 1948). These authors even remarked the coincidence with an ionospheric radio fadeout, caused by solar flare soft X-ray emission. Covington (1951) noticed occasional high circular polarization of bursts at 10.7 cm. De Feiter, Fokker, and Roosen (1959) reported an association of 545 MHz bursts with Hα flares increasing with flare importance. First spectral observations in the decimeter range were reported to consist to 90% of a 'generalized class of fast-drift bursts' (Young *et al.*, 1961; Kundu *et al.*, 1961). Some of these appear to be the decimetric continuation of meterwave type III bursts, but many others had a different nature, occurring in large groups and showing an 'immeasurably' high drift in frequency. Decimetric emission has also been named 'flare associated continuum' (Pick, 1986). Thompson and Maxwell (1962) refer to them as pulsating structures. This notation or simply 'pulsations' has established itself in the literature and will be used here.

The regularity of the pulsations has been noted early. Gotwols (1972) reports a quasi-periodic pulsation over most of the observed band from 600 – 1000 MHz. Remarkably regular pulses at 1.0 s period from 300 – 350 MHz were the basis of the theory of Roberts, Edwin, and Benz (1984) on magnetohydrodynamic (MHD) oscillations in the corona. In their catalogues, Güdel and Benz (1988) and Isliker and Benz (1994) characterize pulsations in the decimeter range between 'almost periodic' and 'irregular' with pulse separations of 0.1 to 1 second. Some complex cases may be the superposition of several pulsations with different periods (Mészárosová, Rybák, and Karlický, 2011). Ultra-rapid pulsations have been reported by Magdalenic *et al.* (2003) and Fleishman, Stepanov, and Yurovsky (1994). Pulsations have better defined upper and lower bounds in frequency than type III bursts and higher drift rates by a factor of 3 on average (Aschwanden and Benz, 1986). Contrary to type III bursts, pulsations are highly circularly polarized, except when occurring near the limb (Aschwanden, 1986, 2006).

The frequency range of pulsations extends from meter to centimeter wavelengths, but their character changes. At meter waves, McLean *et al.* (1971) observed about 50 strikingly regular pulses with periods increasing from 2.5 s to 2.7 s in time. Pulsations above about 300 MHz are less regular. The highest frequency pulsations reported extend beyond 4 GHz (Saint-Hilaire and Benz, 2003; Tan *et al.*, 2010) and consist of irregular pulses.

The emission process of pulsations is unclear. It is often associated with some velocity space instability of non-thermal electrons, such as a loss-cone instability or plasma emission by beams (Benz, 1980; Fleishman, Stepanov, and Yurovsky, 1994). The frequency of such emissions is at the plasma frequency ν_p, the electron gyrofrequency ν_e, the upper hybrid frequency $(\nu_\mathrm{p}^2 + \nu_\mathrm{e}^2)^{1/2}$, or at twice these characteristic frequencies (review by Benz, 2002). For the driver of such a pulsating instability, electrons trapped in flare loops have been evoked initially (Aschwanden and Benz, 1988). More recently, Kliem, Karlický, and Benz (2000) proposed that pulsations originate in large-scale current sheets instable to tearing mode reconnection leading to magnetic islands. Their coalescence into a continuously growing plasmoid causes quasi-periodic acceleration. Fleishman, Bastian, and Gary (2008) find observational evidence for this hypothesis. It is also supported by the observed relation to plasmoid ejection observed in X-rays (Khan *et al.*, 2002; Bárta, Karlický, and Žemlička, 2008; Aurass, Landini, and Poletto, 2009). However, Karlický, Zlobec, and Mészárosová (2010) conclude that MHD oscillations are more likely in the case of observed sub-second periods. As possible signatures of coronal MHD oscillations, decimetric pulsations have received recent attention (Nakariakov *et al.*, 2010). An alternative origin for short pulses is a periodic self-organizing system of loss-cone instability and particle escape described by Lotka–Volterra coupled equations (Aschwanden and Benz, 1988).

Bremsstrahlung of non-thermal flare electrons is readily observed in hard X-rays (HXR) and is therefore also a signature of particle acceleration. HXR observed in coronal sources (review by Krucker *et al.*, 2008) are generally assumed to originate close to the acceleration site. Thus, originating both from super-thermal electrons, the question of the relation of decimetric emissions to HXR emission is immediate. Radio sources at 450 MHz and lower are often observed far from the coronal source. This is well known for type III bursts (*e.g.* Vilmer *et al.*, 2002). Battaglia and Benz (2009) reported large spatial separations between decimetric spike emissions and the associated, but not time-correlated, HXR coronal source. On the other hand, Saint-Hilaire and Benz (2003) observed that pulsations occurring nearly simultaneously with the HXR peaks are located close to the HXR source ($\approx 10''$). Their flare was at 0.8 solar radii from the center of the disk, but projection effects cannot be excluded. Later in the flare, pulsations became stronger, but drifted to lower frequency and occurred far from the HXR source.

Decimetric pulsations are well associated with HXR. In a survey between 150 MHz and 4000 MHz, Benz *et al.* (2005) find 160 pulsations in 201 HXR flares observed by the *Ramaty High Energy Solar Spectroscopic Imager* (RHESSI; Lin *et al.*, 2002). This percentage of 80% is higher than for other decimetric emissions such as spikes (14%) and fine structures like intermediate drift bursts and parallel drifting bands (1.5%). One third of the pulsations are correlated in time with HXR flux in some detail (Dabrowski and Benz, 2009).

Here we compare for the first time imaging observations of pulsations and HXR events that have previously been found to correlate in time. We report on the location of decimetric pulsations relative to the coronal HXR source in two well observed limb events. The ultimate question is: What can decimetric pulsations tell us about flare energy release?

2. Observations, Selection, and Data Analysis

Solar radio burst types are classified by their characteristics in spectrum and time. Data from the Phoenix-2 spectrometer (Messmer, Benz, and Monstein, 1999) were used. It operated near Bleien, Switzerland (8°6′44″ E, 47°20′26″ N), during the time of the selected events. Its broad spectrograms were essential to identify pulsations. A parabolic dish with a diameter of 7 m allows surveying the full Sun at frequencies from 100 MHz to 4 GHz from sunrise to sunset. The 4000 measurements available per second were distributed into 200 channels, yielding a sampling time of 0.1 seconds in each channel. The data were calibrated and cleaned from terrestrial interference using standard routines.

The Nançay Radioheliograph (Kerdraon and Delouis, 1997) images the radio emission and was used to locate the position of pulsations. It observes the Sun daily for 7.5 hours centered around 12 UT. The interferometer is operated at discrete frequencies, and we use here data at 237, 327, 410.5, and 432 MHz at a time resolution of 0.15 seconds. For the June event selected, the half-power beam width (major axis of lobe) is 77″, 58″, 44″, and 42″, respectively. For the selected event in December, it is 192″, 144″, 115″, and 109″, respectively. Comparisons with the VLA have shown agreement on the order of 20″ for a typical day (Benz *et al.*, 2005).

The RHESSI satellite is used here for imaging thermal and non-thermal X-rays. It was launched on 5 February 2002, and observes X-rays in the range from 3 keV to 17 MeV with an energy resolution of about 1 keV. Spectral resolution allows differentiating between thermal and non-thermal emissions and selecting the energy range of non-thermal photons. RHESSI rotates to modulate the incoming X-ray flux detected behind the shadowing grids. This allows reconstructing images with high spatial resolution (Hurford *et al.*, 2002). Combined with the high energy resolution, thermal and non-thermal sources can be imaged and identified in different energy ranges (*e.g.* Emslie *et al.*, 2003; Battaglia and Benz, 2006).

SOHO/EIT and GOES/SXI data were used for complementary information about hot plasma. EIT is a normal-incidence, multi-layered mirror instrument (Delaboudinière *et al.*, 1995). It imaged a $7' \times 7'$ area with a pixel size of $2.62'' \times 2.62''$. The 195 Å wavelength band was used, including the emission line of Fe XII with diagnostic capabilities for temperatures in the range of $1.1-1.9\times10^6$ K. GOES/SXI images the Sun in soft X-rays from 2.1 – 20.7 keV (Hill *et al.*, 2005). The spatial resolution is approximately 10″ FWHM. Images consist of 512 × 512 pixels with 5″ pixel size. GOES/SXI produces full-disk solar images at a 1 minute cadence.

2.1. Flare Selection

We selected events from all Phoenix-2 data classified as DCIM and published in Solar and Geophysical Data, starting from RHESSI launch in 2002 until the end of 2007. DCIM stands for ‘decimetric events’ that differ in spectrograms from burst types I to V at meter waves. DCIM have the subclasses ‘pulsations’, ‘spikes’, and ‘continuum’. Using Phoenix-2 quicklook images, these subclasses can be distinguished. We followed the selection of Dabrowski and Benz (2009), who found 870 DCIM events and selected those with simultaneous RHESSI observations. With the requirement of at least 75% temporal coverage by RHESSI, they found 107 decimetric pulsations. From this set of simultaneous events they excluded those that were obviously not correlated (delays of more than 20 seconds in cross-correlation). Dabrowski and Benz (2009) finally list 33 pulsations correlating with hard X-rays in detail. Using their list, we selected events that were in the Nançay time window and observable at the Nançay frequencies. In addition we limited the selection to flares that occurred at radial distances larger than 700″ from disk center to simplify geometrical interpretation.

Finally, the events of 10 June 2003 14:07 – 14:09 UT and 5 December 2006 11:15 – 11:18 UT remained, for which Dabrowski and Benz (2009) report correlation delays relative to X-rays of −2.43 s and 4.07 s, respectively (fitting the cross-correlation coefficient by a Gaussian in delay time). Thus, the events selected here are well time-correlated pulsations at the limb observed by RHESSI.

The data are presented in Figures 1 and 2 in spectrum and time. The X-ray time bins are 4 s, about the RHESSI satellite rotation period.

2.2. Data Analysis

RHESSI X-ray images were constructed for 1 minute time intervals for four energy bands in the thermal and non-thermal range (10 – 14, 14 – 18, 18 – 25, 25 – 50 keV). The image

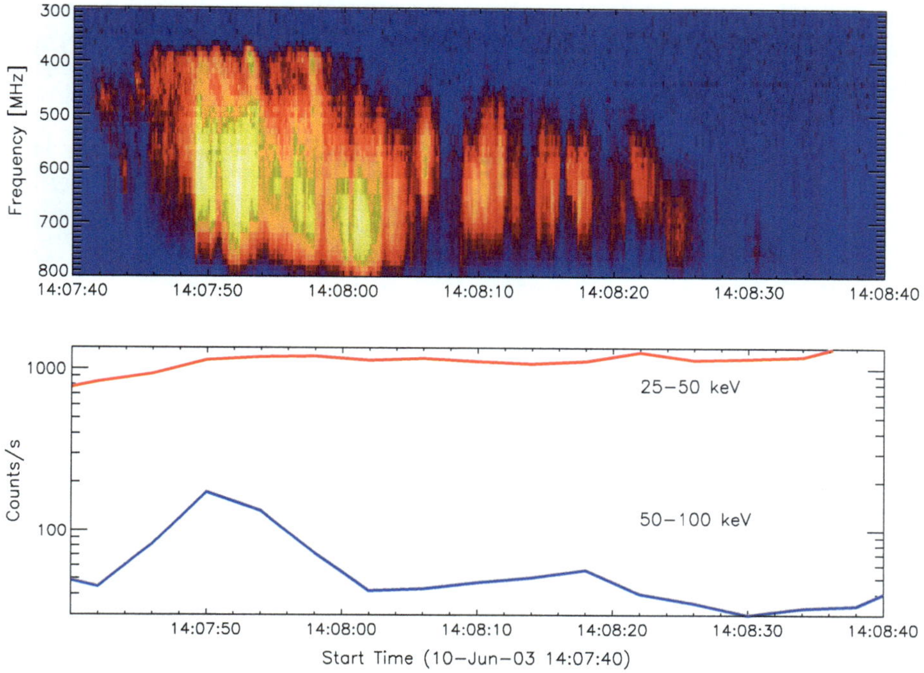

Figure 1 Temporal correlation of radio and X-ray observations of the 10 June 2003 flare. Top: Extract from Phoenix-2 spectrum showing decimetric pulsations. Bottom: Simultaneous time profile of hard X-rays observed by RHESSI between 25 – 50 keV (red) and 50 – 100 keV (blue).

integrations started at 14:07 UT for the 10 June 2003 event and at 11:15 UT for the 5 December 2006 event. Spectral analysis was possible in the photon energy range 6 – 40 keV. The low energies are well fitted by a single temperature thermal component. The higher energies can be fitted by a power-law or a second, much hotter thermal component. Evidence from larger flares suggests the non-thermal interpretation. The spectral and spatial analysis indicates thermal emission by the coronal source for the 10 – 14 keV range and non-thermal footpoint emission in the 25 – 50 keV band, while the 14 – 18 keV and 18 – 25 keV bands might include some non-thermal emission from the coronal source.

In the 10 June 2003 flare an EIT image taken at 14:12:10 UT with 12.56 s integration time was available for complimentary information about the thermal emission at lower temperatures. For the 5 December 2006 flare a GOES/SXI image taken at 11:16:49 UT with 3 s integration was used.

The source parameters in Nançay Radioheliograph images were determined at each frequency and in each time bin, fitting automatically a two-dimensional Gaussian as described by Battaglia and Benz (2009). Peak location (centroid position) in solar coordinates and peak flux are taken from the fit. Solar radio images are generally composed of a practically constant background and a temporary burst component. Only peak fluxes above average flux in the interval are attributed to the pulsating emission and used in the following for plots of individual positions and averages. The standard deviation of the corresponding positions yields an upper limit for the accuracy of the fitted positions (error bars in Figures 3 and 5).

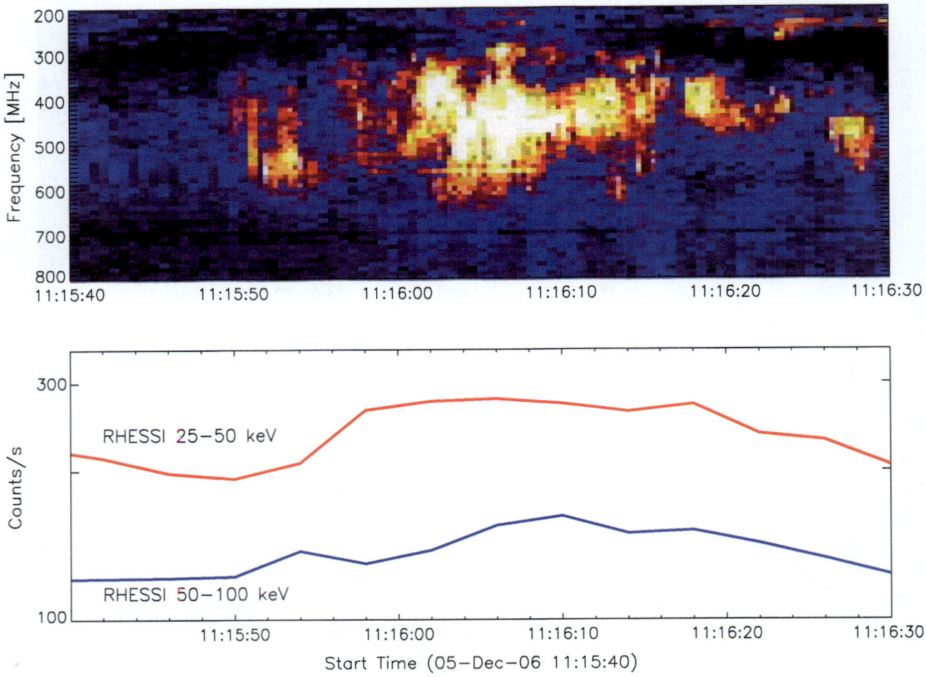

Figure 2 Temporal correlation of radio and X-ray observations of the 5 December 2006 flare. Top: Extract from Phoenix-2 spectrum showing decimetric pulsations. The observed flux density is logarithmically compressed and indicated by colors from blue to white. Bottom: Simultaneous time profile of hard X-rays observed by RHESSI between 25 – 50 keV (red) and 50 – 100 keV (blue).

3. Results

Figure 3 shows centroid positions of enhanced radio flux for the 10 June 2003 event. They differ considerably between the different frequencies. The average altitude clearly increases with decreasing frequency. The 432 and 410 MHz average positions are approximately aligned with the direction to the active region, visible in the EUV image, and preferentially with the hardest X-ray source (blue).

The radio sources of the 10 June 2003 flare originated 250″ – 320″ (or at least 0.3 solar radii) above the X-ray emission of the coronal source. In view of this large separation, the connection between X-ray and radio sources is not clear. Considering the close temporal correlation between X-ray and radio emission, this is a surprising result.

The EIT image indicates a complex active region with possibly two loop systems, where the southern footpoint of the northern system is close to the northern footpoint of the southern system. For this event, Hα observations from Kanzelhöhe were available. The two EIT footpoints in the middle coincide with the Hα peak position within the positional accuracy of the Hα peak, indicated by the error bar. This common footpoint is not visible in X-rays. The total X-ray spectrum can be fitted by a purely thermal electron distribution. The prominent X-ray source at low energies is associated with the southern loop system. It is apparently a coronal source. The harder X-ray source (blue) is a hot spot and possibly also a coronal source related to the northern loop system. The EIT image shows a protrusion from this spot approximately in the direction to the radio sources.

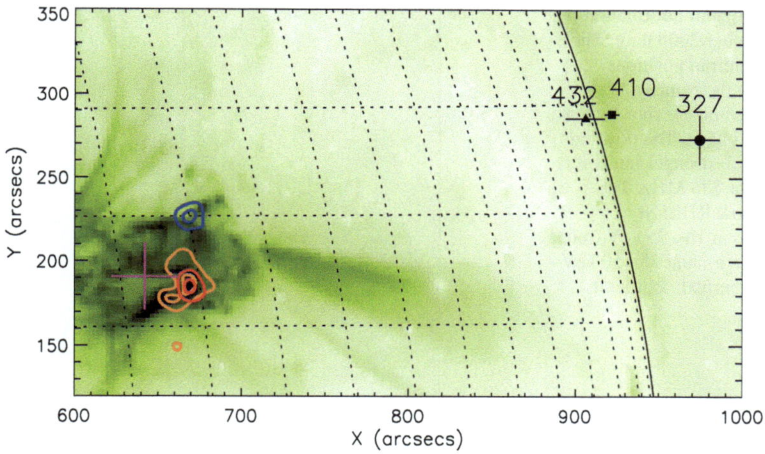

Figure 3 Centroid positions of the radio pulsations in the 10 June 2003 flare measured by the Nançay Radioheliograph overlaid over an EIT image at 195 Å. The time-averaged centroid positions are shown with error bars indicating the full width at half-power of the distribution of the individual measurements. The observed frequencies in MHz are indicated. The contours indicate RHESSI observations at 10 – 14 keV (red, thermal) and 18 – 25 keV (dark yellow, mostly thermal) of the coronal source. The origin of the 25 – 50 keV (blue, non-thermal) emission is at a different location. The cross marks the Hα centroid position and positional accuracy reported by the Kanzelhöhe Observatory (courtesy A. Veronig). The solar coordinate grid and the photospheric limb are indicated for orientation.

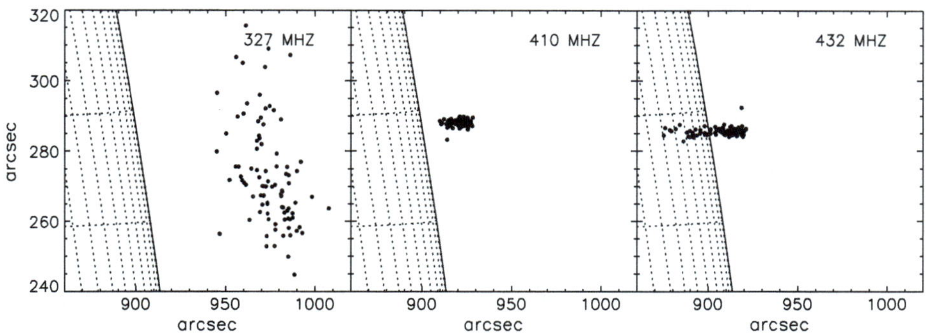

Figure 4 Positions of individual radio pulsation centroids in the 10 June 2003 flare measured by the Nançay Radioheliograph. Only the measurements with flux above the average peak flux are shown. The solar coordinate grid and the photospheric limb are indicated for orientation. The coronal X-ray source(s) (see Figure 3) are far off the maps to the lower left.

Figure 4 presents the individual radio positions at 327, 410, and 432 MHz of time bins with flux above average, thus above background. The centroid positions at 432 and 410 MHz are scattered along a line in x-direction. The flare positions in X-rays are far off the picture to the lower left. The angle of the line from the radio positions to the harder X-ray source (blue in Figure 3) and the x-direction is $28° \pm 2°$. The angle to the softer source (red in Figure 3) is $37° \pm 2°$. The time evolution of this scatter was studied, but no systematic motion was found. The 327 MHz positions are different. They scatter parallel to the limb, but the pulsations are very weak or absent (Figure 1). This lowest frequency may not be relevant for the pulsation as suggested also by other evidence presented later.

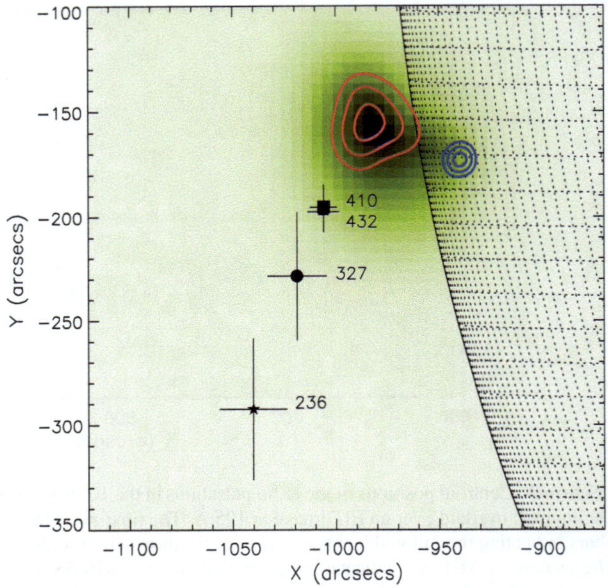

Figure 5 Spatial information on the 5 December 2006 flare. The averaged centroid positions observed by the Nançay Radioheliograph are shown on a GOES/SXI image. The position of the second, stronger interval is displayed for 236 MHz. The contours show RHESSI observations at 18 – 25 keV (red, coronal source), and 25 – 50 keV (blue, non-thermal, footpoint source).

The geometrical situation is even more suggestive in the 5 December 2006 flare. The line of the centroid positions points straight to the coronal X-ray source (red in Figure 5). The highest frequency source, at 432 MHz, is still 40″ (thus 30000 km in projection) above the peak of the soft X-ray emission. Emission at 236 MHz occurs in two intervals, weak at 11:16:02-08 UT and stronger at 11:16:22-27 UT (Figure 2). It originates from different locations. Only the second interval is shown in Figure 5. The first interval is emitted at about twice the distance (see Figure 6). We have investigated the Phoenix-2 spectrogram and polarigram in detail concerning the classification of the 236 MHz emissions. The two intervals were found consistent with being part of the pulsations at higher frequencies. There is no indication in spectrum or polarization for another classification, such as Type I, II, or III. No apparent motion is detected at the other frequencies. The X-ray image shows only one 25 – 50 keV footpoint. The soft X-ray image is consistent with a limb crossing loop and with the hypothesis that the other footpoint is occulted. This suggests that projection effects are minimal.

In the 5 December 2006 flare (Figure 5), the average projected distance s measured from the center of the coronal X-ray source to the radio source can be fitted by the linear relation

$$s \approx -386\,\nu + 1.96 \times 10^5 \text{ [km]}, \tag{1}$$

where ν is the observing frequency in MHz. If Equation (1) is extrapolated, it suggests that the position of the 510±10 MHz emission coincides with the peak of the coronal source ($s = 0$). This frequency is within the range over which pulsations are observed in the spectrogram (Figure 2).

In both events the radio frequency extrapolated to the position of coronal X-ray source (Equation (1)) coincides approximately with the upper limit of the observed pulsation. It can be compared to the plasma frequency derived from the thermal X-ray source density, the square root of the emission measure divided by the volume. For the 10 June 2003 flare, a plasma frequency of 560 MHz results. It is higher if the filling factor is less than unity. However, for the 5 December 2006 event the X-ray determined plasma frequency is 1170 MHz,

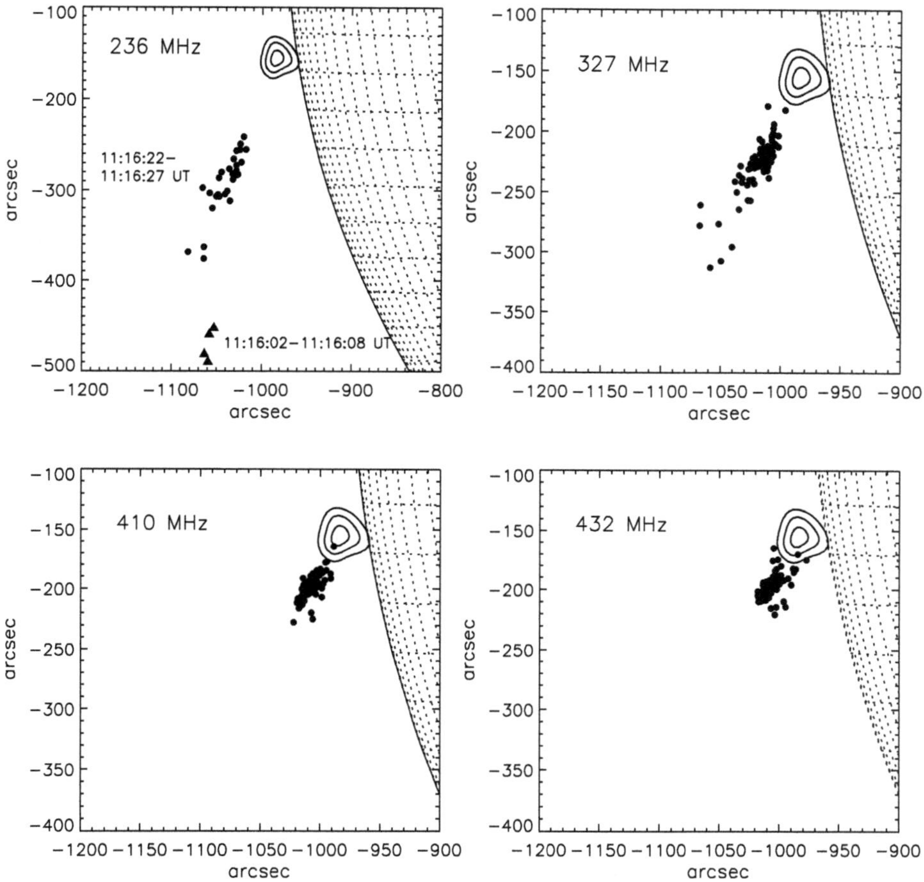

Figure 6 Positions of individual radio pulsation centroids in the 5 December 2006 flare measured by the Nançay Radioheliograph. Only the measurements with flux above the average peak flux are shown. Contours at 50%, 70% and 90% of RHESSI Clean images in the 18 – 25 keV energy band (same as yellow curves in Figure 5) indicate the position and extent of the coronal X-ray source. Note the different scale in the 236 MHz picture.

well above the upper limits of the observed radiation. Thus, the radio emission could not propagate and its origin cannot be in the coronal X-ray source.

A similar extrapolation can be made for the 10 June 2003 flare, although the observed radio sources are much farther away from the X-ray sources. The result is the same; extrapolating the 410 MHz and 432 MHz positions, the location of the highest observed frequency at which pulsations are observed, 830 MHz, would be in or near the coronal X-ray source (blue in Figure 3). The scaling factor in Equation (1) is a factor of 5 larger for the 10 June 2003 event.

Figure 6 shows that the individual centroid positions are scattered along the same line for all frequencies. Again, the lower the frequency, the higher in the corona the radio emission originates. The 236 MHz emission needs a special discussion. During the first interval, 11:16:02 – 08 UT, the radio emission was observed at a projected distance of 360″ from the coronal X-ray source and shifted to the right of the line from the coronal X-ray sources to the other radio sources. The radio emission during the second time interval, 11:16:22 – 27 UT, was much stronger, originated closer to the X-ray source and straight on the above line. The centroid positions at 236 MHz do not scatter excessively and appear to be related to the pulsations at higher frequency.

4. Discussion

The time-correlating radio and X-ray sources are clearly separated in space. The average radio positions are at higher altitude, increasing to lower frequency. Nevertheless, a remarkable relation between the two emissions was found. In the event of 5 December 2006 (Figure 5), the average positions at different frequencies are on a line pointing straight toward the coronal X-ray source. They are at larger apparent distance in the 10 June 2003 event, and point approximately to the X-ray source at higher X-ray energies (Figure 3).

Even more surprising, the scatter of individual centroid positions is not random in both flares. The individual positions are preferentially aligned out on the above line in the direction to the coronal X-ray flare site in the 5 December 2006 (Figure 6). For the 10 June 2003 event, the alignment is at an angle of some 30° from this line (Figure 4).

Scattering of a source position at a given frequency may have several reasons. First, we discuss the possibility of ionospheric scattering. Variable refraction would move the positions of nearby frequencies, measured simultaneously, in the same direction. Thus, the excursions in both x-direction and y-direction would be similar, along a line, and correlated for adjacent frequencies. No correlation between the positions at different frequencies is observed (Figure 7). A plot of the x-positions and y-positions *vs.* time, however, reveals occasional correlation of the baselines at 410 and 432 MHz (Figure 8). Thus, the relatively low cross-correlation value and the scatter in Figure 7 result from uncorrelated excursions, which cannot originate from ionospheric scattering. The scatter of the 10 June 2003 radio positions has the same property.

A second cause for linear scatter could be a seesaw of the centroid between two positions. The positions of the bins below and above average flux density were compared. Their distributions in x and y are similar. The average positions of low and high flux are the same in both events and at all frequencies. Thus, there seems to be only one source moving randomly in position.

In the absence of contrary evidence, we suggest that the linear spreads of the centroid positions in the 5 December 2006 event in all four frequencies and in the 10 June 2003 event at 410 and 432 MHz are real. This is supported by Figure 5 showing a linear alignment of the average sources in decreasing frequency in line with the coronal X-ray source. Figure 3 indicates the same behavior for 432 and 410 MHz. Only the scatter of the 327 MHz positions on 10 June 2003 deviates from this general behavior. As noted before, its flux is very low such that the emission cannot be classified in the spectrogram (Figure 1). Throughout the day, the Nançay Radioheliograph has observed a low intensity noise storm (importance I) at 327 MHz reported at $955'' \pm 10''$ W, $246'' \pm 10''$ N (Solar-Geophysical Data, 2003). This position is only 2 standard deviations southeast of the average position measured here. Thus, the 327 MHz source is most likely not pulsating emission.

The scatter in pulsation position appears to have two components: *i*) rapid fluctuations that are not correlated in frequency even as close as 410 and 423 MHz, and *ii*) correlated changes in positions of less than $\pm 10''$ that appear as slow variations of the baseline in time (Figure 8). The latter seems to be caused by a systematic motion of the pulsations' centroid at both frequencies.

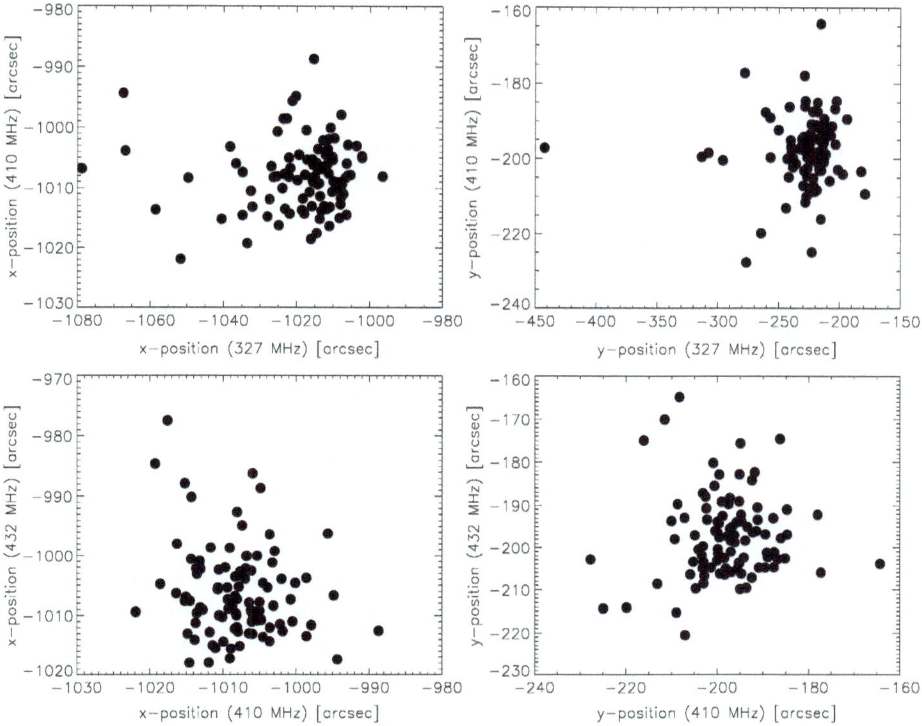

Figure 7 Comparison of simultaneous x-positions (left) and y-positions (right) at different frequencies for the 5 December 2006 flare. Only fluxes above average in both frequencies are shown.

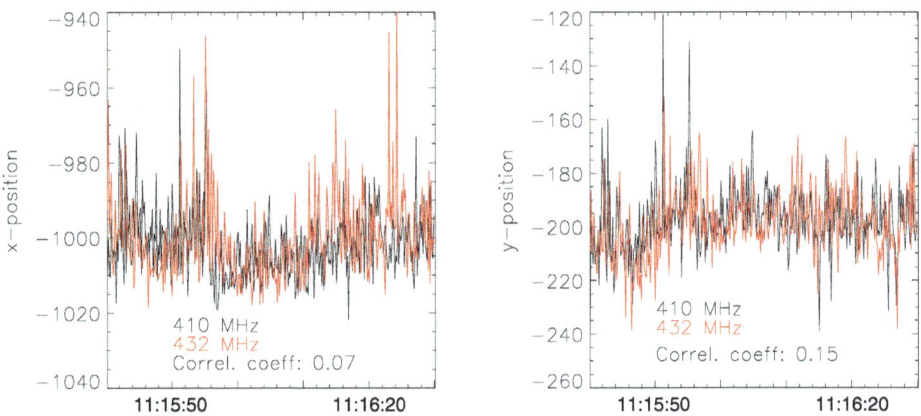

Figure 8 Comparison of positions in adjacent frequencies for the 5 December 2006 flare. All time bins are shown.

5. Conclusions

The positions of radio pulsations and soft and hard X-ray sources are found far apart despite the good correlation in time. The projected difference is at least 30 000 km. This is far beyond the distance an Alfvén wave would travel within the delay time measured by cross-correlation. It is therefore likely that the trigger signal is propagated by energetic particles.

The radio emission originates consistently above the coronal X-ray source. Evoking the loss-cone instability of electrons trapped in such loops would predict the radio sources near the mirroring magnetic fields, thus near the footpoints of the loops and below the coronal X-ray region. This clearly contradicts the old scenario proposing decimetric pulsations to be caused by electrons trapped in flare loops (see Section 1).

We cannot distinguish the cause of the pulsed modulation between two remaining scenarios: pulsating acceleration or particles accelerated elsewhere radiating in an oscillating MHD structure. As the radio sources are displaced from the main acceleration site (generally assumed to be close to the coronal X-ray source), the pulsating acceleration scenario would suggest additional energy release, although less energetic, in a larger volume including the radio sources. This can be envisaged as a current sheet extending upward from the flaring loops possibly behind an ejected plasmoid (see Section 1). We have not been able to find ancillary observations for the two selected events from coronagraphs, or other X-ray or EUV instruments to confirm this hypothesis. If true, decimetric pulsations would trace the coronal current sheet.

The comparison of positions indicates that the radio sources at different frequencies are aligned on a structure pointing upward from the coronal X-ray source in both flares. The scatter of positions at a given frequency is along the same line in one case, and at an angle of some 30° in the other case. This suggests that the radio emission is emitted at a frequency related to the local density or magnetic field (or both) which decrease with altitude. The scatter at a given frequency then could be the result of the source occurring on individual field lines with different density or magnetic field.

In the 5 December 2006 event, emission was also observed at 236 MHz. This source moved 220″ inward within 20 s (Figure 4), thus exceeding the commonly assumed values of the Alfvén velocity by more than an order of magnitude. The motion indicates that the conditions (density and/or magnetic field) at the first source were the same as 20 s later in the second source. Most likely, the emission site moved from a relatively dense region high in the corona to a lower region having the same density. At the higher frequencies and closer to the coronal X-ray source, no motion was observed within the accuracy of the observation.

Due to instrumental limitation, radio pulsations at low frequencies were selected. In addition, the positional information refers to the lower end of the frequency range of the selected events. We cannot exclude that at higher decimetric frequencies radio pulsations occur near or in the coronal X-ray source, as suggested by Equation (1). The confirmation of such a hypothesis would require a new radio interferometer at frequencies above 500 MHz.

Acknowledgements We thank Christian Monstein and Andreas James for constantly improving and operating the Bleien radio spectrometers. The construction of the spectrometers was financially supported by the Swiss National Science Foundation (grants 20-113556 and 200020-121676). The NRH is funded by the French Ministry of Education, the CNES and the Région Centre in France.

References

Aschwanden, M.J.: 1986, *Solar Phys.* **104**, 57.
Aschwanden, M.J.: 2006, *Physics of the Solar Corona*, Praxis, Chichester.
Aschwanden, M.J., Benz, A.O.: 1986, *Astron. Astrophys.* **158**, 102.
Aschwanden, M.J., Benz, A.O.: 1988, *Astrophys. J.* **332**, 466.
Aurass, H., Landini, F., Poletto, G.: 2009, *Astron. Astrophys.* **506**, 901.
Bárta, M., Karlický, M., Žemlička, R.: 2008, *Solar Phys.* **281**, 173.

Battaglia, M., Benz, A.O.: 2006, *Astron. Astrophys.* **456**, 751.
Battaglia, M., Benz, A.O.: 2009, *Astron. Astrophys.* **499**, L33.
Benz, A.O.: 1980, *Astrophys. J.* **240**, 892.
Benz, A.O.: 2002, *Plasma Astrophysics: Kinetic Processes in Solar and Stellar Coronae*, 2nd edn., Kluwer, Dordrecht.
Benz, A.O.: 2009, Landolt-Boernstein VI/4B, Springer, 4.1.2.8.
Benz, A.O., Wentzel, D.G.: 1981, *Solar Phys.* **94**, 100.
Benz, A.O., Grigis, P.C., Csillaghy, A., Saint-Hilaire, P.: 2005, *Solar Phys.* **226**, 121.
Covington, A.E.: 1951, *J. Roy. Astron. Soc. Can.* **45**, 157.
Dabrowski, B.P., Benz, A.O.: 2009, *Astron. Astrophys.* **504**, 565.
De Feiter, L.D., Fokker, A.D., Roosen, J.: 1959, *Nature* **184**, 805.
Delaboudinière, J.-P., Artzner, G.E., Brunaud, J., Gabriel, A.H., Hochedez, J.F., Millier, F., *et al.*: 1995, *Solar Phys.* **162**, 291.
Emslie, A.G., Kontar, E.P., Krucker, S., Lin, R.P.: 2003, *Astrophys. J.* **595**, L107.
Fleishman, G.D., Stepanov, A.V., Yurovsky, Yu.F.: 1994, *Solar Phys.* **153**, 403.
Fleishman, G.D., Bastian, T.S., Gary, D.E.: 2008, *Astrophys. J.* **684**, 1433.
Gotwols, B.L.: 1972, *Solar Phys.* **25**, 232.
Güdel, M., Benz, A.O.: 1988, *Astron. Astrophys. Suppl. Ser.* **75**, 243.
Hill, S.M., Pizzo, V.J., Balch, C.C., Biesecker, D.A., Bornmann, P., Hildner, E., *et al.*: 2005, *Solar Phys.* **226**, 255.
Hurford, G.J., Schmahl, E.J., Schwartz, R.A., Conway, A.J., Aschwanden, M.J., Csillaghy, A., *et al.*: 2002, *Solar Phys.* **210**, 61.
Isliker, H., Benz, A.O.: 1994, *Astron. Astrophys. Suppl. Ser.* **104**, 145.
Karlický, M., Bárta, M.: 2005, *Space Sci. Rev.* **121**, 153.
Karlický, M., Zlobec, P., Mészárosová, H.: 2010, *Solar Phys.* **261**, 281.
Kerdraon, A., Delouis, J.-M.: 1997, In: Trottet, G. (ed.) *Coronal Physics from Radio and Space Observations*, *Lect. Not. Phys.* **483**, 192.
Khan, M., Vilmer, N., Saint-Hilaire, P., Benz, A.O.: 2002, *Astron. Astrophys.* **388**, 363.
Kliem, B., Karlický, M., Benz, A.O.: 2000, *Astron. Astrophys.* **360**, 715.
Krucker, S., Battaglia, M., Cargill, P.J., Fletcher, L., Hudson, H.S., MacKinnon, A., *et al.*: 2008, *Astron. Astrophys. Rev.* **16**, 155.
Kundu, M.R., Roberts, J.A., Spencer, C.L., Kuiper, J.W.: 1961, *Astrophys. J.* **133**, 255.
Lehany, F.J., Yabsley, D.E.: 1948, *Nature* **161**, 645.
Li, Y., Fleishman, G.D.: 2009, *Astrophys. J. Lett.* **701**, L52.
Lin, R.P., Dennis, B.R., Hurford, G.J., Smith, D.M., Zehnder, A., Harvey, P.R., *et al.*: 2002, *Solar Phys.* **210**, 3.
Magdalenic, J., Zlobec, P., Vrsnak, B., Messerotti, M., Aurass, H., Temmer, M.: 2003, *ESA SP* **535**, 619.
McLean, D.J., Sheridan, K.V., Stewart, R.T., Wild, J.P.: 1971, *Nature* **234**, 140.
Messmer, P., Benz, A.O., Monstein, C.: 1999, *Solar Phys.* **187**, 335.
Mészárosová, H., Rybák, J., Karlický, M.: 2011, *Astron. Astrophys.* **525**, A88.
Nakariakov, V.M., Foullon, C., Myagkova, I.N., Inglis, A.R.: 2010, *Astrophys. J.* **708**, L47.
Nindos, A., Aurass, H., Klein, K.-L., Trottet, G.: 2008, *Solar Phys.* **253**, 3.
Pick, M.: 1986, *Solar Phys.* **104**, 19.
Roberts, B., Edwin, P.M., Benz, A.O.: 1984, *Astrophys. J.* **279**, 857.
Saint-Hilaire, P., Benz, A.O.: 2003, *Solar Phys.* **216**, 205.
Tan, B., Zhang, Y., Tan, C., Liu, Y.: 2010, *Astrophys. J.* **723**, 25.
Thompson, A.R., Maxwell, A.: 1962, *Astrophys. J.* **136**, 546.
Vilmer, N., Krucker, S., Lin, R.P., RHESSI team: 2002, *Solar Phys.* **210**, 261.
Young, C.W., Spencer, C.L., Moreton, G.E., Roberts, J.A.: 1961, *Astrophys. J.* **133**, 243.

Solar Phys (2011) 273:377–392
DOI 10.1007/s11207-011-9756-z

Millisecond Radio Spikes in the Decimetric Band

B.P. Dąbrowski · P. Rudawy · M. Karlický

Received: 29 November 2010 / Accepted: 21 March 2011 / Published online: 27 April 2011

Abstract We present the results of the analysis of thirteen events consisting of dm-spikes observed in Toruń between 15 March 2000 and 30 October 2001. The events were obtained with a very high time resolution (80 microseconds) radio spectrograph in the 1352–1490 MHz range. These data were complemented with observations from the radio spectrograph at Ondřejov in the 0.8–2.0 GHz band. We evaluated the basic characteristics of the individual spikes (duration, spectral width, and frequency drifts), as well as their groups and chains, the location of their emission sources, and the temporal correlations of the emissions with various phases of the associated solar flares. We found that the mean duration and spectral width of the radio spikes are equal to 0.036 s and 9.96 MHz, respectively. Distributions of the duration and spectral widths of the spikes have positive skewness for all investigated events. Each spike shows positive or negative frequency drift. The mean negative and positive drifts of the investigated spikes are equal to $-776\ \mathrm{MHz\,s^{-1}}$ and $1608\ \mathrm{MHz\,s^{-1}}$, respectively. The emission sources of the dm-spikes are located mainly at disk center. We have noticed two kinds of chains, with and without frequency drifts. The mean durations of the chains vary between 0.067 s and 0.509 s, while their spectral widths vary between 7.2 MHz and 17.25 MHz. The mean duration of an individual spike observed in a chain was equal to 0.03 s. While we found some agreement between the global characteristics of the groups of spikes recorded with the two instruments located in Toruń and Ondřejov, we did not find any one-to-one relation between individual spikes.

Keywords Solar corona · Flares · Radio bursts

Energy Storage and Release through the Solar Activity Cycle – Models Meet Radio Observations
Guest Editors: Christophe Marqué and Alexander Nindos.

B.P. Dąbrowski (✉) · M. Karlický
Astronomical Institute, Academy of Sciences of the Czech Republic, 251 65 Ondřejov, Czech Republic
e-mail: bdabrow@asu.cas.cz

M. Karlický
e-mail: karlicky@asu.cas.cz

P. Rudawy
Astronomical Institute, University of Wrocław, ul. Kopernika 11, 51-622 Wrocław, Poland
e-mail: rudawy@astro.uni.wroc.pl

1. Introduction

Decimetric radio spikes (hereafter called dm-spikes) are short-lived individual electromagnetic emissions usually of duration not longer than 0.1 s, recorded in a very narrow bandwidth of 0.2 – 2.0% of the center frequency (Benz, 2002). The dm-spikes usually appear in multi-element groups, typically of up to several thousands of individual events each (Benz, 1985). Sometimes one can distinguish columns and chains of spikes on radio spectrograms (Guedel and Benz, 1988; Dąbrowski *et al.*, 2005). A chain of spikes is a series of spikes emitted at about nearly the same frequency. The dm-spikes occur often during DCIM (decimeter, complex, highly structured radio emission) solar events. They can be observed effectively with high time resolution due to their very high flux densities (of the order of 100 solar flux units, 1 sfu $= 10^{-19}$ erg s^{-1} cm^{-2} Hz^{-1}) and very high brightness temperature, greater than $10^{13} - 10^{15}$ K (Benz, 1986, 2002). The intense emission of decimetric radiation is generally considered to originate from nonthermal electron velocity distributions. These distributions can be unstable to maser emission or plasma waves that couple coherently with radio waves (Dąbrowski and Benz, 2009).

The relative spectral widths of the dm-spikes are of the order of a few percent and can vary significantly both from one event to another and within the same event (Benz, 1986; Fleishman and Melnikov, 1999). Csillaghy and Benz (1993) analyzed two events recorded in the 1.08 – 1.6 GHz and 1.0 – 1.35 GHz bands; they found that the average relative spectral width of the spikes was 17 MHz and 49.50 MHz, respectively. Messmer and Benz (2000) observed dm-spikes in the 0.87 – 1.0 GHz band and found that their spectral widths were equal to 7.5 MHz. Table 1 presents a review of the spike durations and spectral widths reported in various papers. Both parameters were evaluated at half height of their maxima, and their mean values were estimated for at least ten or more spikes.

The dm-spikes reveal very fast drifts in frequency, with a wide scatter of the drifts over positive and negative values (a negative drift is defined as the time change of an observed frequency from higher to lower values), but usually the negative drifts dominate. The drifts lie between -1 and -4 GHz s^{-1} at 770 MHz (Guedel and Benz, 1990), while their absolute drift lies between 2.2 to 4.3 GHz s^{-1} in the 2.7 – 3.8 GHz band (Wang, Yan, and Fu, 2002).

The dm-spikes have the highest association rate with flares (95%) of all coherent radio emissions. By contrast, only 2% of all hard X-ray (HXR) events are associated with dm-spikes (Aschwanden and Guedel, 1992). The emission of the dm-spikes is often correlated in time with the HXR emission recorded during impulsive phases of solar flares (Benz, 1986; Aschwanden and Guedel, 1992; Dąbrowski and Benz, 2009). Unfortunately, due to an insufficient temporal resolution of the presently available X-ray observations, it is not possible to investigate temporal correlations of the individual radio spikes with short-term changes of the HXR flux.

Detailed reviews considering solar dm-spikes can be found in numerous papers, *e.g.*, Benz (1986), Fleishman and Melnikov (1998), and Dąbrowski *et al.* (2005).

In this paper we present an analysis of thirteen events in the dm range observed with the Toruń radio spectrograph having very high time resolution (80 microseconds). Various kinds of co-temporal radio events (two of them with dm-spikes), observed by the Ondřejov radio spectrograph, are also analyzed. The main characteristics of the observed spikes were given in the paper by Dąbrowski *et al.* (2005). Here we present new results concerning their statistical properties and evaluate the physical parameters of their emission sources. We also compare the dm-spikes observed simultaneously by the Toruń and Ondřejov radio spectrographs.

Table 1 Durations and spectral widths of the radio spikes observed in 0.3 GHz to 7.5 GHz band.

Freq.* [GHz]	Duration [ms]	Time res. [ms]	Width [MHz]	Freq. res. [MHz]	Reference
0.33	–	–	9.9	1 – 10	(Benz, Csillaghy, and Aschwanden, 1996)
0.35	–	–	4.8	1	(Messmer and Benz, 2000)
0.36	–	–	7.32	1	(Csillaghy and Benz, 1993)
0.36	73	2	–	1	(Guedel and Benz, 1990)
0.40**	4 – 30	1 & 10	< 20	–	(Magdalenić *et al.*, 2006)
0.47	41	0.5	–	3	(Guedel and Benz, 1990)
0.48	–	–	12.6	6	(Csillaghy and Benz, 1993)
0.60	–	100	< 2	0.061	(Benz *et al.*, 2009)
0.73	20	10	–	10	(Guedel and Benz, 1990)
0.77	20	2	–	10	(Guedel and Benz, 1990)
0.83	–	–	7.04	1	(Csillaghy and Benz, 1993)
0.87	19	10	–	10	(Guedel and Benz, 1990)
0.94	–	–	7.5	1	(Messmer and Benz, 2000)
1.01	17	10	–	10	(Guedel and Benz, 1990)
1.16	–	–	49.5	14	(Csillaghy and Benz, 1993)
1.34	–	–	17.0	10	(Csillaghy and Benz, 1993)
1.42	2.0	1	–	–	(Wang and Xie, 1999)
1.42	9.0	1 & 10	–	–	(Mészárosová *et al.*, 2003)
1.69	–	–	32.3	10	(Csillaghy and Benz, 1993)
2.00	1.3	1	–	–	(Wang and Xie, 1999)
2.70	5.1	1 & 10	–	–	(Mészárosová *et al.*, 2003)
2.80	–	–	10.8	10	(Csillaghy and Benz, 1993)
2.84	13.2	1	–	–	(Wang and Xie, 1999)
3.20	14	8	–	10	(Chernov *et al.*, 2001)
3.20	42.7	8	117.3	10	(Wang and Xie, 1999)
3.20	8 – 36	8	20 – 110	10	(Wang *et al.*, 2008)
5.25	–	5	∼ 30	10	(Rozhansky, Fleishman, and Huang, 2008)
7.30	∼ 100	100	120	10	(Benz *et al.*, 1992b)
7.40	–	–	48.7	11	(Csillaghy and Benz, 1993)

*Center of the frequency band.

**Spike-like structures, observed at 237, 327, 408, 610 MHz and in the range 270 – 450 MHz.

2. Instruments and Observational Data

Thirteen events with dm-spikes were observed in Toruń between 15 March 2000 and 30 October 2001, during nearly 2000 hours of observations (Dąbrowski *et al.*, 2005; Dąbrowski and Kus, 2007) and co-temporal various kinds of radio events were observed in Ondřejov (Table 2).

During the observations of the dm-spikes the 15 m radio telescope in Toruń (located in Piwnice) was connected to the fast multi-channel radio spectrograph Penn State Pulsar Machine II – PSPM II (Cadwell, 1997). Radio spectrograms of the spikes were recorded in

Table 2 Thirteen events of dm-spikes observed in Toruń between 15 March 2000 and 30 October 2001 (some events consisted of two groups of spikes), and corresponding events observed in Ondřejov.

Date	Toruń event Time interval [UT]	Ondřejov event	Class GOES	Position	Flare phase GOES
15.03.2000	12:20:06 – 12:21:08	continuum and spikes	C5.0	S18W19	rise
26.03.2000	10:37:44 – 10:38:04	spikes (800 – 1000 MHz)	C2.0	S18E09	maximum
26.04.2000	14:09:55 – 14:10:35	–	B7.0	S15E10	rise
17.05.2000	15:42:21 – 15:42:26	–	C2.2	N20E20	rise
	15:43:45 – 15:43:52	–	C2.2	N20E20	rise
21.05.2000	09:39:26 – 09:39:31	–	C8.2	S18W47	rise
	09:39:44 – 09:39:45	–	C8.2	S18W47	rise
21.06.2000	09:24:59 – 09:26:05	continuum and weak spikes	M1.3	N24W34	rise
11.07.2000	13:16:03 – 13:16:44	strong continuum	X1.0	N18E27	decay
	13:26:30 – 13:27:37	spikes	X1.0	N18E27 or S17W39	decay
11.07.2000	14:41:10 – 14:48:10	spikes	X1.0	N18E27	decay
11.07.2000	14:57:49 – 15:00:37	–	X1.0	N18E27	decay
22.03.2001	13:10:37 – 13:10:50	–	M1.0	N14E60	rise
	13:11:17 – 13:11:34	pulsations	M1.0	N14E60	rise
23.04.2001	10:09:53 – 10:10:03	continuum	C9.1	N17E26	rise
	10:14:45 – 10:15:05	continuum	C9.1	N17E26	rise
14.05.2001	12:47:25 – 12:47:39	–	C2.3	S15W15	decay
30.10.2001	12:20:14 – 12:20:40	continuum	C7.4	N12E10	maximum

the 1352 – 1490 MHz frequency band split into 46 channels, each 3 MHz wide. The time resolution of the collected data is equal to 80 microseconds (Dąbrowski *et al.*, 2005).

The investigated radio events were also observed by the 10 m radio telescope in Ondřejov with a 0.8 – 2.0 GHz swept-frequency radio spectrograph. The 0.8 – 2.0 GHz band was divided into 256 frequency channels with bandwidths of 3 MHz, set 4.69 MHz apart. The temporal resolution of the radio spectra was equal to 0.1 s (Jiřička *et al.*, 1993; Karlický, Jiřička, and Sobotka, 2000).

3. Data Evaluation and Data Analysis Methods

The radio data collected with the radio spectrographs in Toruń and Ondřejov were processed in order to make the sensitivities of the individual channels internally homogeneous, but were not calibrated in absolute flux units or equalized between both telescopes.

For subsequent automatic evaluation of the main observational properties of the individual spikes, we adopted a working definition of a spike as a short-lived increase of the flux with a peak signal more than three times higher than the local background, in accordance with our previous paper (Dąbrowski *et al.*, 2005). The temporal resolution of the data collected in Toruń and Ondřejov was different; then, to compare the spectra, the temporal resolution of the Toruń radio spectrogram was numerically decreased, but the statistical properties of the spikes observed with both instruments were calculated using original data.

Each individual radio spike observed in Toruń was fitted with a two-dimensional elliptical Gaussian (using a nonlinear least-square method), and the basic parameters were measured at an intensity level equal to half of the maximum for duration, spectral width, and frequency drift. All spurious structures (like local peaks) were rejected. A detailed description of this method was given by Dąbrowski *et al.* (2005).

Due to the limited temporal resolution of the data collected with the Ondřejov radio spectrograph, we only analyze the frequency profiles of the individual spikes. An automatic code evaluated the frequency width of the individual spikes by fitting its frequency profile with a one-dimensional Gaussian having four free parameters: height, center frequency, width (standard deviation), and signal level.

4. Results

4.1. Basic Characteristics of the Individual dm-Spikes

We have evaluated the main observational parameters of the 5199 individual dm-spikes (lasting no longer than 0.1 s) observed in Toruń. The mean duration of the spikes was equal to 0.036 s. The measured mean spectral width of the spikes was equal to 9.96 MHz, but this value can be recognized as an upper limit due to the limited spectral resolution of the radio spectrograph, equal to 3 MHz (Benz *et al.*, 2009). For a particular event, the spectral widths of the spikes are usually comparable.

Each spike shows positive (from low to high) or negative (from high to low) frequency drift. The negative drifts range from $-63\ \mathrm{MHz\,s^{-1}}$ to $-13\,600\ \mathrm{MHz\,s^{-1}}$, while the positive ones range from $51\ \mathrm{MHz\,s^{-1}}$ to $13\,400\ \mathrm{MHz\,s^{-1}}$. The mean negative and positive drifts of the investigated spikes are equal to $-776\ \mathrm{MHz\,s^{-1}}$ and $1608\ \mathrm{MHz\,s^{-1}}$, respectively.

Histograms of the basic parameters of the dm-spikes (for all investigated spikes), *i.e.*, duration, spectral width, and frequency drifts, are presented in Figure 1. All investigated parameters have positive skewness distributions (compare with Csillaghy and Benz, 1993).

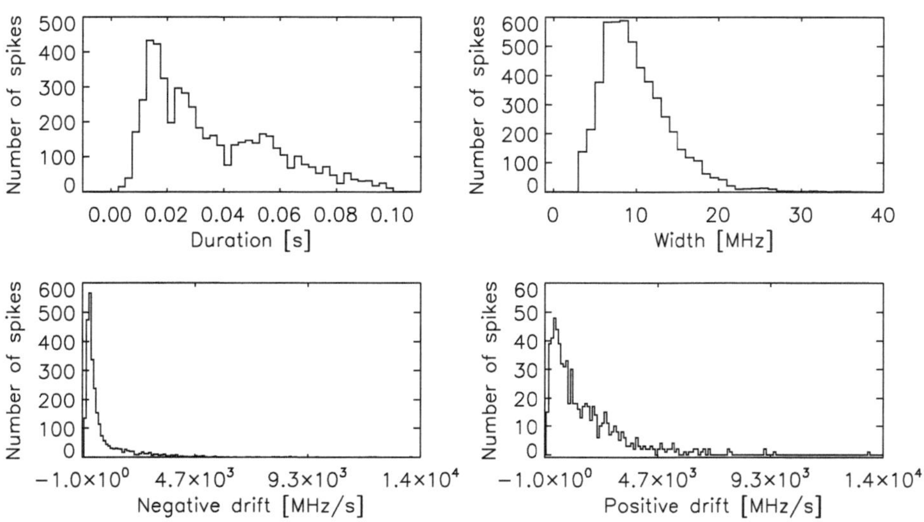

Figure 1 Histograms of the basic parameters of the dm-spikes (for all investigated spikes): duration, spectral width, and frequency drifts.

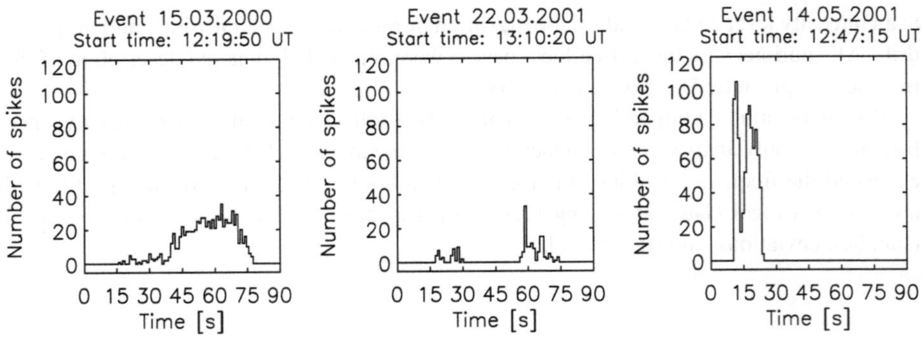

Figure 2 Spike rates (number of spikes per second) in radio events.

4.2. Groups and Rates of dm-Spikes

For the thirteen investigated events we detected eighteen individual groups of dm-spikes (Table 2). The groups of spikes lasted from 1 s up to 420 s; the mean duration was equal to 55.7 s. Each group consists of a dozen or so up to several hundreds of individual spikes.

Spike rates (number of spikes per second) are unique for each individual event (Figure 2). There are: *i*) events that have a gradual increase and next an abrupt drop of the numbers of observed spikes, *ii*) events with spikes clustered into well-distinguishable groups of spikes, and *iii*) events with the highest numbers of spikes just at the beginning of the whole event.

4.3. Chains of the dm-Spikes

We have noticed two kinds of well-outlined complex structures formed by numerous individual spikes on the radio spectrograms of the investigated events: columns and chains of spikes. While the columns were studied by Dąbrowski *et al.* (2005), here we analyze the chains in detail. We define a chain of spikes as a group of at least three individual emissions (spikes) having the same (chains without frequency drift) or similar (chains with frequency drift) frequency bands and appearing close in time.

Two kinds of chains, with and without frequency drifts, are presented in Figures 3 and 4, respectively. The chains of spikes were observed during the following events: 15 March 2000, 26 March 2000, 26 April 2000 (only two chains of spikes were observed), and 30 October 2001. We recorded a total of 29 chains of spikes. The mean durations of the chains, evaluated for each individual event separately, vary between 0.067 s and 0.509 s, while their spectral widths vary between 7.2 MHz and 17.25 MHz. The mean duration of an individual spike observed in the chain was 0.03 s. The main parameters of the chains are presented in Table 3. The drift, spectral widths, and durations of the chains were evaluated using the same fitting method as used for individual spikes.

Histograms of the basic parameters of the chains (for all investigated chains), *i.e.*, duration, spectral width, and frequency drifts, are presented in Figure 5.

4.4. Decimetric Spikes *Versus* Phase of the Solar Flare

Emissions of the investigated dm-spikes were observed during various phases of solar flares. Similar results were published by Benz *et al.* (2005). Half of the dm-spike events were recorded during the rise phases of the solar flares (54%). The dm-spikes were emitted mostly (54%) during C-class flares; similar results were published by Jiřička *et al.* (2001), who observed the dm-spikes in the bandwidth 0.8 – 2.0 GHz between 1992 and 2000.

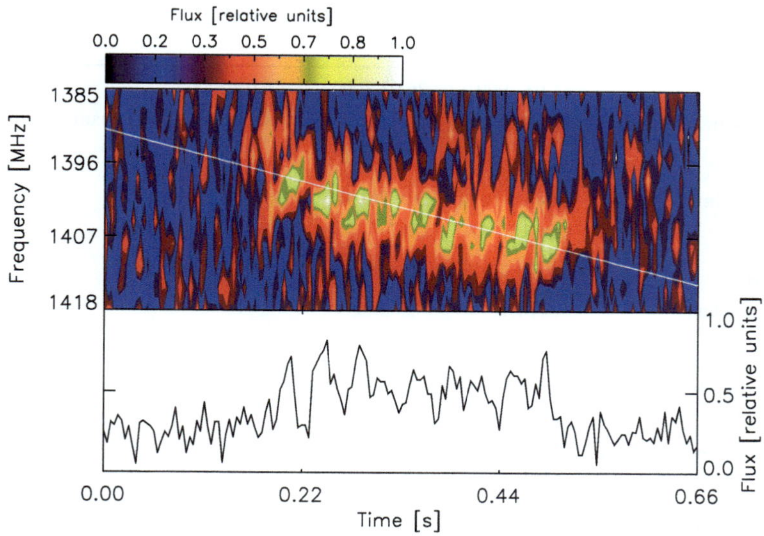

Figure 3 Chain of spikes with a drift in frequency recorded on 26 March 2000 (beginning at 10:38:00.6 UT). The upper panel presents the radio spectrogram. The lower panel shows the changes of signal intensity (along the inclined line in the upper part).

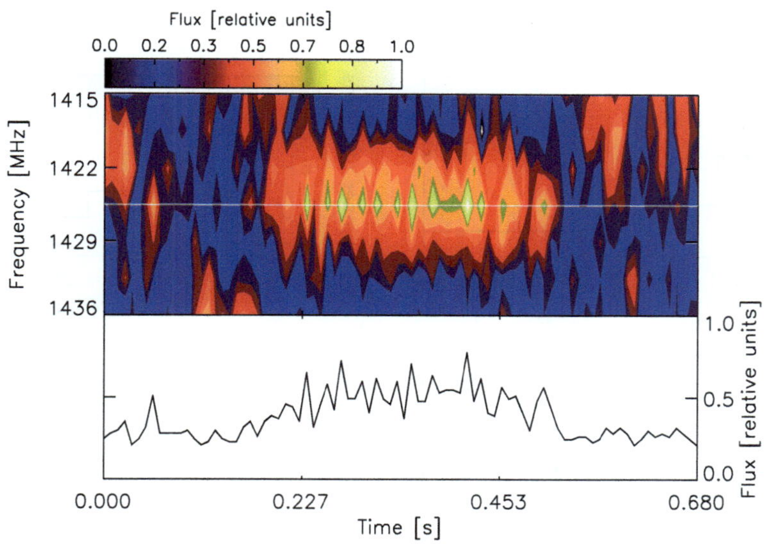

Figure 4 Chain of spikes without a drift in frequency recorded on 26 April 2000 (beginning at 14:10:13.96 UT). The upper panel presents the radio spectrogram. The lower panel shows the changes of signal intensity (at the frequency where we observed the highest signal for this chain of spikes).

4.5. Locations of the Emission Sources of the Radio Spikes

Most of the active regions which host sources of radio spikes ($\sim 77\%$) were located close to the center of the solar disk, inside a circle having a radius of 35° (Figure 6). Due to the low number (thirteen) of dm-spike events observed in Toruń we also analyzed dm-spikes

Table 3 Parameters of the chains of spikes observed in four radio events: 15 March 2000, 26 March 2000, 26 April 2000, and 30 October 2001.

Events	15.03.2000	26.03.2000	26.04.2000*		30.10.2001
Number of chains	10	9	1st chain	2nd chain	8
Duration [s]					
minimum/maximum	0.048/0.10	0.099/0.34	–	–	0.061/0.159
mean	0.067	0.17	0.275	0.509	0.10
Width [MHz]					
minimum/maximum	5.6/14.4	7.5/15.0	–	–	14.4/25.1
mean	8.27	11.2	7.2	7.8	17.25
Negative drifts [$\mathrm{MHz\,s^{-1}}$]					
number of chains	7	3	–	–	8
minimum/maximum	180/587	52/340	–	–	59/401
mean	304	179	–	–	218
Positive drifts [$\mathrm{MHz\,s^{-1}}$]					
number of chains	–	5	–	–	–
minimum/maximum	–	15/267	–	–	–
mean	–	80	–	–	–

*Only two chains of spikes were observed in this event.

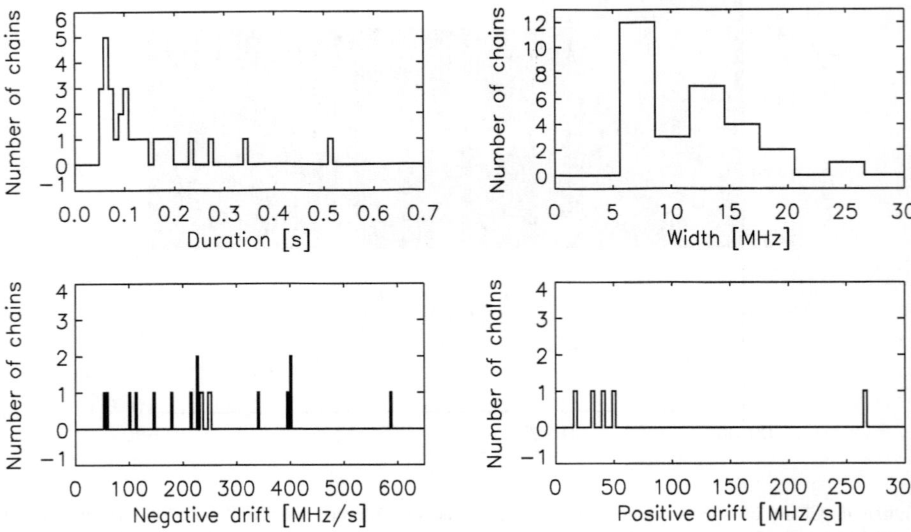

Figure 5 Histograms of the basic parameters of the chains (for all investigated chains): duration, spectral width, and frequency drifts.

observed in Ondřejov in the 0.8 – 2.0 GHz band, which have been described by Karlický, Sobotka, and Jiřička (1996), Karlický, Jiřička, and Sobotka (2000), and Fárník and Karlický (2005). Thus, the total number of dm-spike events having well-established emission sources increased to twenty-one.

Figure 6 Distribution of active regions associated with the dm-spike emission on the solar disk. Those corresponding to observations from Toruń are marked with asterisks. The event observed on 11 July 2000 in Toruń is marked with a square, but the exact location of the radio source is problematic (two co-temporal solar flares in two distinct active regions were observed). The active regions related with observations from Ondřejov are marked with diamonds. The concentric circles are drawn at 50° and 90° from disk center.

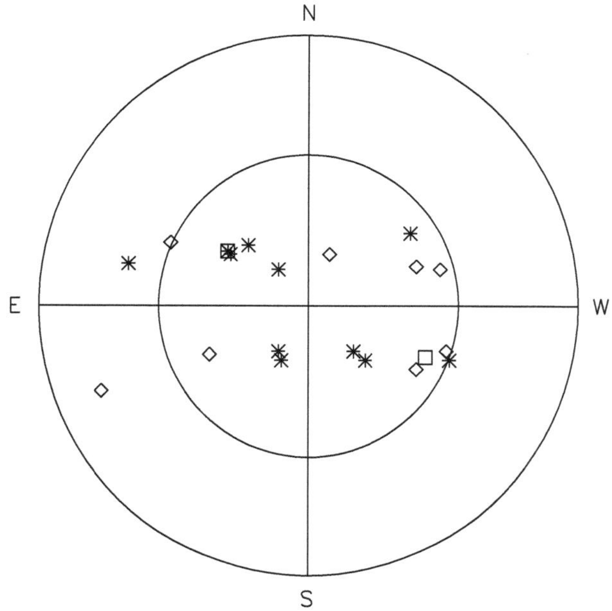

The emission sources of the dm-spikes are determined from Hα, *Yohkoh*, TRACE, and EIT/SOHO observations and are located mainly inside a circle with a radius of 50°. The reasons for this apparent preference in location are unclear, and this topic needs future investigation.

4.6. Radio Events Observed Co-temporal in Toruń and Ondřejov

The radio spectrographs of both observatories have complementary frequency bands (1352 – 1490 MHz in Toruń and 0.8 – 2.0 GHz in Ondřejov), but different time and spectral resolutions.

In Table 2 we present a list of the dm-spike events observed with the Toruń radio telescope and the relevant events observed with the radio telescope in Ondřejov: spikes, continuum emissions, and pulsations. However, for some groups of dm-spikes observed in Toruń there were no specific radio events in Ondřejov. Four dm-spike groups observed in Toruń were simultaneous with events in Ondřejov in the same frequency band, while one dm-spike event observed on 26 March 2000 in Toruń was also observed in Ondřejov, but in a different frequency band (800 – 1000 MHz).

Table 4 presents the main parameters of the spikes observed on 11 July 2000 in the same frequency band in Toruń and Ondřejov (1352 – 1490 MHz). Table 5 presents additional parameters of the dm-spikes observed in Ondřejov in the 0.8 – 2.0 GHz band.

4.6.1. First Event with dm-Spikes on 11 July 2000

The event, which consisted of two groups of spikes, was observed in Toruń between 12:30:10 UT and 13:30:10 UT. The first group of spikes was observed between 13:16:03 UT and 13:16:44 UT, the second one between 13:26:30 UT and 13:27:37 UT. The spikes of both groups were recorded in the entire available frequency range (1352 – 1490 MHz), and no internal structure of the individual spikes was detected.

Table 4 Parameters of the dm-spikes observed on 11 July 2000 in Toruń and Ondřejov in the 1352 – 1490 MHz band.

	Toruń	Ondřejov
Time interval [UT]	13:26:30 – 13:27:37	13:25:40 – 13:27:40
Number of spikes	688	39
Duration [s]		
minimum/maximum	0.01/0.99	–
mean/median	0.03/0.03	–
skewness	1.8	–
Width [MHz]		
minimum/maximum	3.1/31.4	4.8/14.7
mean/median	10.3/9.4	7.6/7.8
skewness	1.1	0.6
Negative drifts [$MHz\,s^{-1}$]		
minimum/maximum	64/3850	–
mean/median	1153/817	–
skewness	0.8	–
Positive drifts [$MHz\,s^{-1}$]		
minimum/maximum	149/3853	–
mean/median	1260/1115	–
skewness	1.0	–

Table 5 Parameters of the dm-spikes observed on 11 July 2000 in Ondřejov in the 0.8 – 2.0 GHz band.

	1st event	2nd event
Time interval [UT]	13:25:40 – 13:27:40	14:43:00 – 14:44:00
Number of spikes	261	22
Width [MHz]		
minimum/maximum	4.7/23.9	4.7/15.9
mean/median	7.9/7.1	7.0/6.2
skewness	1.7	1.9

The first group of spikes was associated with strong continuum emission observed in Ondřejov. The second group of spikes was also observed in Ondřejov between 13:25:40 UT and 13:27:40 UT in the frequency band 1.0 – 1.8 GHz. The dm-spikes recorded in Ondřejov were organized into two branches (bandwidth ∼300 MHz), slowly drifting toward lower frequencies (Karlický *et al.*, 2001) – the bandwidth of the Toruń radio spectrograph is much narrower than the bandwidth of the Ondřejov instrument (Figure 7). The frequency drift of the branches was around $-6\ MHz\,s^{-1}$. Due to electromagnetic interference the data recorded in Ondřejov in the 1391 – 1278 MHz band were not used. The global spectra of the dm-spikes observed in both locations within the frequency band 1352 MHz to 1490 MHz seem to be similar, but we did not find any one-to-one relation between individual spikes (Figures 8 and 9). The correlation coefficient of the total radio fluxes recorded with both

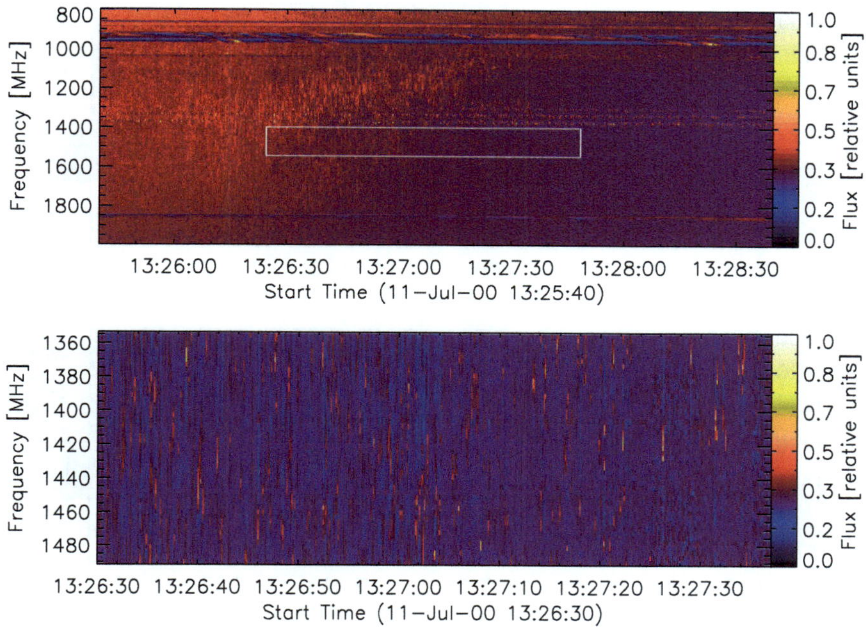

Figure 7 The dm-spikes recorded on 11 July 2000 in Ondřejov (upper panel) and Toruń (lower panel). White rectangle on Ondřejov radio data corresponds to the time and frequency range of Toruń data.

instruments during the same time and frequency range was equal to 0.2 (Figure 8). For two selected sections of the spectra, taken from 13:27:10 to 13:27:12 UT and from 13:27:13 to 13:27:15 UT (Figure 9), the correlation coefficients of the total radio fluxes were equal to 0.45 and 0.02, respectively.

The spikes observed between 13:26:30 UT to 13:27:37 UT in the frequency band 1352 MHz to 1490 MHz have mean spectral widths (FWHM) equal to 10.3 MHz and 7.6 MHz in Toruń and Ondřejov, respectively. Distributions of the spectral widths of the spikes observed in both observatories have positive skewness, in accordance with the results presented by Csillaghy and Benz (1993).

4.6.2. Second Event with dm-Spikes on 11 July 2000

The radio event was observed in Toruń between 14:34:50 UT and 14:51:30 UT. During the event a single group of spikes was identified between 14:41:10 UT and 14:48:10 UT in the frequency band 1352 – 1490 MHz. The spikes were recorded in the entire available frequency range; no internal structure of the individual spikes was detected. The dm-spikes were also observed in Ondřejov between 14:36:10 UT and 14:44:50 UT in the frequency band 0.8 – 1.8 GHz. For statistical analysis, we selected a one minute long period from 14:43:00 UT to 14:44:00 UT; this period consisted of 22 spikes (Table 5). As in the previous case, no one-to-one correlation between the individual spikes observed in Toruń and Ondřejov was found.

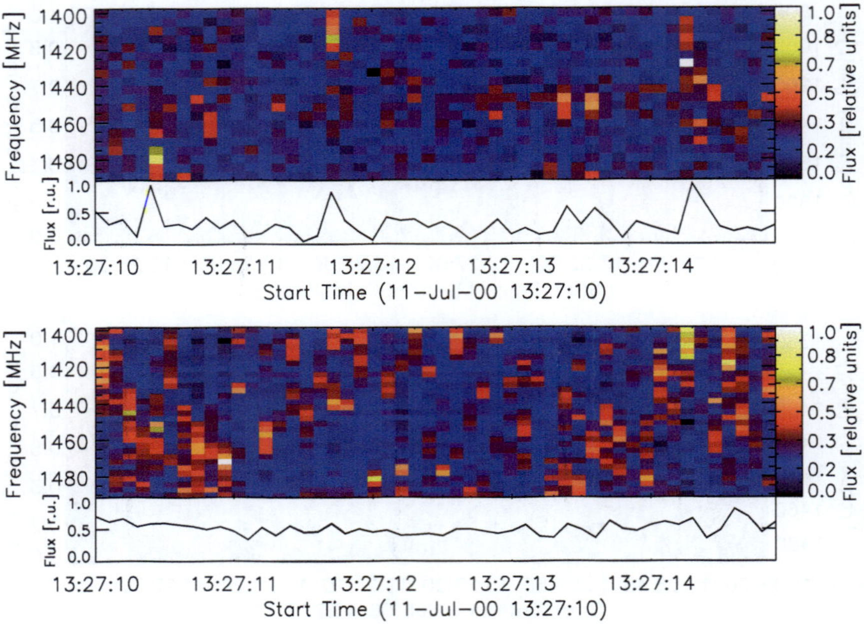

Figure 8 The dm-spikes recorded on 11 July 2000, between 13:27:10 UT and 13:27:15 UT in the band 1400 – 1490 MHz with 0.1 s time resolution in Ondřejov (upper panel) and Toruń (lower panel). Total radio fluxes (in relative units) are presented below the corresponding radio spectrogram. The correlation coefficient of the total radio fluxes is equal to 0.2.

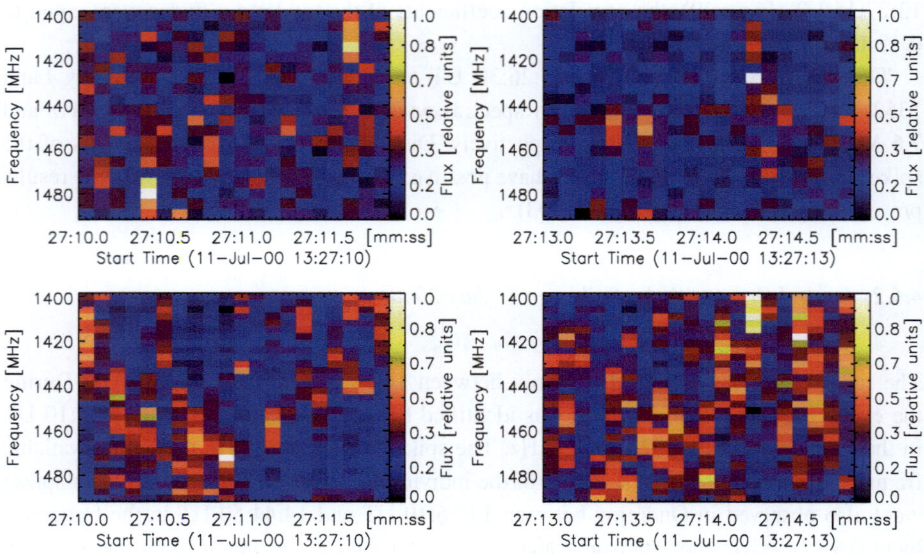

Figure 9 Enlarged sections of the radio spectrograms presented in Figure 8. Upper panels: dm-spikes recorded in Ondřejov. Lower panels: dm-spikes recorded in Toruń. The correlation coefficients of the total radio fluxes of the spectra are equal to 0.45 and 0.02 for left and right columns, respectively.

5. Discussion and Conclusions

We investigated thirteen events having dm-spikes observed in Toruń between 15 March 2000 and 30 October 2001, and complementary radio events observed in Ondřejov: spikes, continuum emissions, and pulsations.

The mean spectral width of the spikes observed in the 1352 – 1490 MHz band was equal to 9.96 MHz; this is nearly three times smaller than the spectral width (32.2 MHz) estimated by Csillaghy and Benz (1993) for spikes observed in the 1400 – 1970 MHz band. However, the mean spectral width measured at half height of the maximum intensity in 1420 MHz using the relation presented in the paper by Csillaghy and Benz (1993) is equal to 13.9 MHz. This value is similar to the one that we obtained. Distributions of the spectral widths of the spikes have positive skewness for all investigated events, in accordance with the results obtained by Csillaghy and Benz (1993). For a particular event the spectral widths of the spikes are usually comparable. This could be attributed to stable or marginally variable conditions in the emission source, but it could also be an observational artifact due to the narrow frequency band of the observations (138 MHz).

The knowledge of the frequency drifts allows us to estimate the bulk velocities of the electron beams exciting the dm-spikes as proposed by Benz *et al.* (1992a). Assuming plasma emission at the fundamental plasma frequency for a $-776\ \mathrm{MHz\,s^{-1}}$ drift (the mean value for all dm-spikes with negative drifts observed in Toruń), the relevant velocity is $\vartheta_b = 1.1 \times 10^9\ \mathrm{cm\,s^{-1}}$, and the kinetic energy of the electrons is equal to 0.344 keV. For a drift of $1608\ \mathrm{MHz\,s^{-1}}$ (the mean value for all dm-spikes with positive drifts observed in Toruń), the bulk velocity is $\vartheta_b = 2.3 \times 10^9\ \mathrm{cm\,s^{-1}}$, and the kinetic energy of the electrons is equal to 1.504 keV, assuming a density scale length of $\sim 10^9$ cm (Benz, 2002) and an emission frequency $\nu = 1420$ MHz. We did not find any correlation between the emission frequency and the frequency drift. However, in some cases the frequency drift was equal or nearly equal for various emission frequencies. In such cases, the electron beams traveled with constant velocities through the whole observed region.

The frequency drifts of the spikes are scattered in a broad range. In the entire set of investigated emissions, we have found some having very low negative and positive frequency drifts. We found eighteen spikes (0.3% of the whole population) with negative drifts lower than $-100\ \mathrm{MHz\,s^{-1}}$ and four spikes (0.08%) with positive drifts lower than $100\ \mathrm{MHz\,s^{-1}}$, respectively. These values of the frequency drifts are much lower than the average for all spikes and the average for type III bursts. Assuming that all the spikes are generated by electron beams, such a broad scattering of their frequency drifts can be explained by a broad range of density gradients in their sources. Namely, the travel distances of these beams are of the order of 1000 km due to the very short duration of the spikes. The density gradients on such short spatial scales can differ essentially from gradients in the mean solar atmosphere, especially in turbulent reconnection outflows, which were proposed as sources of the spikes by Bárta and Karlický (2001).

The spikes are spread uniformly across the radio spectrograms. The only exception consists of chains and columns of spikes, but their number is relatively low. This means that the individual spikes are emitted from various altitudes.

In some events (for example, on 17 May 2000) there are no spikes between the groups of spikes.

The main parameters of the spikes emitted in groups for a single event are stable. It is very likely that the emissions of the whole group occurred in a single region with stable physical conditions. The sizes of the emission sources of the investigated spikes were estimated using the relations described by Benz (2002). Assuming the plasma mechanism of emission, the mean diameters of the emission sources are of the order of 100 km.

For the dm-spikes emitted in 1420 MHz, the relevant electron density is equal to 2.5×10^{10} cm^{-3} (from the relation of the plasma frequency and electron density). According to the solar corona model evaluated by Aschwanden (2002), a resulting density of 2.5×10^{10} cm^{-3} must be present at an altitude of $\sim 10^4$ km above the solar surface. Assuming an active region mean magnetic field of the order of 200 G (Aschwanden, 2004) and a magnetic dipole depth of 3.5×10^4 km (Takakura and Scalise, 1970), the emission source observed at a gyromagnetic frequency of 1420 MHz (central frequency of the analyzed band of frequencies) will be located at an altitude of 5.7×10^3 km (Zhao, 1995). This altitude corresponds to the lower solar corona.

We have found chains of spikes in the analyzed spectrograms. Assuming the plasma emission mechanism, the chains of spikes could be excited by MHD shock waves similarly to the case of the chain of type I radio bursts (Spicer, Benz, and Huba, 1981). These chains thus support the idea that dm-spikes are generated in the turbulent reconnection plasma outflows (Bárta and Karlický, 2001).

While we found some agreement between global characteristics of the groups of spikes recorded with the two instruments located in Toruń and Ondřejov, we did not find any one-to-one relation between individual spikes (Figures 7 – 9). To explain this finding, we considered two possibilities: *i*) high directivity of spike emission and *ii*) differences in measurements at Toruń and Ondřejov. Concerning the first possibility, for the distance between Toruń and Ondřejov observatories (450 km), the angle of the directivity emission cone, explaining this low correlation, should be lower than ~ 1 arcsec. This is much lower than the one (1 arcmin) estimated for spikes by Fleishman (private communication). Therefore, we decided to check the second possibility by making the following test. We generated a one minute artificial spectrum covering the frequency range from 1400 MHz to 1700 MHz; then, we added 1000 spikes randomly distributed in this spectrum. The duration and bandwidth of individual spikes were selected in the observed ranges. Considering the parameters of the radio spectrographs at Toruń and Ondřejov, we generated matrixes that mimic real observations. Then, we computed "radio fluxes" at the grid points of these matrixes. Thus, we obtained two records of the same set of spikes, which represented the spectra from Toruń and Ondřejov. For comparison, we smoothed these spectra to the same time and frequency resolution and then computed their correlation coefficient. We made several computations for different values of the duration and spectral width of spikes. We found that the correlation coefficient increases with the increase of these parameters. For a duration of individual spikes of 0.2 s (the mean value) and a bandwidth of 2 MHz (the value estimated as the real one by Benz *et al.*, 2009), the correlation coefficient was 0.4. Based on these tests, we think that the low correlation between individual spikes observed at Toruń and Ondřejov is due to differences in time and spectral resolutions, and the sensitivity of both instruments.

Acknowledgements The authors are very grateful to the anonymous referees for constructive comments and suggestions, which have proved to be very helpful in improving the manuscript. This paper was supported by Grant 300030701 of the Grant Agency of the Academy of Sciences of the Czech Republic, research project AV0Z10030501, and the Centre for Theoretical Astrophysics, Prague.

References

Aschwanden, M.J.: 2002, Particle acceleration and kinematics in solar flares – a synthesis of recent observations and theoretical concepts. *Space Sci. Rev.* **101**, 1 – 227.

Aschwanden, M.J.: 2004, *Physics of the Solar Corona. An Introduction*. Springer, Berlin.

Aschwanden, M.J., Guedel, M.: 1992, The coevolution of decimetric millisecond spikes and hard X-ray emission during solar flares. *Astrophys. J.* **401**, 736 – 753.

Bárta, M., Karlický, M.: 2001, Turbulent plasma model of the narrowband dm-spikes. *Astron. Astrophys.* **379**, 1045 – 1051.
Benz, A.O.: 1985, Radio spikes and the fragmentation of flare energy release. *Solar Phys.* **96**, 357 – 370.
Benz, A.O.: 1986, Millisecond radio spikes. *Solar Phys.* **104**, 99 – 110.
Benz, A.O.: 2002, *Plasma Astrophysics: Kinetic Processes in Solar and Stellar Coronae*, **279**, 2nd edn., Kluwer Academic, Berlin.
Benz, A.O., Csillaghy, A., Aschwanden, M.J.: 1996, Metric spikes and electron acceleration in the solar corona. *Astron. Astrophys.* **309**, 291 – 300.
Benz, A.O., Magun, A., Stehling, W., Su, H.: 1992a, Electron beams in the low corona. *Solar Phys.* **141**, 335 – 346.
Benz, A.O., Su, H., Magun, A., Stehling, W.: 1992b, Millisecond microwave spikes at 8 GHz during solar flares. *Astron. Astrophys. Suppl.* **93**, 539 – 544.
Benz, A.O., Grigis, P.C., Csillaghy, A., Saint-Hilaire, P.: 2005, Survey on solar X-ray flares and associated coherent radio emissions. *Solar Phys.* **226**, 121 – 142.
Benz, A.O., Monstein, C., Beverland, M., Meyer, H., Stuber, B.: 2009, High spectral resolution observation of decimetric radio spikes emitted by solar flares – first results of the Phoenix-3 spectrometer. *Solar Phys.* **260**, 375 – 388.
Cadwell, B.J.: 1997, Pulsar survey and timing with the Penn State Pulsar Machines. Ph.D. thesis.
Chernov, G.P., Fu, Q.J., Lao, D.B., Hanaoka, Y.: 2001, Ion-sound model of microwave spikes with fast shocks in the reconnection region. *Solar Phys.* **201**, 153 – 180.
Csillaghy, A., Benz, A.O.: 1993, The bandwidth of millisecond radio spikes in solar flares. *Astron. Astrophys.* **274**, 487 – 496.
Dąbrowski, B.P., Benz, A.O.: 2009, Correlation between decimetric radio emission and hard X-rays in solar flares. *Astron. Astrophys.* **504**, 565 – 573.
Dąbrowski, B.P., Kus, A.J.: 2007, Millisecond solar radio spikes observed at 1420 MHz. *Mem. Soc. Astron. Ital.* **78**, 264 – 267.
Dąbrowski, B.P., Rudawy, P., Falewicz, R., Siarkowski, M., Kus, A.J.: 2005, Millisecond radio spikes in the decimetre band and their related active solar phenomena. *Astron. Astrophys.* **434**, 1139 – 1153.
Fárník, F., Karlický, M.: 2005, Narrowband dm-spikes in the frequency range above 1 GHz and hard X-ray emission. *Adv. Space Res.* **35**, 1799 – 1804.
Fleishman, G.D., Melnikov, V.F.: 1998, Millisecond solar radio spikes. *Sov. Phys. Usp.* **41**, 1157 – 1189.
Fleishman, G.D., Melnikov, V.F.: 1999, Solar millisecond radio spikes: recent progress, current state, problems for future. In: A. Wilson *et al.* (eds.) *Magnetic Fields and Solar Processes, ESA Special Publication* **448**, 1247 – 1251.
Guedel, M., Benz, A.O.: 1988, A catalogue of decimetric solar flare radio emission. *Astron. Astrophys. Suppl.* **75**, 243 – 259.
Guedel, M., Benz, A.O.: 1990, Time profiles of solar radio spikes. *Astron. Astrophys.* **231**, 202 – 212.
Jiřička, K., Karlický, M., Kepka, O., Tlamicha, A.: 1993, Fast drift burst observations with the new Ondrejov radiospectrograph. *Solar Phys.* **147**, 203 – 206.
Jiřička, K., Karlický, M., Mészárosová, H., Snížek, V.: 2001, Global statistics of 0.8 – 2.0 GHz radio bursts and fine structures observed during 1992 – 2000 by the Ondřejov radiospectrograph. *Astron. Astrophys.* **375**, 243 – 250.
Karlický, M., Sobotka, M., Jiřička, K.: 1996, Narrowband dm-spikes in the 2 GHz frequency range and MHD cascading waves in reconnection outflows. *Solar Phys.* **168**, 375 – 383.
Karlický, M., Jiřička, K., Sobotka, M.: 2000, Power-law spectra of 1 – 2 GHz narrowband dm-spikes. *Solar Phys.* **195**, 165 – 174.
Karlický, M., Yan, Y., Fu, Q., Wang, S., Jiřička, K., Mészárosová, H., Liu, Y.: 2001, Drifting radio bursts and fine structures in the 0.8 – 7.6 GHz frequency range observed in the NOAA 9077 AR (10 – 14 July 2000) solar flares. *Astron. Astrophys.* **369**, 1104 – 1111.
Magdalenić, J., Vršnak, B., Zlobec, P., Hillaris, A., Messerotti, M.: 2006, Classification and properties of supershort solar radio bursts. *Astrophys. J.* **642**, L77 – L80.
Messmer, P., Benz, A.O.: 2000, The minimum bandwidth of narrowband spikes in solar flare decimetric radio waves. *Astron. Astrophys.* **354**, 287 – 295.
Mészárosová, H., Veronig, A., Zlobec, P., Karlický, M.: 2003, Analysis of solar narrow band dm-spikes observed at 1420 and 2695 MHz. *Astron. Astrophys.* **407**, 1115 – 1125.
Rozhansky, I.V., Fleishman, G.D., Huang, G.L.: 2008, Millisecond microwave spikes: statistical study and application for plasma diagnostics. *Astrophys. J.* **681**, 1688 – 1697.
Spicer, D.S., Benz, A.O., Huba, J.D.: 1981, Solar type I noise storms and newly emerging magnetic flux. *Astron. Astrophys.* **105**, 221 – 228.
Takakura, T., Scalise, E.: 1970, Gyro-synchrotron emission in a magnetic dipole field for the application to the center-to-limb variation of microwave impulsive bursts. *Solar Phys.* **11**, 434 – 455.

Wang, M., Xie, R.X.: 1999, Millisecond spikes in a short decimetric solar radio burst. *Solar Phys.* **185**, 351 – 360.
Wang, S.J., Yan, Y.H., Fu, Q.J.: 2002, Spikes in solar radio bursts at frequencies near 3 GHz. *Solar Phys.* **209**, 185 – 193.
Wang, S.J., Yan, Y.H., Liu, Y.Y., Fu, Q.J., Tan, B.L., Zhang, Y.: 2008, Solar radio spikes in 2.6 – 3.8 GHz during the 13 December 2006 event. *Solar Phys.* **253**, 133 – 141.
Zhao, R.: 1995, A model of solar (radio) active regions. *Astrophys. Space Sci.* **234**, 125 – 137.

Solar Phys (2011) 273:393–402
DOI 10.1007/s11207-011-9794-6

Magnetoacoustic Wave Trains in the 11 July 2005 Radio Event with Fiber Bursts

H. Mészárosová · M. Karlický · J. Rybák

Received: 29 November 2010 / Accepted: 12 May 2011 / Published online: 1 July 2011

Abstract A dm-radio emission with fiber bursts observed on 11 July 2005 was analyzed using wavelet filtration and spectral methods. In filtered radio spectra we found structures with different characteristic period P and frequency drift FD: *i*) fiber substructures (composed of dot emissions) with $P_1 \approx 0.5$ s, $\mathrm{FD}_1 = -87\ \mathrm{MHz\,s^{-1}}$ on average, *ii*) fiber structures with $P_2 \approx 1.9$ s, and *iii*) drifting structures with $P_3 \approx 81.4$ s, $\mathrm{FD}_2 = -8.7, +98.5$, and $-21.8\ \mathrm{MHz\,s^{-1}}$. In the wavelet spectra we recognized patterns having the form of tadpoles. They were detected with the same characteristic periods P as found for the filtered structures. The frequency drift of the tadpole heads is found to be equal to the frequency drift of some groups of fibers for the long-period wavelet tadpoles (P_3) and to the frequency drift of individual fibers for the short-period tadpoles (P_2). Considering these wavelet tadpoles as signatures of propagating magnetoacoustic wave trains, the results indicate the presence of several wave trains in the fibers' source. While the long-period wave trains trigger or modulate a whole group of fibers, the short-period ones look like being connected with individual fiber bursts. This result supports the model of fibers based on magnetoacoustic waves. Using a density model of the solar atmosphere we derived the velocities of the magnetoacoustic waves, 107 and 562 $\mathrm{km\,s^{-1}}$, and setting them equal to the Alfvén ones we estimated the magnetic field in the source of fiber bursts as 10.7 and 47.8 G.

Keywords Sun: corona · Sun: flares · Sun: radio radiation · Sun: oscillations · Methods: data analysis

Energy Storage and Release through the Solar Activity Cycle–Models Meet Radio Observations
Guest Editors: Christophe Marqué and Alexander Nindos

H. Mészárosová (✉) · M. Karlický
Astronomical Institute, Academy of Sciences of the Czech Republic, 25165 Ondřejov, Czech Republic
e-mail: hana@asu.cas.cz

M. Karlický
e-mail: karlicky@asu.cas.cz

J. Rybák
Astronomical Institute, Slovak Academy of Sciences, 05960 Tatranská Lomnica, Slovak Republic
e-mail: rybak@astro.sk

1. Introduction

It was theoretically predicted by Roberts, Edwin, and Benz (1983, 1984) that impulsively generated signals (produced *e.g.* by solar flares) can generate propagating magnetoacoustic wave trains. These wave trains can be trapped in regions with higher density, *e.g.* in coronal loops. Nakariakov *et al.* (2004) studied these magnetoacoustic wave trains numerically. They showed that these trains in the wavelet spectrum correspond to the pattern where a narrowband tail precedes broadband head. Due to the shape of this pattern (tail plus head) they named these characteristic patterns wavelet tadpoles. (Please note here the tadpoles mean tadpoles in the wavelet spectra (Nakariakov *et al.*, 2004; Aschwanden, 2005), not the tadpole radio bursts (Kuijpers, 1975a; Slottje, 1981).) The wavelet tadpoles of the present type were observed in solar eclipse data (Katsiyannis *et al.*, 2003) and in decimetric radio bursts of both the gyrosychrotron (Mészárosová *et al.*, 2009a) and plasma emissions (Mészárosová *et al.*, 2009b).

Complex decimetric radio emissions are characterized by many radio bursts and fine structures (Jiřička *et al.*, 2001). Among them the fiber bursts (or fibers), which are considered to be a fine structure of type IV bursts, are defined as short-lasting and repeated bursts with a frequency drift of about $-100\ \mathrm{MHz\,s^{-1}}$. Individual fibers can even be composed of distinct (isolated) emission dots (for more details of these substructures, see Mészárosová *et al.*, 2008).

In the present paper we search for both the characteristic radio structures and the magnetoacoustic wave trains (tadpoles in the wavelet spectra) in complex radio spectrum. Analyzing these radio structures and wavelet tadpoles we want to contribute to a better understanding of fiber bursts.

The paper is organized as follows: in Section 2 we present the observations, in Section 3 the wavelet analysis of these data is described and in Section 4 we discuss our results and draw our conclusions.

2. Observations

The 11 July 2005 decimetric radio event (16:31:40 – 16:43:20 UT, 800 – 2000 MHz), associated with fiber bursts (Figure 1, upper panel), was recorded by the Ondřejov radio spectrograph (Jiřička *et al.*, 1993, time resolution = 0.1 s, frequency resolution = 5 MHz). This event classified as C1.1 (GOES X-ray maximum at 16:38 UT) was observed in the active region NOAA AR 10786 after the flare classified as C8.4 (GOES X-ray maximum at 15:08 UT, Hα importance SF).

The whole radio event consists of several groups of fiber bursts and underlying continua with a characteristic intensity of 180 – 700 SFU except for the time interval 16:35:45 – 16:36:17 UT with the characteristic intensity of 1080 – 2070 SFU. Individual groups of fibers have different global frequency drifts, as studied in detail in the following. The most remarkable property is the frequency drift of the individual fiber bursts that ranges from -71 to $-144\ \mathrm{MHz\,s^{-1}}$ ($-105\ \mathrm{MHz\,s^{-1}}$ on average). The fibers in the time interval 16:35:50 – 16:36:40 UT are shown in Figure 2 (upper panel). Most of these fibers are composed of emission dots (Mészárosová *et al.*, 2008).

The fibers with long-period wavelet tadpoles in time interval 16:38 – 16:41 UT were previously studied by Mészárosová *et al.* (2009b). Here, we extended our study to the whole radio event (16:31:40 – 16:43:20 UT); especially to its most complex part (16:35:50 – 16:36:40 UT). Furthermore, we extended also the range of periods studied to shorter ones.

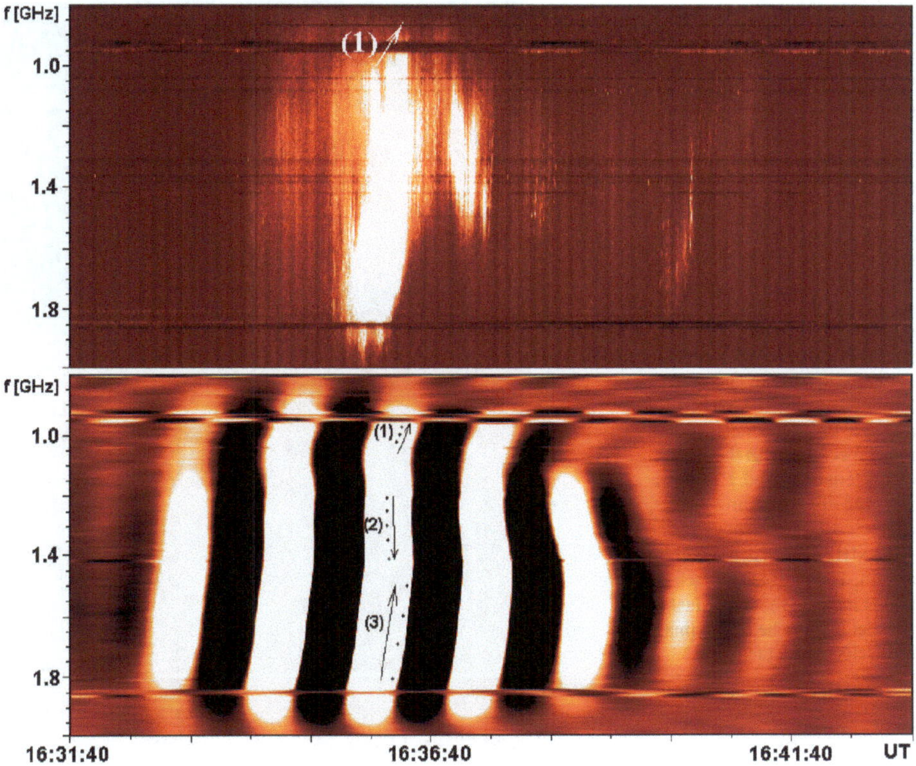

Figure 1 Upper panel: Global view on 11 July 2005 of the dm-radio event recorded by the Ondřejov radio spectrograph (16:31:40 – 16:43:20 UT, 800 – 2000 MHz), which consists of groups of fiber bursts and underlying continuum. Bottom panel: Filtered radio spectrum with the characteristic period $P = 81.4$ s showing structure with three different frequency drifts, marked by black arrows 1, 2, and 3 equal to -8.7, $+98.5$, and $-21.8\ \mathrm{MHz\,s^{-1}}$, respectively. The frequency drift (black arrow 1) is equal to the frequency drift (white arrow 1) in the upper panel. The individual black points indicate the individual wavelet tadpole head maxima in Figures 4 – 6.

3. Wavelet Analysis of the 11 July 2005 Radio Event with Fibers

To find the characteristic properties of the 11 July 2005 radio event with fibers we used two independent methods of wavelet analysis:

i) wavelet time filtration of a complex radio spectrum to separate individual drifting structures which are present in the original complex radio dynamic spectrum;
ii) wavelet power spectra of radio flux time series at individual frequencies to detect characteristic wavelet tadpole patterns as signatures of magnetoacoustic wave trains.

We focused on characteristic periods and frequency drifts of both the separated structures and the corresponding magnetoacoustic wave trains to find possible relations between them.

3.1. Wavelet Time Separation of Drifting Structures in a Complex Radio Spectrum

We used a new method of the separation of drifting structures in a complex radio spectrum developed by Mészárosová, Rybák, and Karlický (2011). It gives us two independent possibilities on how to separate individual structures (bursts) of the original radio spectrum;

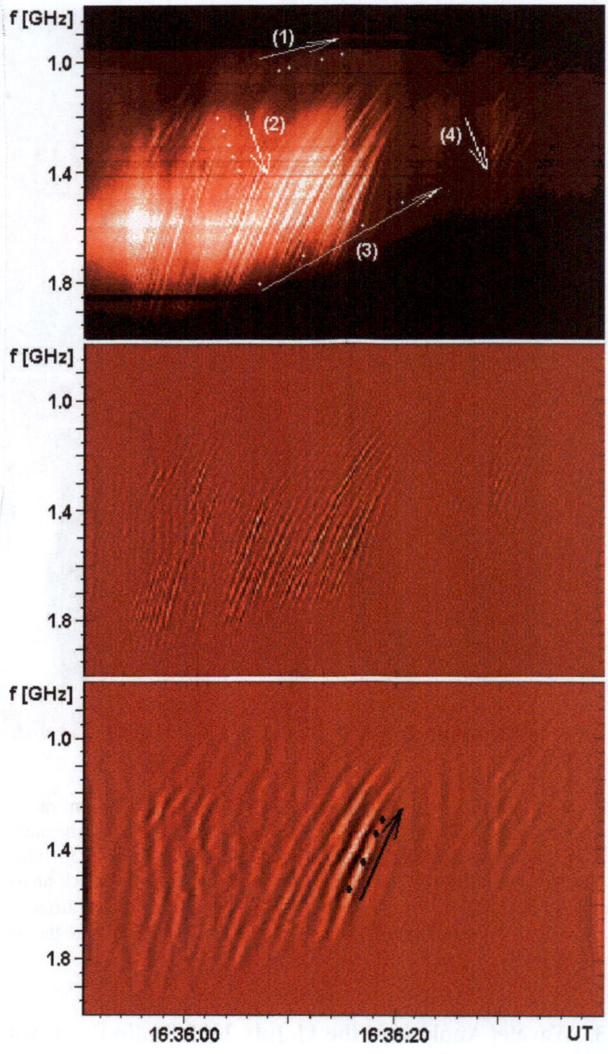

Figure 2 Upper panel: Detailed view on fiber bursts in the time interval 16:35:50 – 16:36:40 UT. The frequency drifts (arrows 1 – 3) equal the frequency drifts obtained by the wavelet method and the individual white points indicate individual wavelet tadpole head maxima in Figures 4 – 6. The frequency drift (arrow 4) equals that of arrow 2. Middle panel: Filtered radio spectrum with the characteristic period $P \approx 0.5$ s showing the fiber substructures (composed of dot emissions) with characteristic frequency drift $-87\ \mathrm{MHz\,s^{-1}}$ on average. Bottom panel: Filtered radio spectrum with the characteristic period $P \approx 1.9$ s showing fiber structures with characteristic frequency drift $-78\ \mathrm{MHz\,s^{-1}}$ on average (black arrow). The individual black points indicate the individual wavelet tadpole head maxima at four selected frequencies.

in time and in frequency. Here, we used the time separation, *i.e.* separation according to characteristic period(s) of the individual structures.

Only one characteristic period $P = 81.4$ s was found to be present at all frequencies during the whole event (16:31:40 – 16:43:20 UT, upper panel in Figure 1). This period P was found in the period range 70 – 90 s, which was determined from the averaged global wavelet spectrum (for more details, see Section 2 in Mészárosová, Rybák, and Karlický, 2011). The filtered radio spectrum is presented in the bottom panel of Figure 1. At least three different frequency drifts in this radio structure were found (black arrows 1, 2, and 3), which are equal to -8.7, $+98.5$, and $-21.8\ \mathrm{MHz\,s^{-1}}$ at frequency ranges about 800 – 1050, 1050 – 1450, 1450 – 2000 MHz, respectively.

Two characteristic periods $P \approx 0.5$ s (computed in the period range 0.3 – 1.0 s) and $P \approx 1.9$ s (computed in the period range 1.0 – 3.0 s) were found to be present during the selected time interval 16:35:50 – 16:36:40 UT (Figure 2, upper panel). The filtered radio spectrum with the characteristic periods $P \approx 0.5$ s is shown in the middle panel of Figure 2.

It presents fiber substructures composed of emission dots. Their frequency drift is -70 down to $-100\ \mathrm{MHz\,s^{-1}}$ ($-87\ \mathrm{MHz\,s^{-1}}$ on average). In addition, the filtered radio spectrum with the characteristic periods $P \approx 1.9$ s is presented in the bottom panel of Figure 2. It consist of fiber structures with frequency drifts ranging from -60 to $-100\ \mathrm{MHz\,s^{-1}}$ ($-78\ \mathrm{MHz\,s^{-1}}$ on average; see also the black arrow, bottom panel).

Thus, the original complex radio spectrum consists of at least three different radio structures.

3.2. Wavelet Power Spectra with Tadpole Patterns

We searched for the wavelet power spectra with tadpole patterns as signatures of the magnetoacoustic wave trains (Nakariakov *et al.*, 2004). We used the wavelet analysis method (Torrence and Compo, 1998) with the Morlet mother function. In these wavelet spectra we determined the characteristic periods of wavelet tadpole patterns, the cone of influence COI (hatched regions where edge effects become important due to finite-length time series) and the confidence level CL (relative to red noise). Only the regions outside the COI with CL above 90% were considered as significant in each time series. The spectra are plotted with the lighter areas indicating greater power.

The original radio spectra, shown in Figure 1 and 2 (upper panels), were divided into 256 radio flux time series at individual frequencies (cuts of the radio spectrum along time) from 800 to 2000 MHz. As a result of this analysis, we found the wavelet tadpole patterns with characteristic periods $P \approx 0.5$ s, $P \approx 1.9$ s and $P = 81.4$ s during time intervals 16:35:50 – 16:36:20, 16:36:10 – 16:36:20, and 16:33:30 – 16:38:20 UT, respectively. Thus, individual wavelet tadpoles have the same characteristic period P as the individual separated radio structures. We search for frequency drifts of the wavelet tadpole head maxima, similarly as in the paper by Mészárosová *et al.* (2009b).

The maxima of wavelet tadpole heads with characteristic period $P \approx 0.5$ s are present rather chaotically during the time interval 16:35:50 – 16:36:20 UT, *i.e.* without a significant frequency drift. A typical example of such a wavelet tadpole as a signature of the magnetoacoustic wave train (plot period *vs.* time) detected at the radio frequency 1391 MHz is shown in Figure 3 (upper panel).

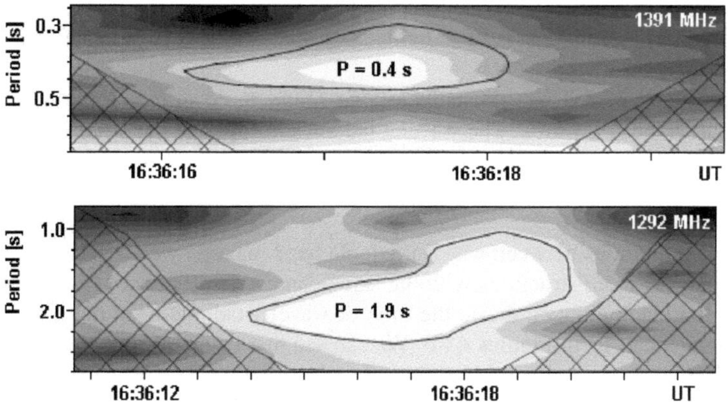

Figure 3 Upper panel: Example of the wavelet power spectrum with the wavelet tadpole pattern and the characteristic period $P = 0.4$ s. Bottom panel: Example of the wavelet power spectrum with the wavelet tadpole pattern and the characteristic period $P = 1.9$ s.

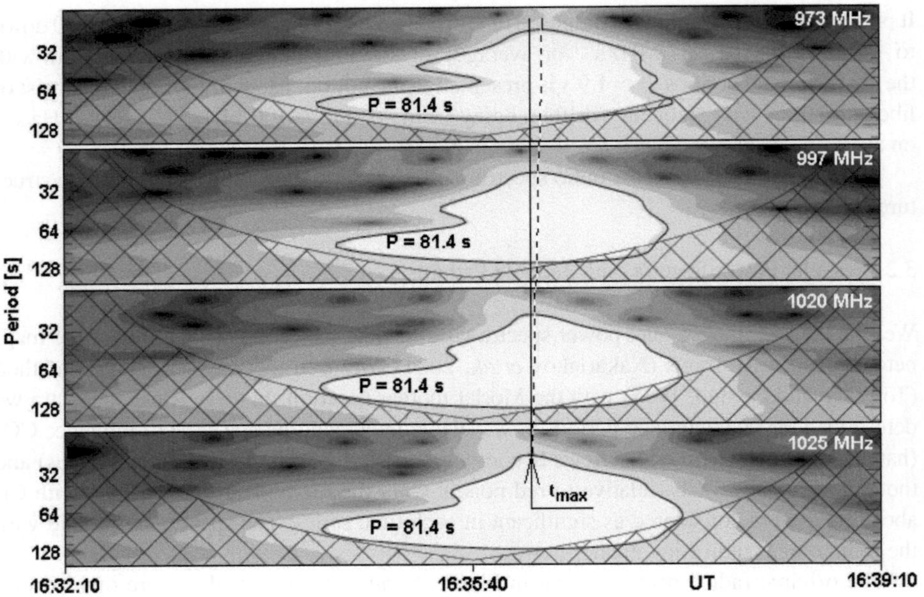

Figure 4 Wavelet tadpoles as signatures of a magnetoacoustic wave train in the 973 – 1025 MHz frequency range. The panels show wavelet power spectra with tadpole patterns and the characteristic period of $P = 81.4$ s. The solid line represents time t_{max} of the tadpole head maximum at the highest frequency and the dashed line shows the negative frequency drift of these head maxima toward lower frequencies ($-8.7\ \mathrm{MHz\,s^{-1}}$; see also arrow 1 in Figure 1).

A typical example of a wavelet tadpole with characteristic period $P \approx 1.9$ s detected at the radio frequency 1292 MHz is shown in the bottom panel (Figure 3). The black points (Figure 2, bottom panel) indicate the time of wavelet tadpole head maxima at four selected frequencies (1292, 1348, 1447, and 1555 MHz). Here, we can see that these maxima show a similar frequency drift as fibers.

Typical examples of the wavelet tadpoles with the period $P = 81.4$ s are shown in Figures 4 – 6. The panels show wavelet power spectra detected at selected radio frequencies. The solid line represents the wavelet tadpole head maximum at the highest frequency and the frequency drift of the heads of individual wavelet tadpoles is indicated by the dashed line.

For these long-period wavelet tadpoles we found three different frequency drifts of wavelet tadpole head maxima. The wavelet tadpoles in Figures 4 and 6 drift toward higher frequencies with the negative frequency drifts -8.7 and $-21.8\ \mathrm{MHz\,s^{-1}}$, respectively. Wavelet tadpoles in Figure 5 drift toward lower frequencies with a positive frequency drift $+98.5\ \mathrm{MHz\,s^{-1}}$.

We studied relations between wavelet tadpole head maxima with a period of $P = 81.4$ s and the separated radio structures with the same characteristic period. In the bottom panel (Figure 1) the black points indicate the time of wavelet tadpole head maxima at the same selected frequencies as in Figures 4 – 6. We can see that these maxima show a similar frequency drift to that of the filtered structures (black arrows 1 – 3). These maxima were found at times of the highest intensity of the whole radio event.

Furthermore, these drifts can also be seen in the original radio dynamic spectrum. The negative frequency drifts -8.7 and $-21.8\ \mathrm{MHz\,s^{-1}}$ are shown at the upper panel of Figure 2 (white arrow 1 and 3), respectively. The first one is also detected at the upper panel of

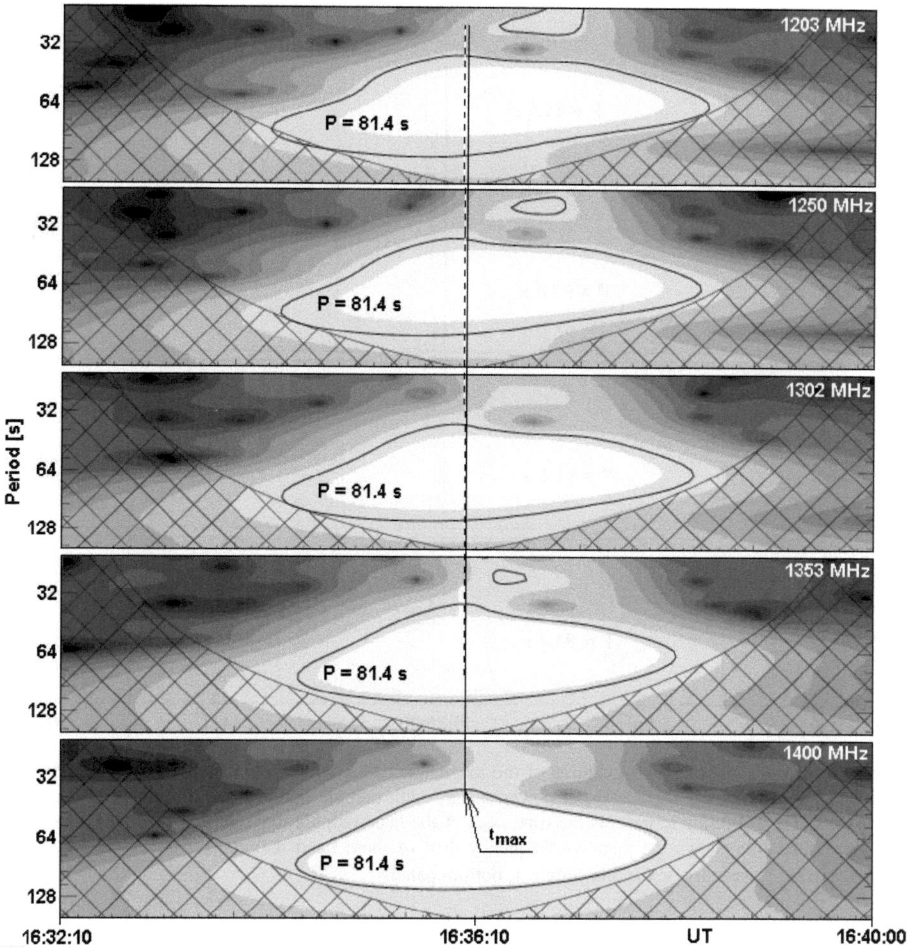

Figure 5 Wavelet tadpoles as signatures of the magnetoacoustic wave train in the 1203 – 1400 MHz frequency range. The panels show wavelet power spectra with tadpole patterns and the characteristic period $P = 81.4$ s. The solid line represents time $t_{\max}$ of the head tadpole maximum at the highest frequency and the dashed line shows the positive frequency drift of these head maxima toward higher frequencies ($+98.5$ MHz s^{-1}; see also arrow 2 in Figure 1, bottom panel).

Figure 1 (arrow 1). The positive frequency drift ($+98.5$ MHz s^{-1}) is shown in the upper panel of Figure 2 (white arrows 2 and 4). The white points (upper panel of Figure 1) indicate the wavelet tadpoles head maxima at selected frequencies in Figures 4 – 6. It is remarkable that the times of the wavelet tadpole head maxima (arrow 3 in Figure 2) correspond to the starting times of fiber bursts. This result is the same as in our previous study of fiber bursts in the interval 16:38 – 16:41 UT (Mészárosová *et al.*, 2009b). The other arrows (1 and 2) do not show such a correspondence. Nevertheless, the drift of arrow 2 is the same as that of starting frequencies of the fibers designated by the arrow 4.

Thus, we found that all three frequency drifts of wavelet tadpoles head maxima with the characteristic period $P = 81.4$ s (Figures 4 – 6) are similar to the ones of separated radio structures (Figure 1, bottom panel) and they correspond to global frequency drifts of the groups of fibers in the original radio spectrum (upper panels in Figures 1 and 2).

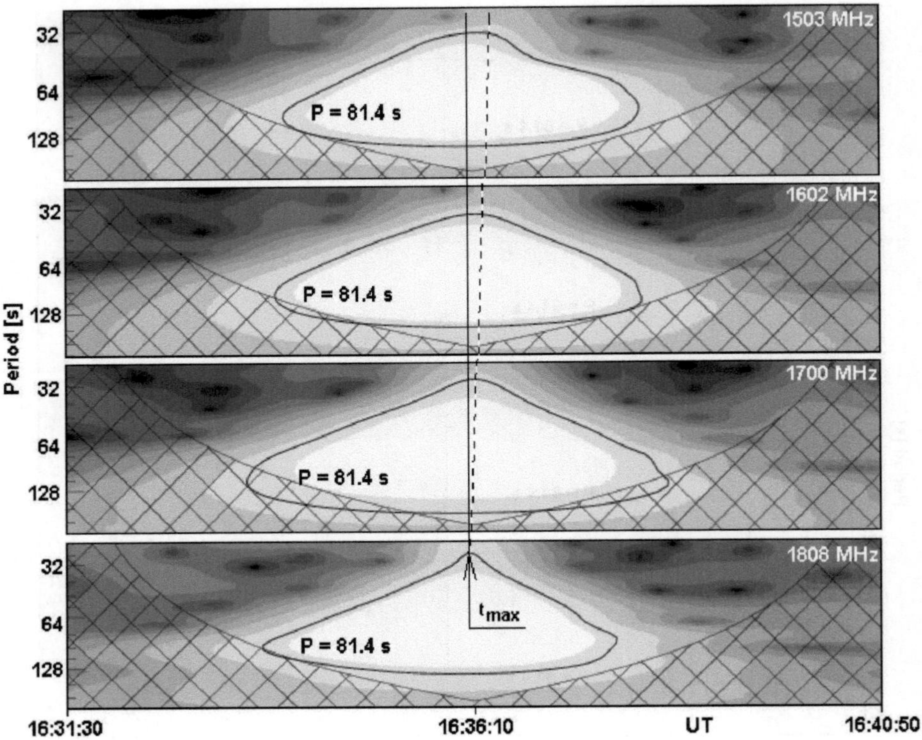

Figure 6 Wavelet tadpoles as signatures of the magnetoacoustic wave train in the 1503 – 1808 MHz frequency range. The panels show wavelet power spectra with tadpole patterns and the characteristic period $P = 81.4$ s. The solid line represents the time $t_{\max}$ of the head tadpole maximum at the highest frequency and the dashed line shows the negative frequency drift of these head maxima toward lower frequencies (-21.8 MHz s^{-1}; see also arrow 3 in Figure 1, bottom panel).

4. Discussion and Conclusions

We have found that the complex radio spectrum observed on 11 July 2005 (upper panels in Figures 1 and 2) consists of at least three different radio structures (middle and bottom panels in Figures 1 and 2) with different characteristic periods. We have also detected the presence of tadpole wavelet patterns; see Figures 3 – 6 with the same characteristic periods as found for the separated (filtered) radio structures. Furthermore, we found that the maxima of the wavelet tadpole heads for the period $P = 81.4$ s were located at the upper frequency boundary of a group of fibers, and the maxima of the tadpole heads for the period $P \approx 1.9$ s drifted along one strong fiber.

We interpret these wavelet tadpole patterns of radio emission fluxes as signatures of magnetoacoustic wave trains in accordance with the papers of Katsiyannis *et al.* (2003), Nakariakov *et al.* (2004), and Mészárosová *et al.* (2009a, 2009b). It means that in the radio source of fibers there are at least three types of magnetoacoustic wave trains. While the long-period wave train (with $P = 81.4$ s) triggers or modulates a whole group of fiber bursts, the short-period one (with $P \approx 1.9$ s) looks like being connected with individual fiber bursts.

Let us discuss these results in the frame of existing models of fiber bursts. Generally, these models can be divided into three groups:

Table 1 Parameters of the long-period (case 3) and short-period magnetoacoustic waves (t_{max} = time of wavelet tadpole head maximum, FD = frequency drift of head maxima, n_e = average electron density, h = mean coronal altitudes, v = wave train velocity, B = magnetic field strength).

Period (s)	Frequency (MHz)	t_{max} (UT)	FD (MHz s^{-1})	n_e (cm^{-3})	h (Mm)	v (km s^{-1})	B (G)
81.4	1503	16:36:21		2.80×10^{10}	10.5		
	1808	16:36:07	−21.8	4.05×10^{10}	8.9	107	10.7
1.9	1292	16:36:19		2.07×10^{10}	11.9		
	1550	16:36:16	−86	2.98×10^{10}	10.2	562	47.8

i) models based on whistler waves (Kuijpers, 1975b; Mann, Karlický, and Motschmann, 1987; Mann *et al.*, 1989),
ii) models considering nonlinear Alfvén solitons (Bernold and Treumann, 1983; Treumann, Güdel, and Benz, 1990), and
iii) models with sausage magnetoacoustic waves (Kuznetsov, 2006).

These models were proposed for the generation of individual fibers. In our case, we found that one fiber drifted simultaneously with the signature of the magnetoacoustic wave train (maxima of the wavelet tadpoles). This fact agrees only with the model of fibers by Kuznetsov (2006). It is a very interesting result; however, it needs to be taken with caution, because it was found by a complex method, and up to now only in the present case. Therefore, further verification of this result for more events is necessary. As regards the long-period ($P = 81.4$ s) magnetoacoustic wave train, which was also present in the source of fibers, it can be considered as the one modulating or triggering the whole group of fibers.

Now, without consideration of any specific fiber bursts model, the frequency drifts of the heads of the magnetoacoustic wave trains can be used for the estimation of the magnetic field.

Let us consider two cases, the long-period magnetoacoustic wave train (case 3, in Figure 2, upper panel) and short-period train (Figure 2, bottom panel). While in the first case the head maxima follow the starting frequencies of the fibers, in the second case the head maxima correspond to the shape of fibers. We assume that the magnetoacoustic wave trains move upwards in the density stratified atmosphere and modulate the radio emission produced by plasma emission mechanism at the plasma frequency. Then using the density model of the solar atmosphere by Aschwanden (2002), we derive the velocities of these wave trains as 107 and 562 km s^{-1} (Table 1). The maximum amplitude of these trains corresponds to the head maximum of the wavelet tadpoles (Nakariakov *et al.*, 2004). Furthermore, at times of the head maximum of wavelet tadpoles, the group velocity of the corresponding magnetoacoustic wave train is close to the Alfvén velocity inside the guiding loop (Roberts, Edwin, and Benz, 1984). Thus, setting these velocities equal to the Alfvén ones we estimated the magnetic field in the mentioned two cases as 10.7 and 47.8 G (Table 1). Furthermore, we can estimate the width of the density structures (loops) in the solar atmosphere guiding these two magnetoacoustic trains as (width $\sim$ period times wave train velocity) 8700 and 1070 km. Because the long-period wave train influences the short-period one, *e.g.* through triggering and modulation of fibers, it indicates that the narrower guiding structure is probably inside or close to the broader one.

Acknowledgements H.M. and M.K. acknowledge support from the Grant IAA300030701 of the Academy of Sciences of the Czech Republic and the research project AVOZ10030501 of the Astronomical Institute AS

CR. The work of J.R. was partly supported by the Slovak Grant Agency VEGA (project 2/0064/09). The program of mobility between the academies of the Czech Republic and Slovakia is also acknowledged. The wavelet analysis was performed using the software based on tools provided by C. Torrence and G.P. Compo at http://paos.colorado.edu/research/wavelets.

References

Aschwanden, M.J.: 2002, *Space Sci. Rev.* **101**, 1.

Aschwanden, M.J.: 2005, In: *Physics of the Solar Corona*, Praxis, Chichester.

Bernold, T., Treumann, R.: 1983, *Astrophys. J.* **264**, 677.

Jiřička, K., Karlický, M., Kepka, O., Tlamicha, A.: 1993, *Solar Phys.* **147**, 203.

Jiřička, K., Karlický, M., Mészárosová, H., Snížek, V.: 2001, *Astron. Astrophys.* **375**, 243.

Katsiyannis, A.C., Williams, D.R., McAteer, R.T.J., Gallagher, P.T., Keenan, F.P., Murtagh, F.: 2003, *Astron. Astrophys.* **406**, 709.

Kuijpers, J.: 1975a, In: *Doctor in de Wiskunde en Natuurwetenschappen Dissertation*, Rijksuniversiteit, Utrecht.

Kuijpers, J.: 1975b, *Solar Phys.* **44**, 173.

Kuznetsov, A.A.: 2006, *Solar Phys.* **237**, 153.

Mann, G., Karlický, M., Motschmann, U.: 1987, *Solar Phys.* **110**, 381.

Mann, G., Baumgartel, K., Chernov, G., Karlický, M.: 1989, *Solar Phys.* **120**, 383.

Mészárosová, H., Rybák, J., Karlický, M.: 2011, *Astron. Astrophys.* **525**, A88.

Mészárosová, H., Karlický, M., Sawant, H.S., Fernandes, F.C.R., Cecatto, J.R., de Andrade, M.C.: 2008, *Astron. Astrophys.* **491**, 555.

Mészárosová, H., Karlický, M., Rybák, J., Jiřička, K.: 2009a, *Astrophys. J. Lett.* **697**, L108.

Mészárosová, H., Karlický, M., Rybák, J., Jiřička, K.: 2009b, *Astron. Astrophys.* **502**, L13.

Nakariakov, V.M., Arber, T.D., Ault, C.E., Katsiyannis, A.C., Williams, D.R., Keenan, F.P.: 2004, *Mon. Not. Roy. Astron. Soc.* **349**, 705.

Roberts, B., Edwin, P.M., Benz, A.O.: 1983, *Nature* **305**, 688.

Roberts, B., Edwin, P.M., Benz, A.O.: 1984, *Astrophys. J.* **279**, 857.

Slottje, C.: 1981, In: *Atlas of Fine Structures of Dynamic Spectra of Solar Type IV-dm and Some Type II Radio Bursts*, Netherlands Foundation for Radio Astronomy, Dwingeloo.

Torrence, C., Compo, G.P.: 1998, *Bull. Am. Meteorol. Soc.* **79**, 61.

Treumann, R., Güdel, M., Benz, A.O.: 1990, *Astron. Astrophys.* **236**, 242.

Solar Phys (2011) 273:403–412
DOI 10.1007/s11207-011-9822-6

Quasi-periodic Oscillations of Solar Active Regions in Connection with Their Flare Activity – NoRH Observations

V.E. Abramov-Maximov · G.B. Gelfreikh · K. Shibasaki

Received: 30 November 2010 / Accepted: 7 July 2011 / Published online: 25 August 2011

Abstract The sunspot-associated sources at the frequency of 17 GHz give information on plasma parameters in the regions of magnetic field about $B = 2000$ G at the level of the chromosphere-corona transition region. The observations of short period (from one to ten minutes) oscillations in sunspots reflect propagation of magnetohydrodynamic (MHD) waves in the magnetic flux tubes of the sunspots. We investigate the oscillation parameters in active regions in connection with their flare activity. We confirm the existence of a link between the oscillation spectrum and flare activity. We find differences in the oscillations between pre-flare and post-flare phases. In particular, we demonstrate a case of powerful three-minute oscillations that start just before the burst. This event is similar to the cases of the precursors investigated by Sych *et al.* (*Astron. Astrophys.* **505**, 791, 2009). We also found well-defined eight-minute oscillations of microwave emission from sunspot. We interpret our observations in terms of a relationship between MHD waves propagating from sunspots and flare processes.

Keywords Oscillations, solar · Radio emission, active regions · Corona, radio emission · Radio bursts, association with flares

1. Introduction

Quasi-periodic oscillations (QPO) are detected practically in all wavelength ranges and in all structures of the solar atmosphere (Lites *et al.*, 1998; Bogdan, 2000; Fludra, 2001;

Energy Storage and Release through the Solar Activity Cycle – Models Meet Radio Observations
Guest Editors: Christophe Marqué and Alexander Nindos.

V.E. Abramov-Maximov (✉) · G.B. Gelfreikh
Central Astronomical Observatory at Pulkovo, Russian Acad. Sci., St. Petersburg 196140, Russia
e-mail: beam@gao.spb.ru

K. Shibasaki
Nobeyama Solar Radio Observatory, Minamisaku, Nagano, Japan
e-mail: shibasaki@nro.nao.ac.jp

Bogdan and Judge, 2006; Kosovichev, 2009). The periods of oscillations typically range from seconds to hours and perhaps days. The dominant periods of the oscillations are three and five minutes. Most of these oscillations are of intermittent nature: both their amplitude and frequency vary with time. The oscillations are often visible as trains of a few periods. The oscillations with different periods have different physical nature. The short-period oscillations are caused by running waves. Their intermittency is probably caused by instability of physical conditions in the regions of generation and propagation.

Studying oscillations can help us understand such fundamental astrophysical problems as accumulation and release of energy, physics of the coronal heating and the origin of flares. The investigation of oscillation processes in the solar corona is a powerful tool for coronal plasma diagnostics (Nakariakov and Erdélyi, 2009).

Observations of QPO of the solar radio emission have been carried out for more than 40 years (Durasova, Kobrin, and Yudin, 1971). The first observations showed that oscillations are mostly produced in solar active regions and their parameters reflect the development of flare activity (Kobrin *et al.*, 1973; Aleshin, Kobrin, and Korshunov, 1973). However, in these studies the spatial and temporal resolution of radio telescopes were not adequate.

Significant progress in this field was achieved with the help of large modern radio telescopes. Especially rich information was gained from the Nobeyama Radioheliograph (NoRH) (Nakajima *et al.*, 1994). The NoRH has been observing the Sun since 1992. The observations are carried out every day (seven – eight hours daily) with spatial resolution of 10 – 20 arcsec and cadence of 1 sec.

The most sensitive analysis of the radio oscillations reflecting the plasma resonance phenomena for MHD oscillations was found for the sunspot-associated sources (Gelfreikh *et al.*, 1999; Shibasaki, 2001; Nindos *et al.*, 2002; Gelfreikh, Nagovitsyn, and Nagovitsyna, 2006). Such oscillations reflect the oscillations of the plasma parameters (temperature and magnetic field) at the level of the chromosphere-corona transition region (CCTR), where the strength of the magnetic field is about $B = 2000$ G. The analysis has shown the presence of regular oscillations of the radio brightness (registered both in I and V Stokes parameters). The periods of oscillations vary from fractions of a minute to several hours. In reality they have a different nature. The nature of oscillations is still in the stage of discussion. However, some mechanisms have been identified. The three- and five-minute oscillations in sunspots are well known from optical observations (Uchida and Sakurai, 1975; Horn, Staude, and Landgraf, 1997) and also analyzed in space-borne data of EUV lines (Brynildsen *et al.*, 1999a; Brynildsen *et al.*, 1999b; De Moortel *et al.*, 2002; Brynildsen *et al.*, 2003; King *et al.*, 2003). It was shown that they usually reflect propagation of MHD waves from the chromosphere toward the corona. In some cases the waves propagate along coronal loops rooted in sunspots (De Moortel *et al.*, 2002).

Recently Sych *et al.* (2009) established the relationship between three-minute oscillations of the microwave emission and quasi-periodic pulsations (QPP) in flares. They found two cases of increase in the power of the three-minute oscillations just before the flares. The oscillations are interpreted as slow magnetoacoustic waves which can cause the QPP.

In many sunspot-associated sources longer periods of QPO (dozens of minutes) are also found. These periods are typical for the coronal loops oscillations (Gelfreikh, Nagovitsyn, and Nagovitsyna, 2006; Bakunina *et al.*, 2009; Chorley *et al.*, 2010). Some of these may begin in a sunspot region reflecting possibly acoustic oscillations in coronal loops (see also Nakariakov *et al.*, 2004). Much longer periods, up to several hours, are also observed for most of the discussed radio sources. Similar long periods were found from the optical observations of sunspots (Efremov, Parfinenko, and Solov'ev, 2010). Global oscillations of the Sun may also be reflected in some cases of sunspot radio emission.

In summary, study of QPO observations of sunspot-associated sources may lead to significant progress in understanding of physical processes in solar plasma. Such observations also contain information on MHD wave propagation in the solar atmosphere. Although so far we have not reached satisfactory interpretation of the physical nature of all observed types of the discussed QPO, there is no doubt that the detailed analysis of their parameters brings new insight concerning the plasma structures in the studied active regions.

The main aim of this work is to study the difference in oscillations observed in pre-flare and post-flare phases and to investigate the reconstruction of the magnetic structure of ARs.

2. Observations

2.1. Methods

In this study we used the NoRH daily (seven – eight hours every day) observations at 1.76 cm (17 GHz). The radio maps of the whole solar disk were synthesized with a cadence of ten seconds and ten seconds averaging. The spatial resolution of the radio maps is about 10 – 15 arcsec.

NoRH provides intensity and circular polarization (I and V, respectively) solar images. At 17 GHz the total intensity and polarization data usually yield similar results for the sunspot-associated sources and this is also the case for our study. This behavior is due to the cyclotron nature of emission. The polarized emission at 1.76 cm is generated at the third harmonic of electron gyrofrequency in a layer where the magnetic field is about $B = 2000$ G (Akhmedov *et al.*, 1982; Shibasaki *et al.*, 1994; Nindos *et al.*, 2000). The pioneer study of oscillations using NoRH data (Gelfreikh *et al.*, 1999) was based on polarization data only. Shibasaki (2001) used time series of the peak brightness temperature of the intensity. Additionally, he used correlation plots for the right and the left circular polarization, but he did not use solar images in polarization. Some authors (Gelfreikh, Nagovitsyn, and Nagovitsyna, 2006; Chorley *et al.*, 2010; Dzhimbeeva, 2011) used total intensity data only. In this study we only present the intensity data.

Figures 1b, 3b, 5b show full disk images and magnified images of the investigated sources as well. All radio sources are compact. The structure of the source in Figure 3b is more complicated. But there is a compact feature in the left bottom part of the magnified image. This feature is brighter. The size of the feature is comparable to the synthesized beam size. So, we can consider all sources as unresolved and use the peak brightness temperature to generate time series.

The basic steps of data processing are:

- interactive extraction of the selected region-of-interest (ROI) from the initial image,
- computation of the new position of the ROI corrected for solar rotation for each image according to the time of the observation,
- extraction of the ROI from all the images,
- computation of the time series of the maximum brightness temperature over each ROI,
- spectral wavelet analysis of the time series.

For the analysis of the time series we used wavelet spectra with Morlet functions of the sixth order as a base function (Torrence and Compo, 1998).

2.2. The Active Region 9608 on 11 September 2001

Figure 1a shows the position of the AR 9608. It was situated at the central meridian but shifted by half a solar radius toward the south pole. The NoRH radio map is presented in Figure 1b. It is clearly visible that the analyzed AR has the largest sunspot and the brightest radio source on the solar disk. The parameters of the source are in good agreement with a gyro-resonance radio source at the level of the low solar corona with the field $B = 2000$ G. Two large bursts have been registered during the four-hour interval of analysis (Figure 1c). So, for analysis of the variations of the spectra in quiet state we should analyze three intervals. These are rather short, and analysis of long periods (tens of minutes *e.g.*) is not possible.

Figure 2 depicts the time intervals of the trend-subtracted time series (a, c, and e) and the corresponding wavelet spectra (b, d, and f). To summarize information on oscillations we have made time-period charts based on the wavelet spectra. Figure 1d shows the time-period chart for AR 9608. We register a short wave-train of weak three-minute oscillations during the first time interval (Figures 1d and 2b). Probably five-minute oscillations are also

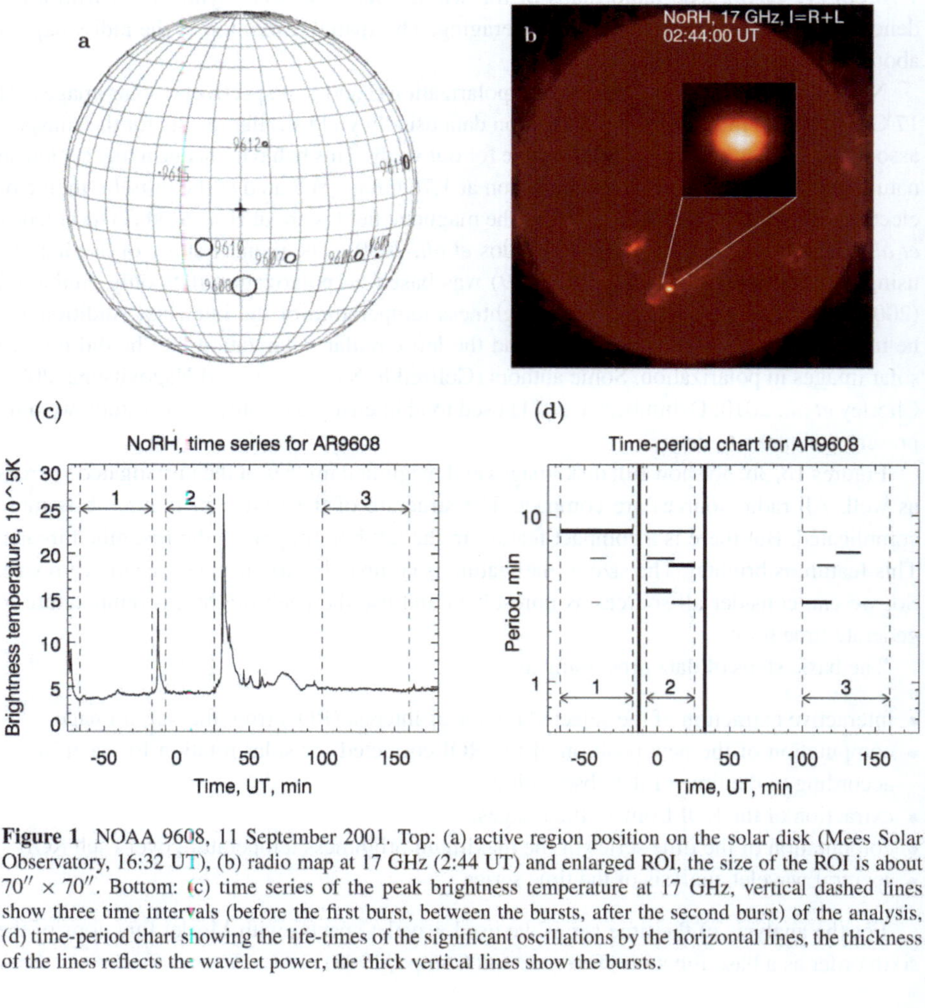

Figure 1 NOAA 9608, 11 September 2001. Top: (a) active region position on the solar disk (Mees Solar Observatory, 16:32 UT), (b) radio map at 17 GHz (2:44 UT) and enlarged ROI, the size of the ROI is about $70'' \times 70''$. Bottom: (c) time series of the peak brightness temperature at 17 GHz, vertical dashed lines show three time intervals (before the first burst, between the bursts, after the second burst) of the analysis, (d) time-period chart showing the life-times of the significant oscillations by the horizontal lines, the thickness of the lines reflects the wavelet power, the thick vertical lines show the bursts.

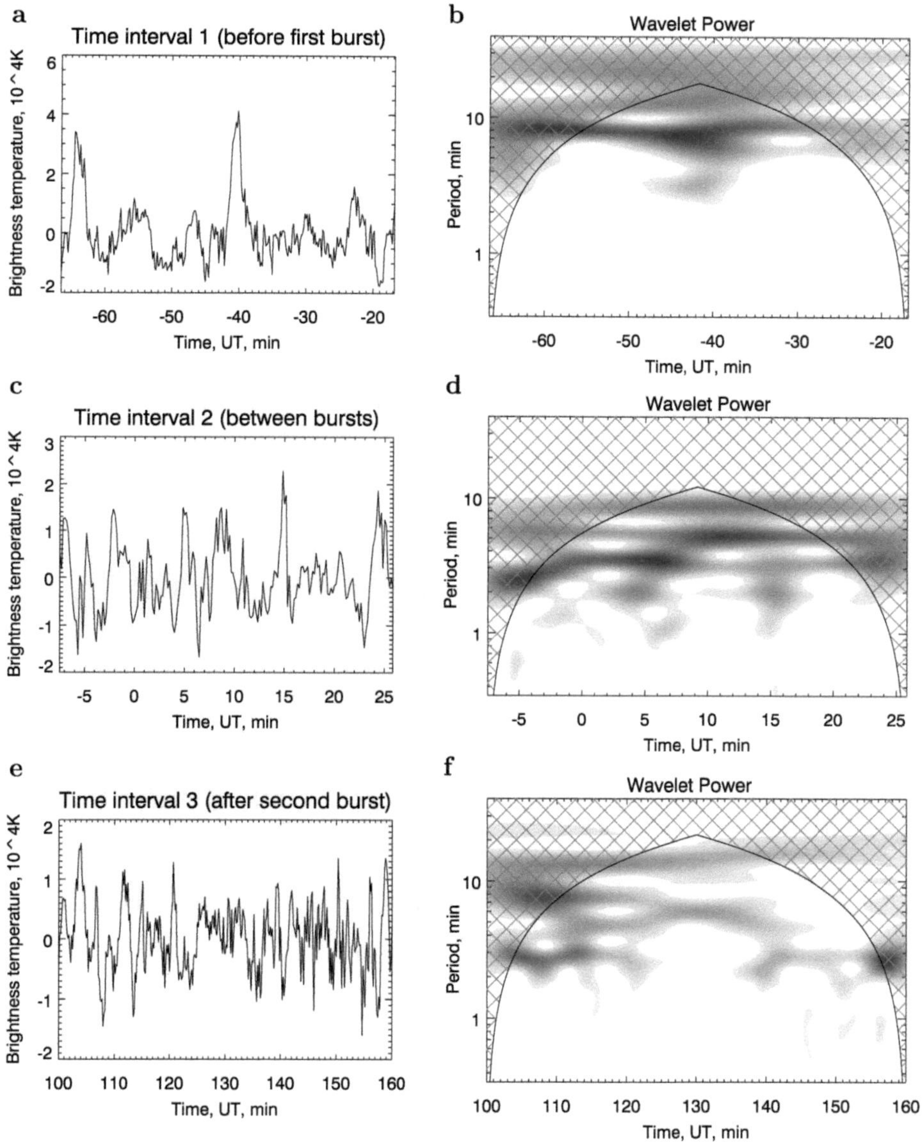

Figure 2 NOAA 9608, 11 September 2001. Top: (a) first time interval of the time series (the trend has been subtracted) of the peak brightness temperature at 17 GHz, (b) wavelet spectrum of the first time interval of the time series, darker regions correspond to higher power, the crosshatched area shows the cone of influence (COI). Middle: (c) second time interval of the time series (the trend has been subtracted) of the peak brightness temperature at 17 GHz, (d) wavelet spectrum of the second time interval of the time series. Bottom: (e) third time interval of the time series (the trend has been subtracted) of the peak brightness temperature at 17 GHz, (f) wavelet spectrum of the third time interval of the time series.

present. We can clearly see the oscillations with period of nearly eight minutes during the whole interval. During the time interval between the bursts (Figures 1d and 2d) the three- and five-minute oscillations are present. The eight-minute oscillations are also seen, but their power is weaker. After the second burst (Figures 1d and 2f) there are unstable wave-trains of the three- and five-minute oscillations and the eight-minute oscillations are practically absent.

It is known that the three-minute oscillations are a typical feature of the sunspot atmosphere (not at all heights, however). So, the appearance or disappearance of oscillations in radio wavelengths reflect the reconstruction of the structure of magnetic flux tube of the large sunspot at the level of CCTR (or lower corona).

2.3. The Active Region 9866 on 14 March 2002

This active region includes a large bipolar sunspot (see Figure 3(a,b)), crossing the central meridian a day later. A large burst was registered at 1:40 UT during 5.3-hour period of the observations (Figure 3c). The flare occurred in the west part of the region, which had a bipolar magnetic structure. Generating the time series, we verified the positions of the peak (X, Y-coordinates of the maximum brightness temperature) to make sure that the peak belongs to the same area. Actually, the peak position varied during the flare because the flare had a complicated structure. However, we did not analyze any details of the time series during the flare. We used time series before and after flare, when the source was stable.

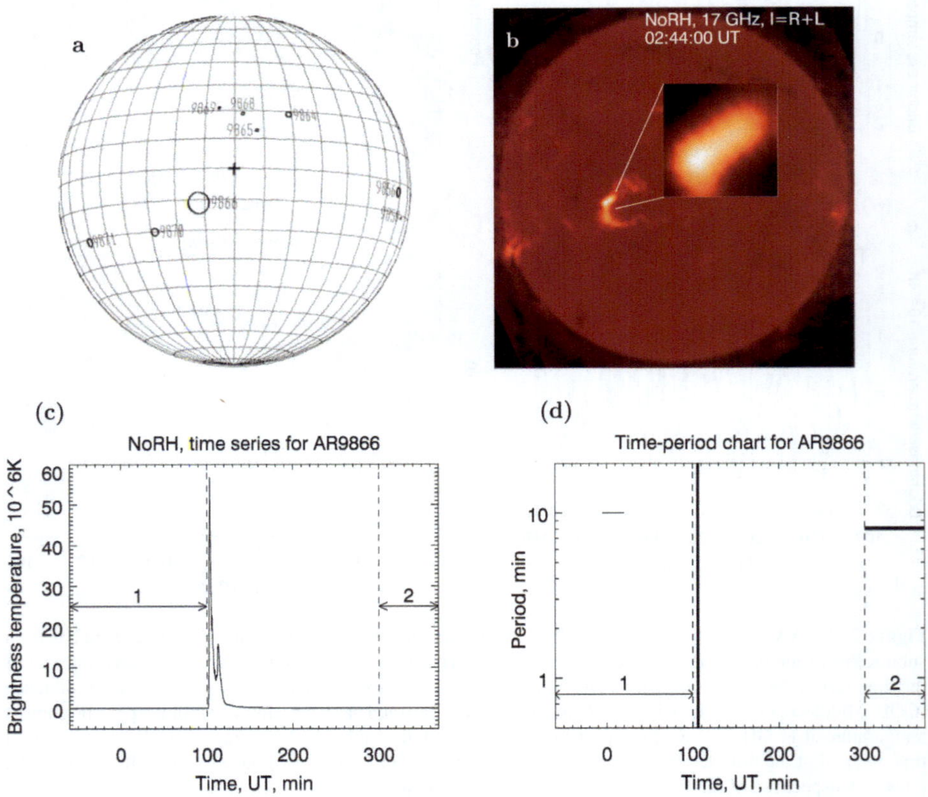

Figure 3 NOAA 9866, 14 March 2002. Top: (a) active region position on the solar disk (Mees Solar Observatory, 16:32 UT), (b) radio map at 17 GHz (2:44 UT) and enlarged ROI, the size of the ROI is about $90'' \times 90''$. Bottom: (c) time series of the peak brightness temperature at 17 GHz, vertical dashed lines show two time intervals (before the burst and after the burst) for the analysis, (d) time-period chart showing the life-times of the significant oscillations by the horizontal lines, the thickness of the lines reflects the wavelet power, the continuous vertical line shows the burst.

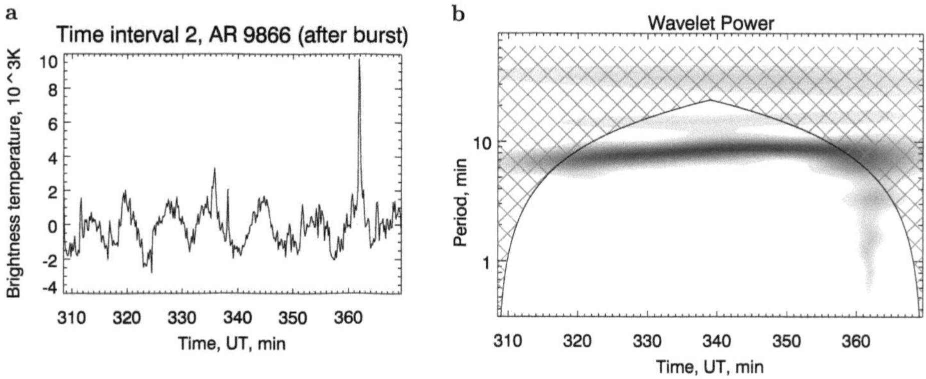

Figure 4 NOAA 9866, 14 March 2001. (a) Second time interval of the time series (the trend has been subtracted) of the peak brightness temperature at 17 GHz, (b) wavelet spectrum of the second time interval of the time series. The eight-minute oscillations are clearly present during the whole time interval.

The time-period chart is shown in Figure 3d. One can notice the appearance of powerful eight-minute oscillations after the burst.

It is interesting to note the presence of powerful well-pronounced eight-minute oscillations (Figure 4). The eight-minute oscillations are less studied than the three- and five-minute oscillations, although they were observed earlier in the radio and optical ranges (Abramov-Maximov *et al.*, 2011). Some authors have found periodicity of about eight minutes in the 171 Å and 195 Å bandpasses in the coronal loops (King *et al.*, 2003; Van Doorsselaere, Birtill, and Evans, 2009).

2.4. The Active Region 10139 on 7 October 2002

The observations were carried out one day before the central-meridian passage (CMP). Figure 5a shows the position of the AR on the solar disk for the date of the analysis. The overall trend of the emission is caused by daily variations of the synthesized beam.

There were some small bursts during the day. Figure 5c shows the time series of the peak brightness temperature at 17 GHz. For our analysis we used five time intervals from the time series between the bursts. The start and end times of the intervals are shown in Figure 5c by vertical dashed lines. The time-period chart is depicted in Figure 5d.

We find that the spectra of the oscillations for the different time intervals differ significantly. The first time interval is noisy. Practically, there are no oscillations visible. There might be oscillations with periods about ten minutes or more but we do not see the three- and five-minute oscillations.

After the bursts, the oscillations appear. During the second time interval we can see a wave-train of five-minute oscillations with duration about 25 minutes. Ten-minute oscillations are also present during the whole time interval.

The third time interval shows the presence of a short wave-train of eight-minute oscillations and beginning of three-minute oscillations at the end of interval. It is noteworthy that the well-defined three-minute oscillations begin 15 – 20 minutes before the burst (Figure 6). Perhaps it is a precursor, similar to the cases investigated by Sych *et al.* (2009).

In the fourth time interval we can see three-minute oscillations and a long wave-train of powerful ten-minute oscillations. In the fifth interval there is a weak short wave-train of three-minute oscillations and a long wave-train of oscillations with periods of nearly eight – ten minutes.

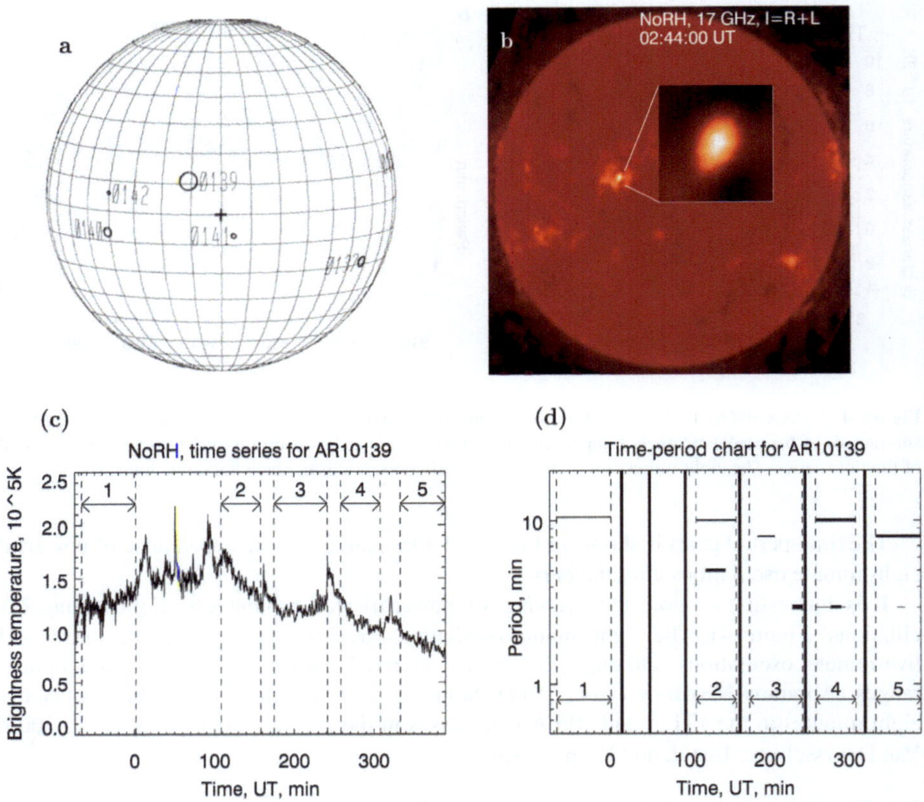

Figure 5 NOAA 10139, 7 October 2002. Top: (a) active region position on the solar disk (Mees Solar Observatory, 16:32 UT), (b) radio map at 17 GHz (2:44 UT) and enlarged ROI, the size of the ROI is about $70'' \times 70''$. Bottom: (c) time series of the peak brightness temperature at 17 GHz, vertical dashed lines show five time intervals of the analysis, (d) time-period chart showing the life-times of the significant oscillations by the horizontal lines, the thickness of the lines reflects the wavelet power, the continuous vertical lines indicates the bursts.

3. Discussion and Conclusions

We have presented three cases that confirm the existence of a link between oscillation spectrum and flare activity.

The analysis of the wavelet spectra of sunspot oscillations at the wavelength of 1.76 cm have confirmed the presence of the variations in the spectra connected with the flare activity. Both disappearance and appearance of some periods were found.

The first case (NOAA 9608, 11 September 2001) shows a decrease of power of the eight-minute oscillations after the first burst and their disappearance after the second, more powerful burst. This case also shows an increase of power of the three- and five-minute oscillations between the bursts. The second case (AR 9866, 14 March 2002) shows the appearance of the eight-minute oscillations after the burst. The most impressive feature in the third case (NOAA 10139, 7 October 2002) is the beginning of well-defined three-minute oscillations 15 – 20 minutes before the burst.

Two of the three cases demonstrate an increase of power of the short-term oscillations before and after the bursts. This result is in a good agreement with the results obtained

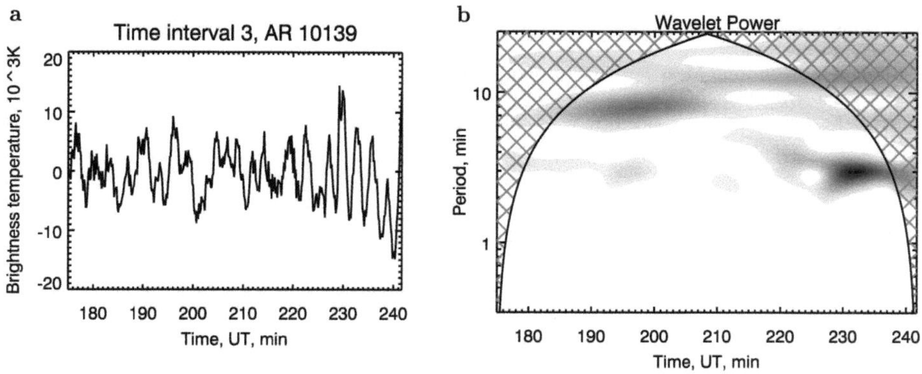

Figure 6 NOAA 10139, 7 October 2002. (a) Third time interval of the time series (the trend has been subtracted) of the peak brightness temperature at 17 GHz, (b) wavelet spectra of the third time interval of the time series. Darker regions correspond to higher power, the crosshatched area shows the COI. The appearance of well-defined three-minute oscillations at the end of the interval before the burst is clearly visible.

by Sych *et al.* (2009). They found similar cases of a gradual increase in the power of the three-minute oscillations before flares in sunspot-associated sources. They proposed that slow magnetoacoustic waves propagate from a sunspot along coronal loops upwards the flare site triggering the energy release. This interpretation is likely to be valid for our cases, too.

In our previous study (Abramov-Maximov *et al.*, 2011) we showed observational evidence that waves traveled from the chromosphere toward the corona along the umbral magnetic flux tube of the sunspot. We found similar wave-trains in the oscillation processes at two different levels of the solar atmosphere and measured time shift between wave-trains. In this study we confirmed the relationship between traveling waves and flare activity. One of the possible interpretation of the presented observational material is that the MHD waves propagating from the sunspot can trigger energy release at the flare site.

The increase of power of the three-minute oscillations before the burst can be used for flare forecasting. Of course, there are some difficulties. First, the detection of the precursor requires a complicated data processing. Second, this effect appears only a few minutes before the flare. So, further studies are needed. Studies of this effect may shed light on the problem of energy release in solar flares. NoRH is a very suitable instrument for such investigations due to its long series of observations (since 1992), high time (one second) and spatial (ten arcseconds) resolution for the full solar disk, and its long daily observing time (up to eight hours per day).

Our conclusions are as follows:

i) We have confirmed the existence of a link between the oscillation spectrum and flare activity.
ii) We found one case of increase of the power of the short-term oscillations before the burst.
iii) We interpret our observations in terms of a relationship between MHD waves propagating from sunspots and flare processes.
iv) The NoRH is a very suitable instrument for such studies.

Acknowledgements This work was partially supported by the Presidium of Russian Academy of Sciences under grant OFN-15.

We are grateful to the anonymous referee and Editors for important remarks, which helped us to improve the paper.

Data used here from Mees Solar Observatory, University of Hawaii, are produced with the support of NASA grant NNG06GE13G.

Wavelet software was provided by C. Torrence and G. Compo, and is available at URL: http://paos.colorado.edu/research/wavelets/.

References

Abramov-Maximov, V.E., Gelfreikh, G.B., Kobanov, N.I., Shibasaki, K., Chupin, S.A.: 2011, *Solar Phys.* **270**, 175.
Akhmedov, Sh.B., Gelfreikh, G.B., Bogod, V.M., Korzhavin, A.N.: 1982, *Solar Phys.* **79**, 41.
Aleshin, V.I., Kobrin, M.M., Korshunov, A.I.: 1973, *Radiophys. Quantum Electron.* **16**, 571.
Bakunina, I.A., Abramov-Maximov, V.E., Lesovoy, S.V., Shibasaki, K., Solov'ev, A.A., Tikhomirov, Y.V.: 2009, In: Gopalswamy, N., Webb, D., Shibata, K. (eds.) *"Universal Heliophysical Processes", IAU Symposium* **257**, 155.
Bogdan, T.J.: 2000, *Solar Phys.* **192**, 373.
Bogdan, T.J., Judge, P.G.: 2006, *Phil. Trans. Roy. Soc. A* **364**, 313.
Brynildsen, N., Kjeldseth-Moe, O., Maltby, P., Wilhelm, K.: 1999a, *Astrophys. J. Lett.* **517**, L159.
Brynildsen, N., Leifsen, T., Kjeldseth-Moe, O., Maltby, P., Wilhelm, K.: 1999b, *Astrophys. J. Lett.* **511**, L121.
Brynildsen, N., Maltby, P., Kjeldseth-Moe, O., Wilhelm, K.: 2003, *Astron. Astrophys.* **398**, L15.
Chorley, N., Hnat, B., Nakariakov, V.M., Inglis, A.R., Bakunina, I.A.: 2010, *Astron. Astrophys.* **513**, 27.
De Moortel, I., Ireland, J., Hood, A.W., Walsh, R.W.: 2002, *Astron. Astrophys.* **387**, L13.
Durasova, M.S., Kobrin, M.M., Yudin, O.I.: 1971, *Nature* **229**, 82.
Dzhimbeeva, L.N.: 2011, *Solar Syst. Res.* **45**(1), 84.
Efremov, V.I., Parfinenko, L.D., Solov'ev, A.A.: 2010, *Solar Phys.* **267**, 279.
Fludra, A.: 2001, *Astron. Astrophys.* **368**, 639.
Gelfreikh, G.B., Grechnev, V.V., Kosugi, T., Shibasaki, K.: 1999, *Solar Phys.* **185**, 177.
Gelfreikh, G.B., Nagovitsyn, Yu.A., Nagovitsyna, E.Yu.: 2006, *Publ. Astron. Soc. Japan* **58**, 29.
Horn, T., Staude, J., Landgraf, V.: 1997, *Solar Phys.* **172**, 69.
King, D.B., Nakariakov, V.M., Deluca, E.E., Golub, L., McClements, K.G.: 2003, *Astron. Astrophys.* **404**, L1.
Kobrin, M.M., Korshunov, A.I., Snegirev, S.D., Timofeyev, B.V.: 1973, *Soln. Dannye* **10**, 79 (in Russian).
Kosovichev, A.G.: 2009, *AIP Conf. Proc.* **1170**, 547.
Lites, B.W., Thomas, J.H., Bogdan, T.J., Cally, P.S.: 1998, *Astrophys. J.* **497**, 464.
Nakariakov, V.M., Tsiklauri, D., Kelly, A., Arber, T.D., Aschwanden, M.J.: 2004, *Astron. Astrophys.* **414**, L25.
Nakariakov, V.M., Erdélyi, R.: 2009, *Space Sci. Rev.* **149**, 1.
Nakajima, H., Nishio, M., Enome, S., Shibasaki, K., Takano, T., Hanaoka, Y., Torii, C., Sekiguchi, H., Bushimata, T., Kawashima, S., Shinohara, N., Irimajiri, Y., Koshiishi, H., Kosugi, T., Shiomi, Y., Sawa, M., Kai, K.: 1994, *Proc. IEEE* **82**, 705.
Nindos, A., Kundu, M.R., White, S.M., Shibasaki, K., Gopalswamy, N.: 2000, *Astrophys. J. Suppl.* **130**, 485.
Nindos, A., Alissandrakis, C.E., Gelfreikh, G.B., Bogod, V.M., Gontikakis, C.: 2002, *Astron. Astrophys.* **386**, 658.
Shibasaki, K., Enome, S., Nakajima, H., Nishio, M., Takano, T., Hanaoka, Y., Torii, C., Sekiguchi, H., Kawashima, S., Bushimata, T., Shinohara, N., Koshiishi, H., Shiomi, Y., Irimajiri, Y., Leka, K.D., Canfield, R.C.: 1994, *Publ. Astron. Soc. Japan* **46**, L17.
Shibasaki, K.: 2001, *Astrophys. J.* **550**, 1113.
Sych, R., Nakariakov, V.M., Karlicky, M., Anfinogentov, S.: 2009, *Astron. Astrophys.* **505**, 791.
Torrence, C., Compo, G.P.: 1998, *Bull. Am. Meteorol. Soc.* **79**, 61.
Uchida, Y., Sakurai, T.: 1975, *Publ. Astron. Soc. Japan* **27**, 259.
Van Doorsselaere, T., Birtill, D.C.C., Evans, G.R.: 2009, *Astron. Astrophys.* **508**, 1485.

Solar Phys (2011) 273:413–419
DOI 10.1007/s11207-011-9735-4

Electron Spikes, Type III Radio Bursts and EUV Jets on 22 February 2010

A. Klassen · R. Gómez-Herrero · B. Heber

Received: 30 November 2010 / Accepted: 18 February 2011 / Published online: 31 March 2011

Abstract The Solar Electron Proton Telescope on board the twin STEREO spacecraft measures electrons and ions in the energy range from 30 to above 400 keV with an energy resolution better than 10%. On 22 February 2010 during a short interval of 100 minutes, a sequence of impulsive energetic electron events in the range below 120 keV was observed with the STEREO-A/SEPT instrument. Each of the four events was associated with a type III radio burst and a narrow EUV jet. All the events show nearly symmetric "spike"-like time profiles with very short durations $\simeq 5$ min. The estimated electron injection time for each individual event shows a small time delay between the electron spike and the corresponding type III radio emission and a close coincidence with an EUV jet. These observations reveal the existence of spike-like electron events showing nearly "scatter-free" propagation from the Sun to STEREO-A. From the time coincidence we infer that the mildly relativistic electrons are accelerated at the same time and at the same location as the accompanying type III emitting electrons and coronal EUV jets. The characteristics of the spikes reflect the injection and acceleration profiles in the corona rather than interplanetary propagation effects.

Keywords Flares, dynamics · Energetic particles, electrons · Radio bursts, type III · Jets

1. Introduction

Solar impulsive near-relativistic (> 30 keV) electron events are usually flare-related and show a strong association with type III radio emission (Lin, 1985; Kahler *et al.*, 2007). It is well established that type III radio bursts are generated by low energy electrons (< 30 keV) propagating through the solar corona and the interplanetary (IP) medium (see, *e.g.*, Suzuki and Dulk, 1985).

Energy Storage and Release through the Solar Activity Cycle – Models Meet Radio Observations
Guest Editors: Christophe Marqué and Alexander Nindos.

A. Klassen (✉) · R. Gómez-Herrero · B. Heber
Institut für Experimentelle und Angewandte Physik, Universität Kiel, 24118 Kiel, Germany
e-mail: klassen@physik.uni-kiel.de

The impulsiveness of electron events is normally defined by their rise time, the time between the onset and the intensity maximum. That time is basically below 10 – 20 minutes. In contrast, the decay time is usually much longer than the rise time. An additional important property of impulsive events is the presence of velocity dispersion. The low energy electrons need more time to arrive at the satellite from the Sun than the higher energy electrons (Krucker *et al.*, 1999).

Haggerty and Roelof (2009) introduced three broad categories of impulsive beam-like electron events using their time profiles: *i*) Spikes (rapid and equal rise and decay), *ii*) Pulses (rapid rise, slower decay); *iii*) Ramps (rapid rise followed by a plateau). This classification could be useful for a better understanding of the relationship between energetic particle events and the related solar activity phenomena: flares, CMEs, shocks, X-ray and EUV jets *etc.* It should be mentioned that the first spike-like particle event observation was already reported in the 1970s (see, *e.g.*, Lin, 1974).

Aurass, Klein, and Martens (1994) and Raulin *et al.* (1996) using radiospectrographic and imaging observations with the Nançay Radioheliograph found close temporal and spatial coincidences between type III bursts and X-ray jets.

The close association between impulsive electron events and type III bursts is well known since the 1970s; however, the relation between *in-situ* electrons and jets (X-ray, or EUV) has been discussed only since a decade. Wang, Pick, and Mason (2006) and Nitta *et al.* (2006) investigated the association between EUV jets, ^{3}He-rich events and impulsive Solar Energetic Particles (SEP) and they concluded "jets associated with type III bursts to be an important observable for impulsive SEP events. They represent flaring that involves open field lines" (Nitta *et al.*, 2006).

In this study, a sequence of four electron spikes detected during a short time interval of 100 minutes by the Solar Electron and Proton Telescope (SEPT) on the STEREO-A spacecraft is presented. We focus on the relationship between spikes, type III radio bursts and EUV coronal jets and discuss the common origin of all mentioned phenomena.

2. Instruments and Data Selection

The observations were carried out using the Solar Electron and Proton Telescope (SEPT) part of the Solar Energetic Particle (SEP) suite of the STEREO IMPACT investigation. SEPT consists of two dual double-ended magnet/foil particle telescopes which separate and measure electrons in the energy range 30 – 400 keV and ions from 60 to 7000 keV (Müller-Mellin *et al.*, 2008). The energy range is divided into 32 logarithmically spaced channels with an energy resolution of $\simeq 10\%$ in the range ≤ 1200 keV.

Anisotropy information on the non-spinning spacecraft is provided by two separate telescopes: SEPT-E looking in the ecliptic plane along the Parker spiral magnetic field (*i.e.* 45° west of the spacecraft–Sun line) both toward and away from the Sun, and SEPT-NS looking perpendicular to the ecliptic plane toward North and South. The dual set-up refers to two adjacent sensor apertures for each of the four view directions: one for protons, one for electrons. The double-ended set-up refers to the detector stack with view cones in two opposite directions: one side (electron side) is covered by a thin foil, the other side (proton side) is surrounded by a magnet. The foil leaves the electron spectrum essentially unchanged but stops protons up to 400 keV. The magnet sweeps away electrons but lets ions pass. The geometry factor for each electron and proton telescope is 0.13 cm^2 sr and 0.17 cm^2 sr, respectively. The SEPT time resolution is one minute and the field of view (FOV) is 52°.

Complementary solar wind data, radio emission spectra and EUV images were obtained from PLASTIC (Galvin *et al.*, 2008), SWAVES (Bougeret *et al.*, 2008) and SECCHI-EUVI

(Howard *et al.*, 2008) instruments on board STEREO-A. EUVI observes the solar atmosphere in four spectral channels (He II 304 Å: $T \simeq 80\,000$ K, Fe IX 171 Å: $T \simeq 1.3$ MK, Fe XII 195 Å: $T \simeq 1.6$ MK, Fe XV 284 Å: $T \simeq 2.0$ MK). SWAVES provides radio observations in the frequency range between 2.5 kHz and 16 MHz.

3. Observations

On 22 February 2010 STEREO-A was at a distance of 0.96 AU from the Sun and 70° ahead the Earth on its orbit, detecting with SEPT a sequence of four electron spikes during a period of only 100 minutes from 04:50 until 06:30 UT.

Figure 1 shows the SWAVES dynamic radio spectrum in the range 16 – 0.03 MHz with a series of type III bursts (top), and the time profile of a sequence of electron spikes in the range 35 – 65 keV (middle). The bottom panel presents the SEPT dynamic energy spectrum in the range 35 – 1000 keV. The time axis for electrons is shifted so that it gives the arrival time in UT minus 13 min. The 13 min difference removes the prolonged travel time of ($\simeq 21$ min) electrons (35 – 65 keV) from the Sun to the spacecraft along the Parker magnetic field spiral with respect to the propagation time of the type III radio emission (8 min). The length of the nominal Parker spiral is 1.04 AU and was calculated using the solar wind speed of 550 km s^{-1} as observed by the STEREO-A/PLASTIC instrument. The vertical dashed lines pointing at the onset times of type IIIs show the relative delay between the *in-situ* detected electron bursts and the type III radio emission. It is evident that these delays are smaller than six minutes which could be mainly explained if we take into account the intrinsic velocity dispersion in the range 35 – 65 keV and the background level, which can partly mask the true onset. A more extended Parker spiral than the calculated ideal length of 1.04 AU could also contribute to the observed delays listed in Table 1. Note that the particles streaming along the Parker spiral were detected only in the sunward detector indicating very low scattering in the interplanetary medium.

The numbers 1 – 4 in Figure 1 indicate the consecutive spikes. Evidently the spikes with numbers 1 and 3 exhibit double peaks seen in the time profiles as well as in the dynamic spectrum. It means that there are two individual sub-spikes in each event. Indeed, in spike 1 they partly overlap each other, but in spike 3 the double structure appears distinctly in the dynamic energy spectrum (see Figure 1, bottom panel). All spikes were detected only at energies below 120 keV. They exhibit velocity dispersion and durations of full width at half maximum (FWHM) shorter than 5 min. Such short durations are very unusual for most of the particle events observed *in-situ* at 1 AU.

The prompt rising peaks and the intrinsic electron velocity dispersion of about 5 – 6 min in the channel 35 – 65 keV set an upper limit on the duration of the electron injection into the IP medium of ≤ 1 min. That is much shorter than the observed 5 min duration and comparable with the duration of the type III radio bursts.

In Table 1 we summarize the spike characteristics and associated radio and EUV emissions.

Obviously the series of type III bursts observed with SWAVES were one to one coincident with the sequence of electron spikes appearing in the same time interval. Each type III burst was associated with an electron spike. Furthermore, the double type IIIs at 04:46 – 04:51 UT and at 05:32 – 05:37 UT which occur only 5 min apart from each other, were in coincidence with the double electron spikes (numbers 1 and 3 in Figure 1).

Moreover, all individual type III bursts and electron spikes were temporally associated with coronal EUV jets as shown in Figure 2 and Table 1. For each event we present a

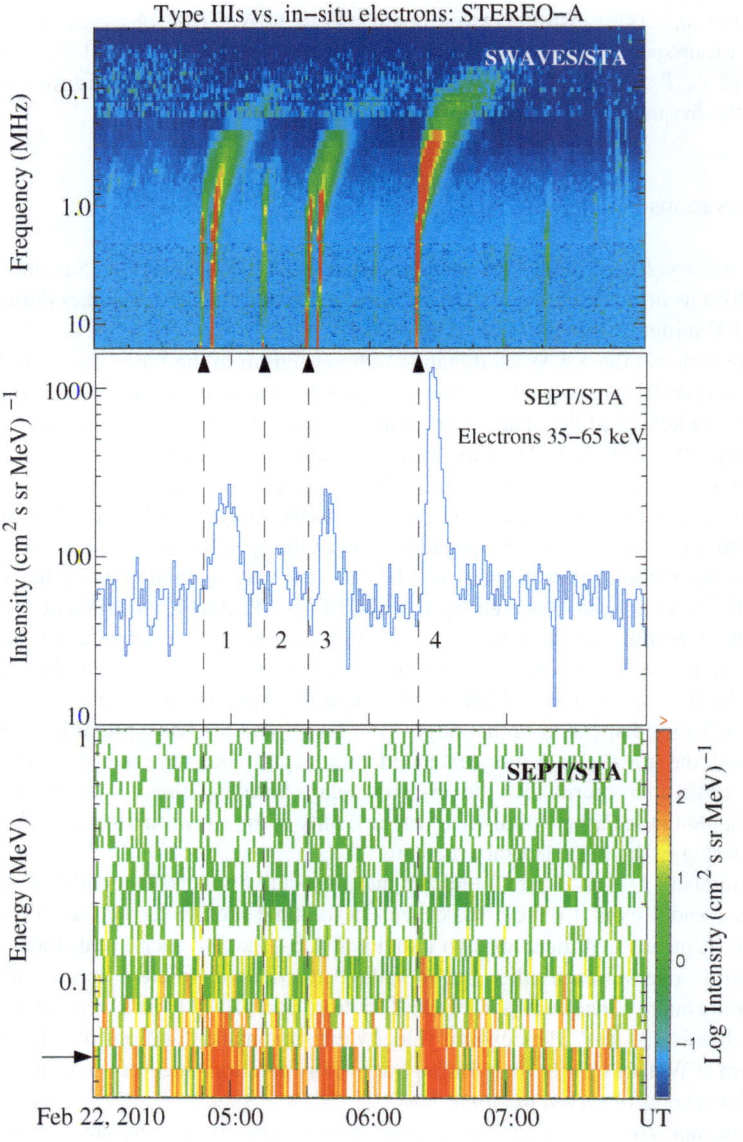

Figure 1 (Top) Dynamic radio spectrum of type III bursts on 22 February 2010 (SWAVES/STA). (Middle) Time profile of the sequence of electron events in the range 35 – 65 keV (SEPT/STA). The vertical lines indicate the onset of type IIIs relative to *in-situ* electron spikes. (Bottom) The dynamic energy spectrum of electrons in the range 35 – 1000 keV (SEPT/STA). The time axis for electrons is shifted (middle and bottom panels) and gives the arrival time minus 13 min. The 13 min difference removes the prolonged travel time of electrons along the IP magnetic field with respect to the propagation time of the type III radio emission. The arrow on the left indicates the 35 – 65 keV energy channel.

difference image taken during the jet and a few minutes earlier using the EUVI lines He II (304 Å) or Fe IX (171 Å). The image cadence for the jets associated with spikes Nos. 1 – 3 was 1.5 – 2.5 min. Only for event 4 the images were taken 10 min apart. It is evident that each type III burst and the accompanying electron spike occur together with a jet.

Table 1 Spike characteristics.

Event	1 (double spike)	2	3 (double spike)	4
Type III onset, UT	04:46 & 04:51	05:12	05:32 & 05:37	06:20
Jet onset, UT	04:46 & 04:51	05:11	05:33 & 05:37	06:16 – 06:26
Spike onset[a]	04:52 & 04:55	05:17	05:37 & 05:40	06:23
– duration[b], min	11[c] (2 × 5.5)	3	7[c] (2 × 3.5)	4
– velocity dispersion	poor	poor	yes	yes
– energy range, keV	35 – 75	35 – 75	35 – 120	35 – 120
– time delay *vs.* type III, min	6 & 4	5	5 & 3	3

[a]Onset time = observation time UT – propagation time + 8 min.

[b]Duration of FWHM in energy range 35 – 65 keV.

[c]Total duration of both spikes together.

Figure 2 (leftmost top image) shows direct 171 Å image of the active region and the location where all jets appear (indicated by an arrow). The other three images in the top panel show running difference images in 304 Å corresponding to the times of spikes number 1, 2, and 4. For the double spike 1 only one jet at 04:51:15 UT is presented. The jet for the first sub-spike at 04:46 UT (Table 1) was very faint and difficult to discern; for this reason it has not been included in Figure 2. In the bottom panel, we present a sequence of 171 Å runing difference images between 05:33:30 and 05:38:30 UT during the times of the second double type III burst and the corresponding double spike 3. The two pairs of images, in which two different recurrent jets can be identified, clearly show the appearance, the decay and the propagation of both jets from the same location toward the west limb.

It is obvious that all very collimated jets originate at the same position and propagate in the same direction, suggesting a recurrent ejection along open magnetic field lines. The measured jet propagation speed in projection, determined only for the very distinct jet 3 at 05:37:15 – 05:38:30 UT, is about 500 $\mathrm{km\,s^{-1}}$. For the other events the image time cadence was not good enough to estimate the jets speed. The lengths of the jets were in the range 40 000 – 70 000 km and only slightly different from event to event.

The EUV jet recurred on the periphery of the AR 11046 at W49N23 close to its eastern boundary and to a small Coronal Hole (CH) or CH-like dark bay (Figure 2, the direct 171 Å image at 05:38:30 UT). Such a CH/bay suggests that there were open magnetic field lines in the vicinity of closed active region magnetic field lines (Rust *et al.*, 2008). The coordinates are given for the STEREO-A field of view. A few days before, on 17 February, the AR 11046 was close to the west solar limb as observed from SOHO, showing only a few small spots and pores. In the following days (18 – 20 February) the spots disappear and the region rotates spotless behind the west limb. Therefore the observations of X-ray emission were not available for the jet times, because the AR was already 25° behind the solar limb as seen from RHESSI and GOES.

4. Summary and Conclusion

We confirm and extend previous observations suggesting very close association between impulsive electron events, type III radio bursts and EUV jets. The investigated sequence of

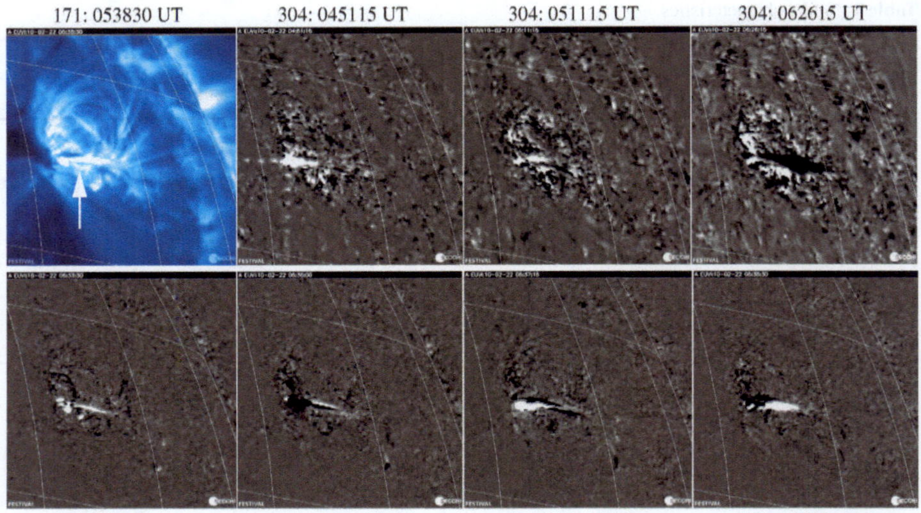

Figure 2 (Top) The leftmost panel shows an EUVI 171 Å image of the active region and the location of all jets (arrow). The rest show running difference images of the jets sequence in 304 Å corresponding to the spikes number 1, 2, and 4. (Bottom) Sequence of EUV 171 Å running difference images showing two consecutive jets in the time interval 05:33 – 05:38 UT coincident with the double type III burst and electron spike (number 3 in Figure 1). North is on top, west is to the right. Regions appear white (dark) when the emission has increased (decreased) between two consecutive images.

short-lived energetic electron spikes is an excellent example showing its close association with the sequence of type III radio bursts and EUV coronal jets. Such a close association suggests the common origin of these phenomena due to recurrent magnetic reconnection process occurring at the same location in the corona (see, *e.g.*, Chifor *et al.*, 2008) or due to electron acceleration in front of propagating coronal jets (Miteva *et al.*, 2007). Moreover, since all spikes show similar profiles, the propagation conditions in the interplanetary medium should be the same during the whole interval of 100 min, revealing nearly scatter-free electron propagation from the Sun to STEREO-A.

The main results are summarized as follows:

i) A sequence of four spikes occur during a short time interval of 100 min and show almost symmetric time profiles with durations below 5 min.

ii) All electron spikes were temporally coincident with type III radio bursts and very collimated recurrent EUV coronal jets, appearing at the same location and showing the same trajectory.

iii) The sharp spike-like profiles and the observed durations between three and five minutes set an upper limit on the duration of the electron injection into the interplanetary medium of ≤ 1 min, comparable with the durations of type III radio bursts.

Acknowledgements The authors acknowledge the FESTIVAL team that developed the browser for manipulation of solar imaging data. We thank the STEREO/SWAVES/SECCHI teams for their open data policy and the anonymous referee for the comments. The STEREO/SEPT project is supported under grant 50 OC 0902 by the German Bundesministerium für Wirtschaft through the Deutsches Zentrum für Luft- und Raumfahrt (DLR).

References

Aurass, H., Klein, K.-L., Martens, P.C.H.: 1994, First detection of correlated electron beams and plasma jets in radio and soft X-ray data 1994. *Solar Phys.* **155**, 203. doi:10.1007/BF00670741.

Bougeret, J.L., Goetz, K., Kaiser, M.L., Bale, S.D., Kellogg, P.J., Maksimovic, M., *et al.*: 2008, S/WAVES: The radio and plasma wave investigation on the STEREO mission. *Space Sci. Rev.* **136**, 487. doi:10.1007/s11214-007-9298-8.

Chifor, C., Isobe, H., Mason, H.E., Hannah, I.G., Young, P.R., Del Zanna, G., *et al.*: 2008, Magnetic flux cancellation associated with a recurring solar jet observed with *Hinode*, RHESSI, and STEREO/EUVI. *Astron. Astrophys.* **491**, 279. doi:10.1051/0004-6361:200810265.

Galvin, A.B., Kistler, L.M., Popecki, M.A., Farrugia, C.J., Simunac, K.D.C., Ellis, L., *et al.*: 2008, The plasma and suprathermal ion composition (PLASTIC) investigation on the STEREO observatories. *Space Sci. Rev.* **136**, 437. doi:10.1007/s11214-007-9296-x.

Haggerty, D.K., Roelof, E.: 2009, Probing SEP acceleration processes with near-relativistic electrons. In: Ao, X., Burrow, R.H., Zank, G.P. (eds.) *Shock Waves in Space and Astrophysical Environments*, AIP, New York, 3 – 10. doi:978-0-7354-0724-4.

Howard, R.A., Moses, J.D., Vourlidas, A., Newmark, J.S., Socker, D.G., Plunkett, S.P., *et al.*: 2008, Sun Earth Connection Coronal and Heliospheric Investigation (SECCHI). *Space Sci. Rev.* **136**, 67. doi:10.1007/s11214-008-9341-4.

Lin, R.P.: 1974, Non-relativistic solar electrons. *Space Sci. Rev.* **16**, 189.

Lin, R.P.: 1985, Energetic solar electrons in the interplanetary medium. *Solar Phys.* **100**, 537. doi:10.1007/BF00158444.

Kahler, S., Aurass, H., Mann, G., Klassen, A.: 2007, Solar radio bursts and solar wind associations with inferred near-relativistic electron injections. *Astrophys. J.* **656**, 567. doi:10.1086/510230.

Krucker, S., Larson, D.E., Lin, R.P., Thompson, B.J.: 1999, On the origin of impulsive electron events observed at 1 AU. *Astrophys. J.* **519**, 864. doi:10.1086/307415.

Miteva, R., Mann, G., Vocks, C., Aurass, H.: 2007, Excitation of electrostatic fluctuations by jets in a flaring plasma. *Astron. Astrophys.* **461**, 1127. doi:10.1051/0004-6361:20053736.

Müller-Mellin, R., Böttcher, S., Falenski, J., Rode, E., Duvet, L., Sanderson, T., *et al.*: 2008, The Solar Electron and Proton Telescope for the STEREO mission. *Space Sci. Rev.* **136**, 363. doi:10.1007/s11214-007-9204-4.

Nitta, N., Reames, V., DeRosa, M.L., Liu, Y., Yashiro, S., Gopalswamy, N.: 2006, Solar sources of impulsive solar energetic particle events and their magnetic field connection to the Earth. *Astrophys. J.* **650**, 438. doi:10.1086/507442.

Rust, D.M., Haggerty, D.K., Georgoulis, M.K., Sheeley, N.R., Wang, Y., De Rosa, M.L., Schrijver, C.J.: 2008, On the solar origin of open magnetic fields in the heliosphere. *Astrophys. J.* **687**, 635. doi:10.1086/592017.

Raulin, J.P., Kundu, M.R., Hudson, H.S., Nitta, N., Raoult, A.: 1996, Metric type III bursts associated with soft X-ray jets. *Astron. Astrophys.* **306**, 299.

Suzuki, S., Dulk, G.A.: 1985, Bursts of type III and type V. In: McLean, D.J., Labrum, N.R. (eds.) *Solar Radiophysics*, Cambridge University Press, Cambridge, 289 – 332.

Wang, Y.-M., Pick, M., Mason, G.M.: 2006, Coronal holes, jets, and the origin of He-rich particle events. *Astrophys. J.* **639**, 495. doi:10.1086/499355.

Solar Phys (2011) 273:421–432
DOI 10.1007/s11207-011-9746-1

Relation Between the 3D-Geometry of the Coronal Wave and Associated CME During the 26 April 2008 Event

M. Temmer · A.M. Veronig · N. Gopalswamy · S. Yashiro

Received: 29 November 2010 / Accepted: 28 February 2011 / Published online: 31 March 2011

Abstract We study the kinematical characteristics and 3D geometry of a large-scale coronal wave that occurred in association with the 26 April 2008 flare-CME event. The wave was observed with the EUVI instruments aboard both STEREO spacecraft (STEREO-A and STEREO-B) with a mean speed of $\sim 240\ \mathrm{km\,s^{-1}}$. The wave is more pronounced in the eastern propagation direction, and is thus, better observable in STEREO-B images. From STEREO-B observations we derive two separate initiation centers for the wave, and their locations fit with the coronal dimming regions. Assuming a simple geometry of the wave we reconstruct its 3D nature from combined STEREO-A and STEREO-B observations. We find that the wave structure is asymmetric with an inclination toward East. The associated CME has a deprojected speed of $\sim 750\pm 50\ \mathrm{km\,s^{-1}}$, and it shows a non-radial outward motion toward the East with respect to the underlying source region location. Applying the forward fitting model developed by Thernisien, Howard, and Vourlidas (Astrophys. J. **652**, 763, 2006), we derive the CME flux rope position on the solar surface to be close to the dimming regions. We conclude that the expanding flanks of the CME most likely drive and shape the coronal wave.

Energy Storage and Release through the Solar Activity Cycle – Models Meet Radio Observations
Guest Editors: Christophe Marqué and Alexander Nindos.

Electronic supplementary material The online version of this article (doi:10.1007/s11207-011-9746-1) contains supplementary material, which is available to authorized users.

M. Temmer (✉) · A.M. Veronig
Kanzelhöhe Observatory-IGAM, Institute of Physics, University of Graz, Universitätsplatz 5, 8010 Graz, Austria
e-mail: manuela.temmer@uni-graz.at

A.M. Veronig
e-mail: asv@igam.uni-graz.at

N. Gopalswamy · S. Yashiro
NASA Goddard Space Flight Center, Code 695, Greenbelt, MD 20771, USA

N. Gopalswamy
e-mail: nat.gopalswamy@nasa.gov

Keywords Shock waves · Coronal mass ejections

1. Introduction

Wave-like disturbances in the solar corona were for the first time imaged by the Extreme-ultraviolet Imaging Telescope (EIT; Delaboudinière *et al.*, 1995) instrument onboard the *Solar and Heliospheric Observatory* (SOHO), thereafter called EIT waves (Moses *et al.*, 1997; Thompson *et al.*, 1998) or, more generally, EUV waves. Their generation mechanism and nature is still an issue of debate. In possible scenarios it is assumed that these disturbances are fast-mode MHD waves which are flare-initiated and/or CME driven (see, *e.g.*, Wills-Davey and Thompson, 1999; Wang, 2000; Warmuth *et al.*, 2001; Long *et al.*, 2008; Veronig, Temmer, and Vršnak, 2008; Gopalswamy *et al.*, 2009a), solitons or slow-mode waves (see, *e.g.*, Wills-Davey, DeForest, and Stenflo, 2007). Other models suppose that these disturbances are non-wave features that occur due to magnetic reconfiguration associated with the erupting CME (see, *e.g.*, Delannée, 2000; Chen *et al.*, 2002; Attrill *et al.*, 2009; Dai *et al.*, 2010). Recently, theoretical models were developed that combine wave and non-wave scenarios within hybrid models (see Zhukov and Auchère, 2004; Cohen *et al.*, 2009; Liu *et al.*, 2010).

The original model developed by Uchida (1968) assumes that the 3D structure of a shock wave in the corona sweeps over the chromosphere which would account for the observations of coronal as well as chromospheric wave signatures (Moreton waves imaged in Hα). Based on combined observations in different wavelengths of a coronal wave, Narukage *et al.* (2002) found from the resulting differences in propagation a 3D structure of the wave. A recent review by Vršnak and Cliver (2008) describes the formation of MHD waves with a 3D piston mechanism driven either by the CME expansion or by a flare-associated pressure pulse. The *Solar Terrestrial Relations Observatory* (STEREO) provides new opportunities to observe coronal waves under different vantage points. Intriguing results giving evidence for the 3D geometry of coronal waves are reported for the STEREO quadrature event from 13 February 2009. For this event the EUV wave could be observed from the lateral direction as well as simultaneously on-disk. From these unique observations Kienreich, Temmer, and Veronig (2009) and Patsourakos and Vourlidas (2009) conclude that the disturbance is a 3D fast-mode MHD wave which is partly driven by the flanks of the associated CME. They further show that the EUV wave is propagating at heights of about 100 Mm above the solar surface which is comparable to the coronal scale height for quiet Sun conditions (Patsourakos *et al.*, 2009). In a recent study by Veronig *et al.* (2010) the full 3D dome of an EUV wave was identified, and interpreted to be freely propagating in the lateral direction after the lateral expansion of the CME stopped, whereas the upper part of the wave dome was permanently driven during the eruption. For recent reviews on the issue of EUV wave initiation and their nature we refer to Wills-Davey and Attrill (2009), Gallagher and Long (2010), and Warmuth (2010).

During the 26 April 2008 CME-flare event a coronal shock wave was imaged with the Extreme Ultraviolet Imager (EUVI) aboard both spacecraft of the STEREO mission, which were separated from each other by 49.5°. In this paper we present a study on the kinematics and geometry of the wave as seen simultaneously from two different vantage points. These observations are combined with a forward fitting model of the erupting flux rope, which enables us to derive geometrical information about the 3D structure of the disturbance.

2. Data and Methods

The EUV wave under study occurred on 26 April 2008 and was associated with a CME and a flare of GOES class B3.8/Hα importance SF from a source region located at N08E08 (Earth view). GOES soft X-ray (SXR) observations reveal that the thermal emission of the flare starts at 13:54 UT with a maximum at 14:08 UT. The event could be observed from both STEREO satellites (STEREO-A and STEREO-B) which were separated by 49.5°. Complementary observations from SOHO/LASCO and EIT provide information from Earth view. We stress that the EUV wave under study was associated with a coronal[1] and interplanetary type II radio burst observed by *Wind*/WAVES[2]. The type II burst in the metric and interplanetary space was studied in (Gopalswamy *et al.*, 2009b).

The EUVI instrument and the coronagraphs COR1 and COR2 are part of the Sun Earth Connection Coronal and Heliospheric Investigation (SECCHI; Howard *et al.*, 2008) instrument suite onboard the STEREO mission (Kaiser *et al.*, 2008), launched in October 2006. The STEREO mission consists of two identical spacecraft, which orbit the Sun ahead (STEREO-A) and behind (STEREO-B) the Earth near the ecliptic plane, and slowly separate from each other by $\sim 44°$ per year. EUVI observes the chromosphere and low corona in four different EUV emission lines out to 1.7 R_s (with R_s the solar radius) (Wuelser *et al.*, 2004; Howard *et al.*, 2008). During the coronal wave event under study, the EUVI imaging cadence was 2.5 min in the 171 Å and 10 min in the 195 Å filter. For the comparative analysis of the early evolution of the associated CME in the low corona we use data from the STEREO/SECCHI inner coronagraph COR1, a classic Lyot internally occulting refractive coronagraph with a field-of-view (FOV) from 1.4 to 4 R_s, and COR2 the externally occulted coronagraph with a FOV of 2 to 15 R_s (Howard *et al.*, 2008). For our study we used polarized brightness COR2 images. In addition, SOHO/LASCO C2 (FOV: 1.5 $R_s - 6$ R_s) and C3 (FOV: 3.5 $R_s - 30$ R_s) data are studied (Brueckner *et al.*, 1995).

The coronal wave is tracked by manually detecting the wave fronts, separately for SECCHI/EUVI-A and -B observations in both the 195 Å and 171 Å passband, respectively (*cf.* Figure 2). From a circular fit to the first wave front, the initiation center of the wave is derived (for more details see Veronig *et al.*, 2006). The distance of the wave to the initiation center is calculated by averaging the measured distance of each point of the tracked wave front from the derived initiation center along the spherical solar surface.

In order to infer the direction of motion of the CME we apply the triangulation method developed by Temmer, Preiss, and Veronig (2009) on the distance–time measurements of the leading edge of the CME observed from LASCO, STEREO-A, and -B. This method uses the information of the spacecraft separation angles and takes the propagation direction of the CME as a free parameter, in order to transform SOHO/LASCO distance–time measurements to STEREO-A and STEREO-B view. By comparing the transformed distances with the distances actually observed from STEREO, the direction of motion, which gives the minimum deviation, results in the best estimation of the true CME propagation direction. To obtain an independent result for the propagation direction of the CME as well as to derive the surface location of the CME flux rope we also use the flux rope forward fitting model developed by Thernisien, Howard, and Vourlidas (2006) and Thernisien, Vourlidas, and Howard (2009). This model is a raytrace simulation method which enables us to compute synthetic total and polarized brightness images using the Thomson scattering formulas

[1] ftp://ftp.ngdc.noaa.gov/STP/SOLAR_DATA/SOLAR_RADIO/BURSTS/.

[2] http://cdaw.gsfc.nasa.gov/CME_list/radio/waves_type2.html.

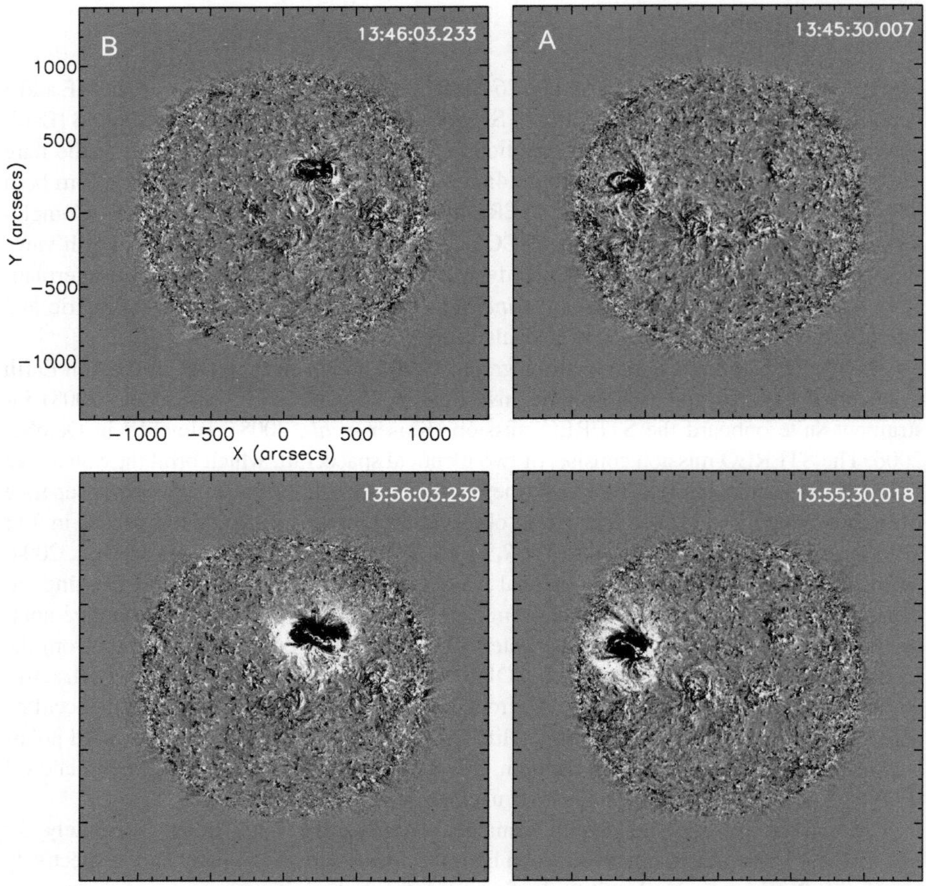

Figure 1 Sequence of running difference images from STEREO-A and -B in the EUVI 195 Å passband. A dome shaped structure observed from STEREO-A is indicated with arrows. See also the online animation.

from an assumed electron density model. Since the graduated cylindrical shell (GCS) model is a reasonable simulation of a flux-rope CME it can be applied to investigate the appearance of a CME. Characteristic CME parameters (width, cone angle, latitude, longitude) are derived by fitting the density model until a best match is found for contemporaneous image pairs from STEREO-A and STEREO-B, which observe the CME from two different vantage points.

3. Results

Figure 1 shows a sequence of EUVI 195 Å running difference images from STEREO-A and -B (see also the online movie). The location of the AR from which the wave is launched is E08 (if not stated otherwise, heliographic coordinates refer to Earth view). Hence, from the vantage point of STEREO-B the region is located less than 15° west from Sun center, whereas from the vantage point of STEREO-A it is more than 35° East. Therefore, measuring the wave fronts from STEREO-B is less affected by projection effects originating

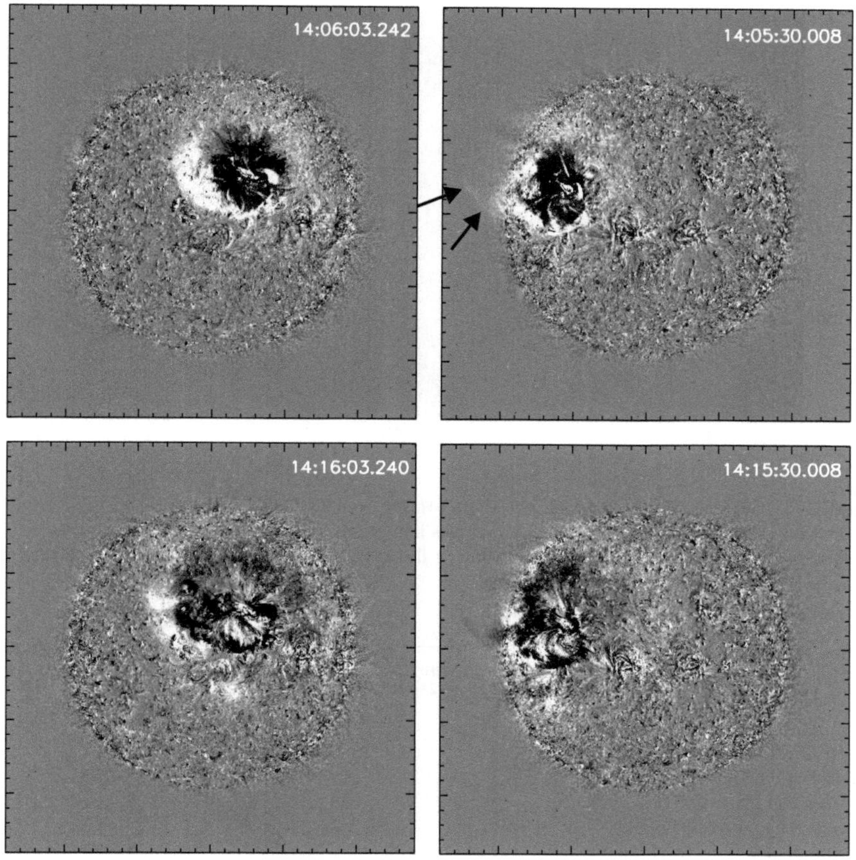

Figure 1 (*Continued.*)

from the three-dimensional nature of the wave (see Kienreich, Temmer, and Veronig, 2009; Patsourakos *et al.*, 2009). Both STEREO-A and -B observations reveal that the wave intensity is higher for the eastern propagation direction. The wave appears to be brighter as well as more diffuse in 195 Å compared to the 171 Å passband. In general, coronal disturbances are found to be better observed in 195 Å than in other wavelengths (Wills-Davey and Thompson, 1999). We also note that a dome shaped structure is identified in STEREO-A observations which images the disturbance from an almost lateral direction.

Figure 2 shows direct images of EUVI-A and EUVI-B in the 195 Å passband together with the wave fronts determined from EUVI 171 Å and 195 Å running difference images. From the vantage point of STEREO-A the AR is close to the eastern limb, therefore, the wave is best observed for the western direction (although less intense) and we focused on the construction of the wave kinematics along this direction. STEREO-B imagery enable us to track the wave in both the eastern and western propagation direction. By fitting circles to the first wave fronts separately for the eastern and western propagation directions observed in STEREO-B, two initiation centers on opposite sides of the AR are derived. This is obtained from EUVI-B 195 Å as well as 171 Å observations.

Figure 3 shows the derived wave kinematics for the western propagation direction obtained from STEREO-A and -B. In the 171 Å passband the wave can be tracked over a

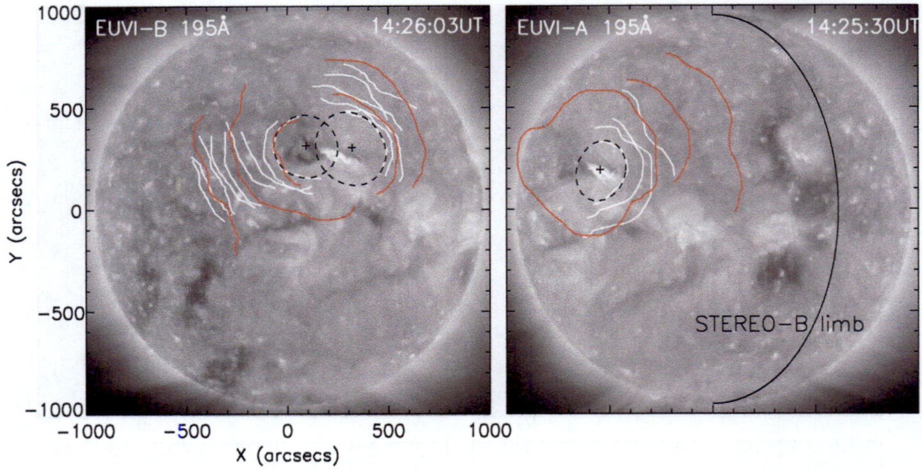

Figure 2 Traced wave fronts observed in STEREO-B (left) and STEREO-A (right) EUVI 195 Å (red lines) and 171 Å (white lines) images. Black dashed circles indicate the fit to the first wave front as seen in 171 Å. Black crosses give the estimated initiation center of the wave. The limb of STEREO-B is indicated in the image of STEREO-A.

Figure 3 Wave kinematics for the western direction as derived from EUVI 171 Å and 195 Å images of STEREO-A and STEREO-B (*cf.* Figure 2). Solid and dashed lines show the quadratic and linear fit, respectively, to the data points. The mean velocity, v, resulting from the linear fit is given in the legend.

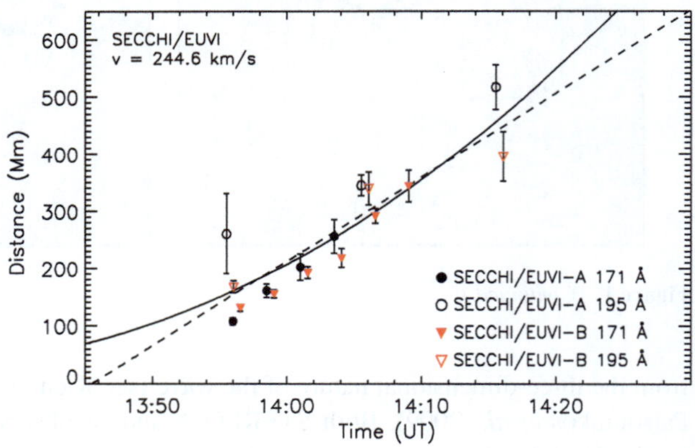

shorter distance than in 195 Å due to the weaker signal. For this propagation direction distances as derived from the 195 Å filter deviate from those measured in 171 Å. This is found from STEREO-A as well as STEREO-B measurements. Especially the first front observed in STEREO-A 195 Å differs clearly from a circular shape since part of the wave is detected off limb. Performing a linear fit over all measurements we obtain a mean velocity of $\sim 245\pm20\ \mathrm{km\,s^{-1}}$. The (point-like) center of initiation from STEREO-A 171 Å observations is derived at $[-544'', 193'']$, from STEREO-B 171 Å at $[318'', 306'']$ from Sun center. A transformation of the coordinates from A to B view gives $[253'', 258'']$, *i.e.* there is a mismatch of $\approx 50''$ when deriving the wave initiation center from the different vantage points. We assume that this difference is due to projection effects, uncertainties in the manual tracking of the wave as well as due to the simplified assumption of a point-like center of initiation (see, *e.g.*, Muhr *et al.*, 2010).

Figure 4 shows the derived wave kinematics from STEREO-B observations for the eastern propagation direction. The mean velocity of the wave is $\sim 240\pm12\ \mathrm{km\,s^{-1}}$ which is

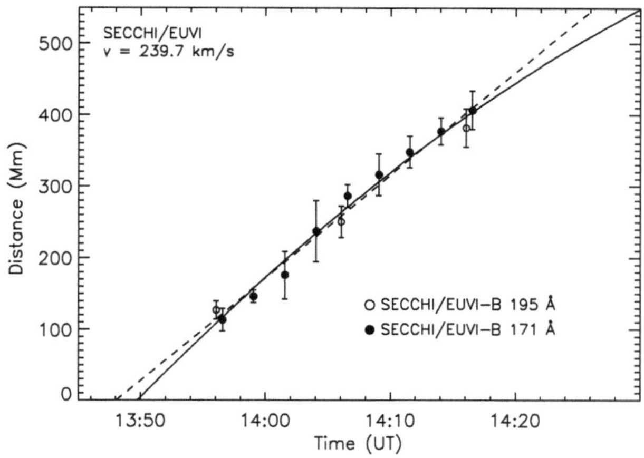

Figure 4 Wave kinematics for the eastern direction as derived from EUVI 171 Å and 195 Å images observed with STEREO-B. Solid and dashed lines show the quadratic and linear fit, respectively, to the data points. The mean velocity, v, resulting from the linear fit to the distance–time measurements is given in the legend.

comparable to the western propagation direction. The wave extracted from 171 Å images can be tracked up to the same distance as in the 195 Å passband. Since for the western propagation direction we observe the wave well in STEREO-A, we can exclude that the wave "disappears" behind the STEREO-B limb. Therefore, we infer an asymmetry in the intensity of the wave, *i.e.* the eastern propagation direction is more intense. In the following we will show that the main direction of the CME propagation as well as the 3D structure of the wave are in accordance with the more intense eastern wave propagation direction.

3.1. Propagation Direction of the Associated CME

In Figure 5 the result of the forward fit to the STEREO-A and -B COR1 image pair recorded at 14:45 UT is shown using the model by Thernisien, Howard, and Vourlidas (2006) and Thernisien, Vourlidas, and Howard (2009). The CME observations from 14:45 UT do not overlap in time with the EUV wave, but are only ~ 20 min after the last detected wave front. COR1 observations from STEREO-A and -B would show the CME in earlier images. However, the more of the CME structure is visible the more reliable is the forward fit. As can be seen in the lower right panel of Figure 5, the structure to the right of the top arrow possibly corresponds to a deflected streamer and not to the CME eruption itself (see also the accompanying movie). Therefore, it is possible that the CME fit, which includes this structure, overestimates the true width of the CME. From the fit we derive a CME propagation direction of E20 – E25 and a width of $\sim 60°$ as well as the position of the flux rope on the solar surface (*cf.* Figure 7). Figure 6 shows a second method from which the CME propagation direction is derived using the triangulation method developed by Temmer, Preiss, and Veronig (2009). From the triangulation of the STEREO-A and -B CME kinematics we obtain a propagation direction of E28 – E35 (see Figure 6) which is consistent with the result derived from the flux rope forward fit. For comparison, the AR from which the CME is launched has a longitudinal position of E08. Using the derived information of the CME propagation direction, we obtain a deprojected CME speed of $\sim 750 \pm 50$ km s^{-1}.

Figure 7 shows a base difference image from STEREO-B EUVI 195 Å and the derived wave fronts over the time range 13:56 UT – 14:16 UT together with the flux rope position on the solar surface obtained with forward fitting applied to the COR1 STEREO-A and -B image pairs at 14:45 UT. The EUVI base difference image reveals coronal dimming regions on opposite sides of the AR. Their location is consistent with the two initiation centers obtained for the wave. The position of the footpoints of the flux rope obtained from the

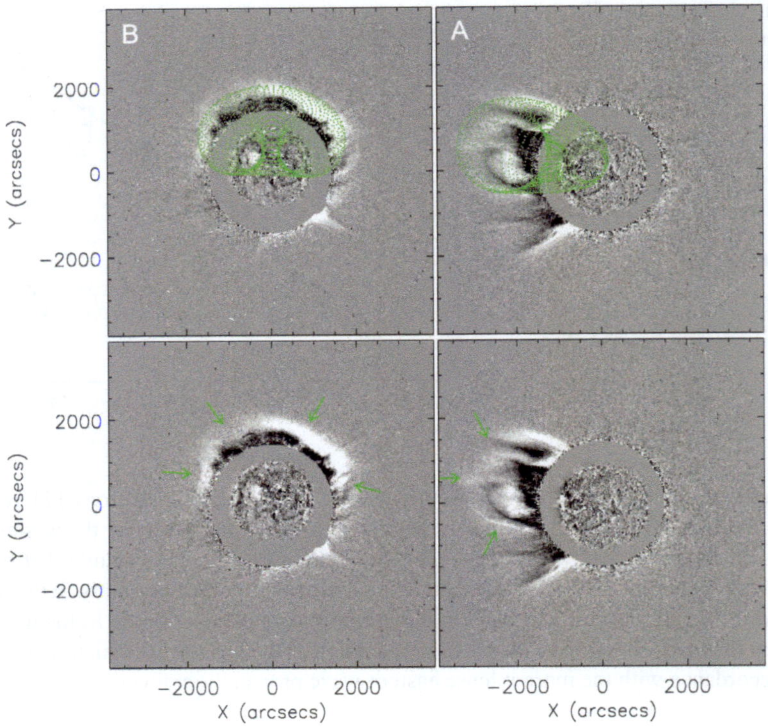

Figure 5 Flux rope (green mesh) resulting from the raytracing-fitting method applied to COR1 observations from STEREO-A (right) and STEREO-B (left) overplotted on composite EUVI 195 Å -COR1 difference images from ∼ 14:45 UT. Arrows outline the shape of the CME to which the flux rope was fitted. See also the online animations of STEREO-A and -B composites from EUVI and COR1 running difference images.

forward fit is also closely located to the dimming regions. The eastern footpoint of the flux rope is positioned to the east from the AR and outside the dimming region. The western footpoint is close to the AR and within the dimming area.

3.2. 3D Geometry of the Wave

Figure 8 (top panels) shows difference images of the wave in EUVI 171 Å as observed from STEREO-A and STEREO-B at 13:59 UT, respectively. For a better comparison the STEREO-B image is rotated to the viewing angle of STEREO-A. Imaging the wave from two different vantage points reveals differences in the morphology of the wave as well as "foreshortening effects". This hints at the 3D nature of the wave, and can be used to infer information of its 3D geometry. For the sake of simplicity we assume that the wave consists of a base and two lateral borders which are inclined with respect to the vertical. As can be seen from the red bars in Figure 8 the distances between the active region and the eastern and western part of the wave are similar observed from STEREO-B, but different observed from STEREO-A. We therefore may assume that STEREO-B observes the base of the wave nearly free from projection effects. Using the observed extension of the eastern and western part of the wave as seen from STEREO-A, and knowing the location of the active region as well as the spacecraft positions, we can simply use trigonometric functions to derive the inclination of the borders of the wave. The results are schematically drawn in the bottom

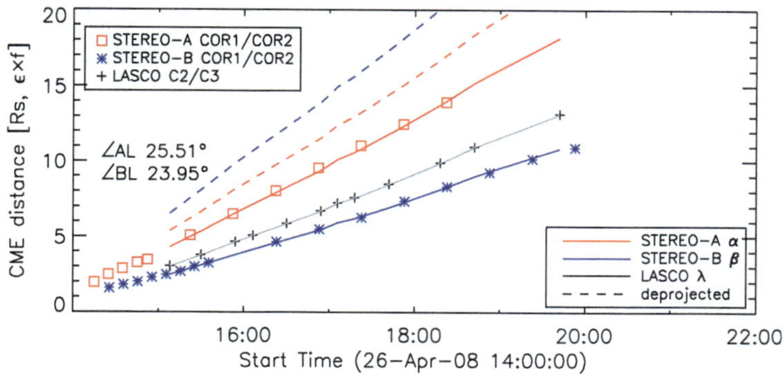

Figure 6 Triangulation method from Temmer, Preiss, and Veronig (2009) applied to the derived CME height–time data from observations of different spacecraft. Solid lines are the modeled distances for STEREO-A view (red), STEREO-B view (blue), and LASCO (gray). The red and blue dashed lines show the resulting de-projected distances separately derived from LASCO/STEREO-A and LASCO/STEREO-B comparison. The separation angle between the STEREO spacecraft with respect to LASCO is labeled (∠AL, ∠BL).

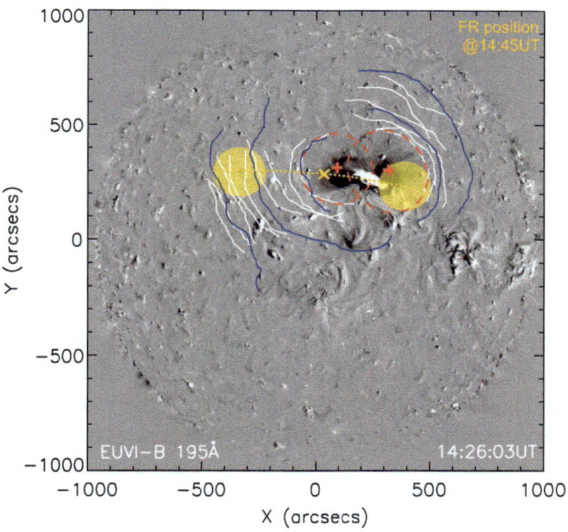

Figure 7 Traced wave fronts (13:56 UT – 14:16 UT) from STEREO-B EUVI 195 Å in blue and 171 Å in white plotted on a base difference image. Red dashed circles give the fit to the first wave front observed in EUVI 171 Å from which the initiation centers (red crosses) of the wave are obtained within the bipolar dimming regions. Results from the raytracing tool derived from COR1 images at 14:45 UT (*cf.* Figure 5) are indicated in yellow: shaded circles mark the position and extension of the footpoints of the flux rope, cross and dashed line give the apex and inclination of the flux rope projected on the solar surface.

panel of Figure 8, revealing that the wave structure is asymmetric, *i.e.* more inclined toward East.

4. Discussion and Conclusion

For the EUV wave that occurred on 26 April 2008 we obtained two initiation centers located on either side of the AR from which the wave was launched. The two initiation centers are located within the coronal dimming regions. A similar result was found by Muhr *et al.* (2010) for the Hα Moreton wave of 28 October 2003. For this event the initial location of the wave center was derived to be located close to the dimming regions well outside of the AR core, which are indicative of the footpoints of the flux rope of a CME (see, *e.g.*, Sterling and Hudson, 1997). Model simulations of fast-mode MHD waves by Wang (2000) showed that the

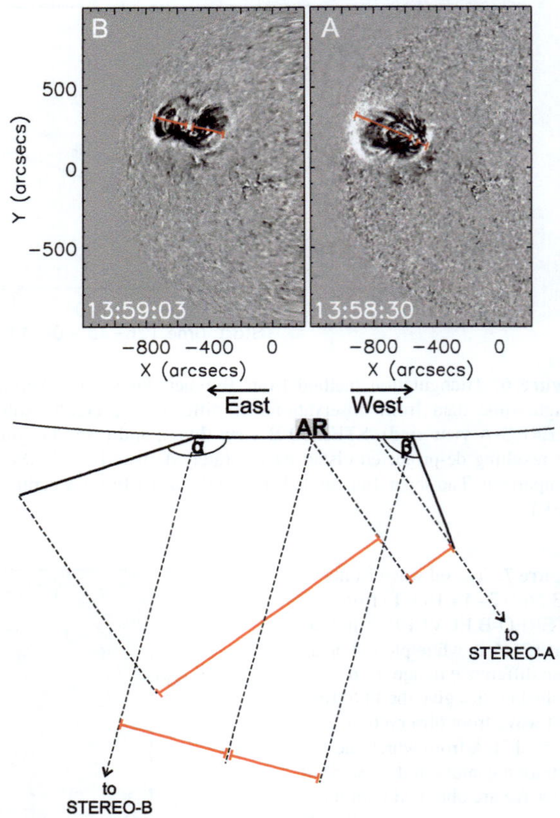

Figure 8 Top: EUVI-A and -B 171 Å difference images of the wave at 13:59 UT. Note that the STEREO-B image (left) is aligned to the viewing direction of STEREO-A (right) in order to make the differences in the morphology of the wave as viewed from two different vantage points clearly visible. The lateral extension of the wave is schematically indicated with red bars. Bottom: Reconstructed 3D geometry (view from solar north pole) with $\alpha \sim 162°$ and $\beta \sim 111°$. AR indicates the location of the active region. For details of the reconstruction, see text.

wave disturbances needed to be launched from the periphery of the AR to expand horizontally over the surface, whereas the initiation in the core of the AR with strong magnetic fields would lead to a vertical upward motion of the wave. These results are in accordance with the observations and the obtained initiation centers on opposite sites of the active region in weak magnetic field. Simultaneous imaging from two different vantage points enabled us to apply the forward fitting model by Thernisien, Howard, and Vourlidas (2006) and Thernisien, Vourlidas, and Howard (2009). From this we derived the flux rope position on the solar surface and found that it is close to the coronal dimming regions and wave initiation centers, respectively. These findings suggest that the CME expanding flanks are initiating the wave. A recent study by Patsourakos, Vourlidas, and Stenborg (2010) showed that CMEs undergo a strong but short lived lateral overexpansion in their early evolution.

The mean speed of the wave of $\sim 240\ \mathrm{km\,s^{-1}}$, which is in the range of the characteristic speed of the ambient quiet solar corona, is derived to be similar for the western and eastern propagation directions as well as from observations of two different vantage points. The wave shows no clear deceleration which is due to its slow initial speed and, as a consequence of the extreme solar minimum, due to the weak magnetic field environment in which the wave is propagating (see, *e.g.*, Kienreich *et al.*, 2011). The derived speed of the wave lies within the velocity range of $210-350\ \mathrm{km\,s^{-1}}$ as expected for fast magnetosonic waves in quiet Sun conditions (see Mann *et al.*, 1999). The radial speed of the upward moving CME is much larger and of $\sim 750\ \mathrm{km\,s^{-1}}$. In this respect it would be interesting to do a comparison to the model by Attrill *et al.* (2007) who interprets EUV waves as magnetic footprints of

CMEs. This model would require a motion of the EUV wave similar to the global average motion defined by the lateral expansion of the CME. However, the CME is observed off-limb well after the end of the EUV wave and therefore it is not possible to compare simultaneous measurements of the wave and CME widths.

For the western propagation direction, the measurements from the 195 Å passband show systematically larger distances as compared to 171 Å. Patsourakos *et al.* (2009) also found differences in the wave appearance between 171 Å and 195 Å, especially at the beginning of the wave evolution. For the eastern propagation direction the wave is observed to be more intense. The transformation of EUVI images from STEREO-B view to STEREO-A view makes foreshortening effects clearly visible (see Figure 8). Reconstructing the 3D structure of the wave by applying simple geometry, we find that the wave is inclined toward East. These results indicate that the 3D structure of the wave is not symmetric. From two independent methods (forward fitting model Thernisien, Howard, and Vourlidas (2006) and triangulation Temmer, Preiss, and Veronig (2009)) we derived that the propagation direction of the associated CME deviates eastward from the AR position. The appearance of the wave, especially its asymmetric intensity and 3D geometry, seems to be closely related to the non-radial evolution of the associated CME. The results from the present study support the wave nature of the phenomenon as well as the close association of the wave structure to the evolutionary characteristics of the expanding CME flanks.

Since in the EUVI 171 Å passband wave features might be confused with expanding loops (Patsourakos and Vourlidas, 2009), we would like to emphasize that the double-initiation center is found also from wave fronts identified in 195 Å images. Moreover, the smooth kinematics observed in the eastern as well as western propagation direction gives no indication that the first fronts identified would not be associated to the propagating wave. In this respect we would also like to note that, if the first measured fronts are expanding loops, we face a very smooth transition between loop expansion and the build up of the wave. We expect that detailed studies with high spatial and temporal resolution SDO data will shed more light on the initiation phase of CMEs and their association to coronal waves. Liu *et al.* (2010) presented the first results from SDO imagery for the EUV wave of 8 April 2010 and proposed a hybrid hypothesis combining both wave and non-wave aspects to explain the observations.

Acknowledgements We thank the referee for helpful comments. M.T. and A.V. gratefully acknowledge the Austrian Science Fund (FWF): P20867-N16.

References

Attrill, G.D.R., Harra, L.K., van Driel-Gesztelyi, L., Démoulin, P.: 2007, *Astrophys. J. Lett.* **656**, 101. doi:10.1086/512854.
Attrill, G.D.R., Engell, A.J., Wills-Davey, M.J., Grigis, P., Testa, P.: 2009, *Astrophys. J.* **704**, 1296. doi:10.1088/0004-637X/704/2/1296.
Brueckner, G.E., Howard, R.A., Koomen, M.J., Korendyke, C.M., Michels, D.J., Moses, J.D., Socker, D.G., Dere, K.P., Lamy, P.L., Llebaria, A., Bout, M.V., Schwenn, R., Simnett, G.M., Bedford, D.K., Eyles, C.J.: 1995, *Solar Phys.* **162**, 357. doi:10.1007/BF00733434.
Chen, P.F., Wu, S.T., Shibata, K., Fang, C.: 2002, *Astrophys. J. Lett.* **572**, 99. doi:10.1086/341486.
Cohen, O., Attrill, G.D.R., Manchester, W.B., Wills-Davey, M.J.: 2009, *Astrophys. J.* **705**, 587. doi:10.1088/0004-637X/705/1/587.
Dai, Y., Auchère, F., Vial, J., Tang, Y.H., Zong, W.G.: 2010, *Astrophys. J.* **708**, 913. doi:10.1088/0004-637X/708/2/913.
Delaboudinière, J.P., Artzner, G.E., Brunaud, J., *et al.*: 1995, *Solar Phys.* **162**, 291. doi:10.1007/BF00733432.
Delannée, C.: 2000, *Astrophys. J.* **545**, 512. doi:10.1086/317777.

Gallagher, P.T., Long, D.M.: 2010, *Space Sci. Rev.* **135**. doi:10.1007/s11214-010-9710-7.
Gopalswamy, N., Yashiro, S., Temmer, M., Davila, J., Thompson, W.T., Jones, S., McAteer, R.T.J., Wuelser, J.P., Freeland, S., Howard, R.A.: 2009a, *Astrophys. J. Lett.* **691**, 123. doi:10.1088/0004-637X/691/2/L123.
Gopalswamy, N., Thompson, W.T., Davila, J.M., Kaiser, M.L., Yashiro, S., Mäkelä, P., Michalek, G., Bougeret, J., Howard, R.A.: 2009b, *Solar Phys.* **259**, 227. doi:10.1007/s11207-009-9382-1.
Howard, R.A., Moses, J.D., Vourlidas, A., Newmark, J.S., Socker, D.G., Plunkett, S.P., *et al.*: 2008, *Space Sci. Rev.* **136**, 67. doi:10.1007/s11214-008-9341-4.
Kaiser, M.L., Kucera, T.A., Davila, J.M., St. Cyr, O.C., Guhathakurta, M., Christian, E.: 2008, *Space Sci. Rev.* **136**, 5. doi:10.1007/s11214-007-9277-0.
Kienreich, I.W., Temmer, M., Veronig, A.M.: 2009, *Astrophys. J. Lett.* **703**, 118. doi:10.1088/0004-637X/703/2/L118.
Kienreich, I.W., Veronig, A.M., Muhr, N., Temmer, M., Vršnak, B., Nitta, N.: 2011, *Astrophys. J. Lett.* **727**, 43. doi:10.1088/2041-8205/727/2/L43.
Liu, W., Nitta, N.V., Schrijver, C.J., Title, A.M., Tarbell, T.D.: 2010, *Astrophys. J. Lett.* **723**, 53. doi:10.1088/2041-8205/723/1/L53.
Long, D.M., Gallagher, P.T., McAteer, R.T.J., Bloomfield, D.S.: 2008, *Astrophys. J. Lett.* **680**, 81. doi:10.1086/589742.
Mann, G., Klassen, A., Estel, C., Thompson, B.J.: 1999, In: Vial, J.C., Kaldeich-Schü, B. (eds.) *8th SOHO Workshop: Plasma Dynamics and Diagnostics in the Solar Transition Region and Corona* **SP-446**, ESA, Noordwijk, 477.
Moses, D., Clette, F., Delaboudinière, J.P., *et al.*: 1997, *Solar Phys.* **175**, 571. doi:10.1023/A:1004902913117.
Muhr, N., Vršnak, B., Temmer, M., Veronig, A.M., Magdalenić, J.: 2010, *Astrophys. J.* **708**, 1639. doi:10.1088/0004-637X/708/2/1639.
Narukage, N., Hudson, H.S., Morimoto, T., Akiyama, S., Kitai, R., Kurokawa, H., Shibata, K.: 2002, *Astrophys. J. Lett.* **572**, 109. doi:10.1086/341599.
Patsourakos, S., Vourlidas, A.: 2009, *Astrophys. J. Lett.* **700**, 182. doi:10.1088/0004-637X/700/2/L182.
Patsourakos, S., Vourlidas, A., Stenborg, G.: 2010, *Astrophys. J. Lett.* **724**, 188. doi:10.1088/2041-8205/724/2/L188.
Patsourakos, S., Vourlidas, A., Wang, Y.M., Stenborg, G., Thernisien, A.: 2009, *Solar Phys.* **259**, 49. doi:10.1007/s11207-009-9386-x.
Sterling, A.C., Hudson, H.S.: 1997, *Astrophys. J. Lett.* **491**, 55. doi:10.1086/311043.
Temmer, M., Preiss, S., Veronig, A.M.: 2009, *Solar Phys.* **256**, 183. doi:10.1007/s11207-009-9336-7.
Thernisien, A.F.R., Howard, R.A., Vourlidas, A.: 2006, *Astrophys. J.* **652**, 763. doi:10.1086/508254.
Thernisien, A., Vourlidas, A., Howard, R.A.: 2009, *Solar Phys.* **256**, 111. doi:10.1007/s11207-009-9346-5.
Thompson, B.J., Plunkett, S.P., Gurman, J.B., Newmark, J.S., St. Cyr, O.C., Michels, D.J.: 1998, *Geophys. Res. Lett.* **25**, 2465. doi:10.1029/98GL50429.
Uchida, Y.: 1968, *Solar Phys.* **4**, 30. doi:10.1007/BF00146996.
Veronig, A.M., Temmer, M., Vršnak, B.: 2008, *Astrophys. J. Lett.* **681**, 113. doi:10.1086/590493.
Veronig, A.M., Temmer, M., Vršnak, B., Thalmann, J.K.: 2006, *Astrophys. J.* **647**, 1466. doi:10.1086/505456.
Veronig, A.M., Muhr, N., Kienreich, I.W., Temmer, M., Vršnak, B.: 2010, *Astrophys. J. Lett.* **716**, 57. doi:10.1088/2041-8205/716/1/L57.
Vršnak, B., Cliver, E.W.: 2008, *Solar Phys.* **253**, 215. doi:10.1007/s11207-008-9241-5.
Wang, Y.: 2000, *Astrophys. J. Lett.* **543**, 89. doi:10.1086/318178.
Warmuth, A.: 2010, *Adv. Space Res.* **45**, 527. doi:10.1016/j.asr.2009.08.022.
Warmuth, A., Vršnak, B., Aurass, H., Hanslmeier, A.: 2001, *Astrophys. J. Lett.* **560**, 105. doi:10.1086/324055.
Wills-Davey, M.J., Attrill, G.D.R.: 2009, *Space Sci. Rev.* **149**, 325.lls. doi:10.1007/s11214-009-9612-8.
Wills-Davey, M.J., DeForest, C.E., Stenflo, J.O.: 2007, *Astrophys. J.* **664**, 556. doi:10.1086/519013.
Wills-Davey, M., Thompson, B.: 1999, *Solar Phys.* **190**, 467. doi:10.1023/A:1005201500675.
Wuelser, J., Lemen, J.R., Tarbell, T.D., *et al.*: 2004, In: Fineschi, S., Gummin, M.A. (eds.) *Society of Photo-Optical Instrumentation Engineers (SPIE) Conference Series* **5171**, 111. doi:10.1117/12.506877.
Zhukov, A.N., Auchère, F.: 2004, *Astron. Astrophys.* **427**, 705. doi:10.1051/0004-6361:20040351.

Solar Phys (2011) 273:433–460
DOI 10.1007/s11207-011-9780-z

ENERGY STORAGE AND RELEASE – MODELS MEET RADIO OBSERVATIONS

Coronal Shock Waves, EUV Waves, and Their Relation to CMEs. I. Reconciliation of "EIT Waves", Type II Radio Bursts, and Leading Edges of CMEs

V.V. Grechnev · A.M. Uralov · I.M. Chertok · I.V. Kuzmenko · A.N. Afanasyev · N.S. Meshalkina · S.S. Kalashnikov · Y. Kubo

Received: 22 November 2010 / Accepted: 19 April 2011 / Published online: 7 June 2011

Abstract We show examples of the excitation of coronal waves by flare-related abrupt eruptions of magnetic rope structures. The waves presumably rapidly steepened into shocks and freely propagated afterwards like decelerating blast waves that showed up as Moreton waves and EUV waves. We propose a simple quantitative description for such shock waves to reconcile their observed propagation with drift rates of metric type II bursts and kinematics of leading edges of coronal mass ejections (CMEs). Taking account of different plasma density falloffs for propagation of a wave up and along the solar surface, we demonstrate a close correspondence between drift rates of type II bursts and speeds of EUV waves, Moreton waves, and CMEs observed in a few previously studied events.

Energy Storage and Release through the Solar Activity Cycle – Models Meet Radio Observations
Guest Editors: Christophe Marqué and Alexander Nindos

Electronic supplementary material The online version of this article (doi:10.1007/s11207-011-9780-z) contains supplementary material, which is available to authorized users.

V.V. Grechnev (✉) · A.M. Uralov · A.N. Afanasyev · N.S. Meshalkina · S.S. Kalashnikov
Institute of Solar-Terrestrial Physics SB RAS, Lermontov St. 126A, Irkutsk 664033, Russia
e-mail: grechnev@iszf.irk.ru

I.M. Chertok
Pushkov Institute of Terrestrial Magnetism, Ionosphere and Radio Wave Propagation (IZMIRAN), Troitsk, Moscow Region 142190, Russia
e-mail: ichertok@izmiran.ru

I.V. Kuzmenko
Ussuriysk Astrophysical Observatory, Solnechnaya St. 21, Primorsky Krai, Gornotaezhnoe 692533, Russia
e-mail: kuzmenko_irina@mail.ru

Y. Kubo
National Institute of Information and Communications Technology, Tokyo, Japan
e-mail: kubo@nict.go.jp

Keywords Coronal mass ejections, low coronal signatures · Coronal mass ejections, initiation and propagation · Prominences, active · Radio bursts, dynamic spectrum · Radio bursts, type II · Waves, shock

1. Introduction

Some solar eruptions are accompanied by large-scale wave-like disturbances visible in various spectral ranges. Moreton waves (Moreton, 1960) observed in the Hα line have been initially proposed by Uchida (1968) to be a chromospheric trail of a coronal fast-mode MHD wave. Observations of the low corona in extreme ultraviolet (EUV) with EIT (Delaboudinière *et al.*, 1995) on SOHO revealed large-scale wave-like disturbances visible as fronts of enhanced (but still low) brightness, either quasi-stationary or propagating over large distances up to the whole disk along the solar surface or expanding above the limb. These transients called "EIT waves" (or "EUV waves") are registered by a number of EUV telescopes – EIT, TRACE, STEREO/EUVI, SDO/AIA, *etc.* (see reviews by Warmuth, 2007; Wills-Davey and Attrill, 2009; Gallagher and Long, 2010).

Analysis of observations and interpretation of such phenomena meet problems. Glaring flare emission hampers detection of faint EUV waves. Having been acquired with SOHO/EIT, the bulk of the observational material is limited by 12-min imaging rate. Faster TRACE observations (Handy *et al.*, 1999) have a small field of view. It is often difficult to reliably identify and trace a moving feature of interest. To overcome these difficulties, special methods are employed, but they might contribute artifacts. Multi-instrument analyses sometimes encounter timing problems. Thus, some results used in interpretation and modeling might not be completely adequate to the observed phenomena.

The nature of EUV waves has been debated starting from their discovery (Thompson *et al.*, 1998). The most popular interpretations of a near-surface EUV wave are *i*) traces of an MHD fast-mode wave in the lower corona (see, *e.g.*, Thompson *et al.*, 1999; Warmuth *et al.*, 2001, 2004b; Khan and Aurass, 2002; Hudson and Warmuth, 2004) and *ii*) plasma compression in bases of coronal loops in their successive stretching by an expanding CME (see, *e.g.*, Delannée, 2000). A numerical 2D MHD simulation of a magnetic flux rope eruption (Chen, Fang, and Shibata, 2005) revealed both the *i*) and *ii*) disturbances. Schmidt and Ofman (2010) presented the first 3D MHD modeling of an "EIT wave" as a disturbance produced by an eruption-driven shock wave. A fast-mode wave detected in the simulation corresponded to an EUV wave observed in the modeled event including reflection from a coronal hole in support of the wave hypothesis.

None of existing models describes all properties of EUV waves. Most likely, this is because the multitude of transients observed as EUV waves actually correspond to different phenomena. This conjecture is supported by a variety of morphologic and dynamic characteristics of observed EUV waves. For example, *i*) their velocities estimated from observations of some events exceeded the coronal fast-mode speed, whereas they were lower in other events (Wills-Davey, DeForest, and Stenflo, 2007; Warmuth, 2010); *ii*) the wave front can be either diffuse or sharp; *iii*) kinematics of an EUV wave can be incompatible with the fast-mode MHD wave model (Zhukov, Rodriguez, and de Patoul, 2009). On the other hand, such properties of EUV waves as deceleration, decay, and broadening the disturbance (Warmuth *et al.*, 2001, 2004a, 2004b; Long *et al.*, 2008; Veronig *et al.*, 2010), bypassing regions of an increased Alfvén velocity – coronal holes and active regions (Thompson *et al.*, 1999), possible reflections (Veronig, Temmer, and Vršnak, 2008; Gopalswamy *et al.*, 2009) appear to correspond to the hypothesis of a coronal MHD wave. Considerations of wave-like transients sometimes observed in EUV to expand above the limb also suggest that different phenomena might be involved (see, *e.g.*, Zhukov and Auchère, 2004; Grechnev *et al.*, 2006b, 2008; Pohjolainen, Hori, and Sakurai, 2008; Meshalkina *et al.*, 2009; Chertok, Grechnev, and Uralov, 2009; Cohen *et al.*, 2009). In Paper III (Grechnev *et al.*, 2011) we consider an event with a two-component "EIT wave"; a propagating component matched properties of a coronal shock wave, while a stationary component was presumably associated with a stretched CME structure.

If some EUV waves are really due to coronal shock waves, then their correspondence with signatures of shocks in higher corona is expected. Type II radio bursts are considered as manifestations of shock waves propagating upwards in the corona (see, *e.g.*, Vršnak and Cliver, 2008). Klassen *et al.* (2000) concluded that almost all metric type II bursts were accompanied by EUV waves but stated no correlation between their speeds; on average, the speeds of EUV waves were three times less than estimates from drift rates of type II bursts. Biesecker *et al.* (2002) found that many EUV waves were not associated with type II bursts.

Sheeley, Hakala, and Wang (2000) proposed that kinks or deflections of coronal rays at the flanks and rear ends of fast CMEs could be signatures of shock waves. An important common property of these flank/rear kinks was their deceleration, while the authors did not reveal deceleration of CME leading edges.[1] A scenario usually considered for the formation of shock waves, which might show up in LASCO images, seems to be incompatible with wave signatures on the solar surface. In this scenario, when the CME speed exceeds the local fast-mode speed at heliocentric distances $\gtrsim 1.5 R_\odot$, a bow shock forms continuously pressed by a fast CME. However, a bow shock followed by a Mach cone can only be formed by a supersonic body of a fixed size, whereas CMEs expand omnidirectionally (with respect to their center). Hence, the conic bow-shock geometry appears to be unlikely for wide CMEs. Since neither mechanisms nor heights of the shock formation have been established, possible association between surface EUV waves and CME components cannot be excluded. Indeed, Veronig *et al.* (2010) found the upper part of an expanding EUV dome to coincide with a white-light CME, while the lower skirt of the dome was a surface EUV wave.

For these reasons it is difficult to expect that all observed properties of EUV waves could be explained by a single mechanism. Based on this assumption, our three companion papers are focused on those EUV waves, which are most likely associated with coronal shock waves. We address a few events, all of which were previously studied. Even for a rather uniform subset of phenomena probably corresponding to similar processes, apparent contradictions between some observational facts might occur. We endeavor to reconcile such discrepancies.

One of challenging issues is reconciliation of EUV wave velocities and drift rates of type II bursts. This is a subject of the present Paper I, the first one of the three companion papers. Assuming a shock-wave nature of the EUV waves under consideration, we try to settle disagreement between different studies. In this paper we use the simplest approximation of a self-similar shock wave that is convenient in comparisons with observations, which often show self-similarity of the wave front expansion. However, this approximation corresponds to the strong shock-wave limit and cannot apply to all stages of events. A weaker shock seems to be more realistic, but its calculations are more complex. Propagation of a weak shock along the solar surface is analytically modeled in Paper II (Afanasyev and Uralov, 2011). Paper III analyzes propagation of a probable shock wave in the 17 January 2010 event using both strong and weak shock approximations.

[1]The SOHO LASCO CME Catalog at http://cdaw.gsfc.nasa.gov/CME_list/ (afterwards 'the CME Catalog'; Yashiro *et al.*, 2004) shows pronounced decelerations for three of the four events considered by the authors.

2. Methodical Issues

Appearance of shock waves in eruptive events seems to be generally accepted, but their sources have not been established. Three possible exciters of shock waves are considered:

i) a pressure pulse produced by a flare,
ii) a super-Alfvénic piston, and
iii) an impulsive piston.

In case *ii*), a bow shock continuously driven by a supersonic CME is expected to appear with kinematics determined by the driver. In case *i*) or *iii*), the shock wave expelled by an impulsive driver propagates afterwards freely like a decelerating blast wave (*cf.* Pomoell, Vainio, and Kissmann, 2008). Observations suggest that shock waves excited by impulsive drivers and freely propagating in the low corona do exist. This is indicated by deceleration of Moreton/EUV waves (see, *e.g.*, Warmuth *et al.*, 2001, 2004a), differences between propagation directions of the wave and a possible driver (Hudson *et al.*, 2003), and very early appearance of type II bursts. We accept this possibility as a working hypothesis and use a description of propagation of a blast shock wave. It is possible to calculate it analytically for two limits. One limit is a strong self-similar wave, whose length along a propagation direction is comparable with the curvature radius of the wave front. The opposite limit is a weak shock wave, whose length is much less than the curvature radius of the front and a typical size of inhomogeneities in the medium. Grechnev *et al.* (2008) found that a formal usage of expressions for propagation of a strong self-similar shock wave excited by a point-like explosion in a gas allowed to fit the speeds and positions of a Moreton wave as well as an EUV wave at the initial stage of the motion.

2.1. Self-similar Shock Wave Approximation

Grechnev *et al.* (2008) used a simple model to describe propagation of such a blast-like wave in plasma with a radial power-law (PL) density falloff δ from an eruption center, $n = n_0(x/h_0)^{-\delta}$ with x being the distance and n_0 the density at a distance of h_0. We use $h_0 \approx 100$ Mm, close to the scale height. (Our notations are different from those used in the papers listed above). Self-similarity of a shock wave is ensured by a presumable large pressure excess inside the volume confined by the shock front over a non-disturbed medium. Propagation of a shock wave in the self-similar approximation is determined by plasma density distribution being almost insensitive to the magnetic fields. Such a wave decelerates when $\delta < 3$ due to a growing mass of swept-up material. Propagation of a strong shock *vs.* time t in plasma with a PL density model is described by an expression

$$x(t) \propto t^{2/(5-\delta)}. \tag{1}$$

The approximation has a singularity at $x \to 0$ (here also the wave velocity $v \to \infty$); however, wave signatures are not observed at small distances. The approximation becomes inaccurate at large distances, being not limited from below by the fast-mode speed. A wave traveling along the solar surface weakens at large distances and propagates, in the first approximation, in a flat-layered atmosphere. Expression (1) describes strong spherical shock waves, which seem to be unrealistic in solar conditions, but its usage within some range of distances can be justified.

i) An enhanced plasma density above an active region falls off both vertically and horizontally. A power-law description of the falloff seems to be acceptable.

ii) The self-similar solution of a strong wave satisfactorily describes damping of a gas-dynamic shock wave up to Mach numbers $M \approx 2$, when the wave is neither strong nor weak.

iii) Applicability of gas-dynamic self-similar solutions to MHD blast shock waves is not obvious, because account of the magnetic field seems to be necessary.

We note the following. With $M \gg 2$, the gas pressure behind the shock front exceeds the magnetic pressure, even if $\beta = C_\mathrm{S}^2/V_\mathrm{A}^2 \ll 1$ in non-disturbed plasma before the shock front; here M is a ratio of the shock speed to the fast-mode speed before the front, C_S and V_A are the sound and Alfvén speeds. That is, the plasma flow behind the shock front has a gas-dynamic character. The role of magnetic fields is also not crucial for medium-intensity shocks ($M \geq 2$), which also strongly heat plasma, thus significantly increasing its pressure. For example, with $M \approx 2$, the plasma pressure behind the front of a wave perpendicular to the magnetic field is equal to the magnetic pressure before the front, even if $\beta \ll 1$ there. For a switch-on shock wave running along the magnetic field this occurs with a Mach number $M \approx 1.5$.

Expression (1) was obtained under an assumption that the δ index was independent of the wave propagation direction ϑ. We will formally use this expression also when $\delta = \delta(\vartheta)$, if variations of δ are small with the change of the direction, *i.e.*, $\pi d\delta \ll d\vartheta$. Note that in a limit of a weak, short shock wave, its propagation is determined by a local value of δ even if this condition is not satisfied. The above considerations lead to a heuristic conclusion about a possibility to use expression (1) for approximate estimates of kinematic characteristics of shock waves of intermediate intensity propagating in medium with $\delta = \delta(\vartheta)$.

It is useful to compare the power-law coronal density model with other popular models. The Newkirk (1961) model ($n_\mathrm{e} = 4.2 \times 10^4 \times 10^{4.32/r}$, r is the heliocentric distance expressed in solar radii) describes the radial plasma density distribution in a coronal streamer. The Saito model (Saito, 1970) describes the density distribution above the quiet Sun depending on the latitude ϕ

$$\frac{n_\mathrm{e}(r,\phi)}{10^8} = \frac{3.09}{r^{16}}(1-0.5\sin\phi) + \frac{1.58}{r^6}(1-0.95\sin\phi) + \frac{0.0251}{r^{2.5}}\left(1-\sin^{0.5}\phi\right). \quad (2)$$

Figure 1a presents the Saito model for $\phi = 0°, 60°$, and the Newkirk model. The PL model can be adjusted to any of these models by varying its parameters. The n_0, h_0 parameters are redundant; we have split them to clarify their physical meaning. The x variable in the radial direction is $x \approx (r-1)R_\odot$. The PL model with $\delta = 2$, $n_0 = 2.8 \times 10^8\ \mathrm{cm}^{-3}$ agrees within $\pm 30\%$ with the Newkirk model at $r = (1.2-9)R_\odot$, which is important in analyses of type II bursts. The parameters of the PL model can be adjusted to the Saito model for various ϕ as well, which is important in analyses of CMEs. A single PL model with a direction-dependent δ provides a convenient alternative to complex involvement of various density models and allows one to account for individual properties of active regions as well as highly disturbed conditions just before the onset of a wave.

2.2. Shock Waves and Type II Bursts

Assuming δ to depend on a propagation direction, we get an approximation for a shock of an intermediate intensity propagating in an anisotropic medium (Figure 1b). A quasi-isotropic shock wave propagating in homogeneous corona can only cause drifting continuum radio emission. A strong narrowband harmonic type II burst can appear if the shock front passes along a lengthy structure like a coronal streamer (see, *e.g.*, Reiner *et al.*, 2003; Mancuso

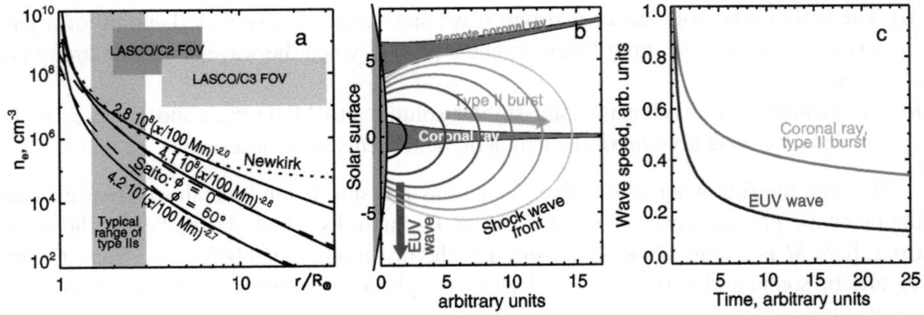

Figure 1 (a) Coronal density models of Newkirk (dotted) and Saito (dashed, for latitudes $\phi = 0°$ and $\phi = 60°$). The solid lines represent the power-law model with different parameters to fit the Newkirk model at $r < 9R_{\odot}$ and the Saito model within the LASCO/C2 & C3 fields of view shown with horizontal shaded bars. The vertical shaded region indicates the typical range of of metrics type II bursts. (b) A cartoon illustrating the relation between a type II burst and an EUV wave. A narrowband type II emission is generated by a shock front propagating in a distinct extended narrow structure like a coronal ray. A near-surface EUV wave runs slower. (c) Speed–time plots of the EUV wave and the type II emission site.

and Raymond, 2004). The cumulation effect increases the density jump in the vicinity of the streamer's current sheet and intensifies radio emission (Uchida, 1974). The situation resembles a flare process running along a coronal ray (Uralova and Uralov, 1994). The difference between the horizontal and vertical directions explains a relation between the speeds of an EUV wave and a type II burst (Figure 1c). The upwards speed is higher than the surface one, because the wave center rises.

Even if a shock wave appears at a zero height, its front rapidly becomes convex and tilted toward the solar surface (Figure 1b). This front shape has actually been observed by Hudson *et al.* (2003) and is consistent with observations discussed by Warmuth *et al.* (2004b). If a convex shock front encounters a remote coronal ray, then the intersection site bifurcates, and its parts move along the ray in opposite directions (*cf.* Mancuso and Abbo, 2004). The contact corresponds to an infinite drift rate followed by bidirectional drifts to resemble a direct bracket "(" in a dynamic spectrum. Note that dynamic spectra present a combination of emissions originating at different sites, so that the intensities are summed.

In Section 3 we reconcile kinematics of "EIT waves" and drift rates of corresponding type II bursts in terms of our approach based on the self-similar shock approximation (Section 2.1). We apply power-law curves with the same onset time to both spectral domains (hereafter 'shock-PL fit'). The density falloff index in a streamer determining the drift rate of a type II burst is expected to be $\delta \sim 2$. Real lateral density falloffs in a streamer should be steeper than along its axis. Otherwise, streamers would not be visible in a homogeneous corona. Thus, real shock fronts should be oblate at medium distances. The density falloff for an EUV wave escaping from an active region can be $0 < \delta < 3$. For possible wave signatures in CMEs, the density falloff index is expected to be close to the Saito model (2), *i.e.*, $\delta \sim 2.6$ at moderate latitudes and steeper at higher latitudes.

2.3. Fit of Presumable Wave Signatures

We fit the drift rate of a type II burst manually. The onset time t_0 of a wave can be approximately estimated from observations. We take parameters of the plasma density model $n_0 = 5.5 \times 10^8$ cm^{-3}, $h_0 = 100$ Mm, and δ according to the considerations in the preceding section. One more input parameter is a reference frequency f_{obs} of a band with a harmonic

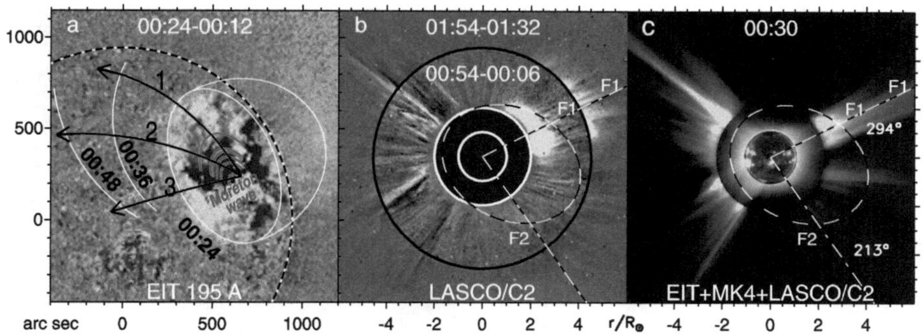

Figure 2 The 13 July 2004 event. (a) The Moreton wave (gray) and EUV wave (white) fronts superposed on the EIT 195 Å difference image. The black arcs 1, 2, and 3 trace the directions of measurements. The dotted circle denotes the limb. (b) The leading part of the CME observed in two LASCO/C2 difference images at 00:54 and 01:54 separated with the black circle. (c) The non-disturbed corona in a combined image composed from non-subtracted EIT 195 Å, Mark4, and C2 images. The dashed ovals in (b) and (c) outline the CME in the 00:54 image. The dash-dotted straight lines in (b) and (c) mark the position angles at which the measurements listed in the CME Catalog were carried out for the fastest features F1 and F2. The axes show the distances from the solar disk center in arc seconds (a) and in solar radii (b and c).

number N_f (usually 1 or 2) actually observed in a dynamic spectrum at a time t_1. The corresponding plasma number density is $n_1 = [f_\mathrm{obs}(t_1)/N_\mathrm{f}/(0.9 \times 10^4)]^2$, and the height is $x_1 = h_0\,(n_0/n_1)^{1/\delta}$. Then the height–time plot of the shock tracer is calculated as $x(t) = x_1\,[(t-t_0)/(t_1-t_0)]^{2/(5-\delta)}$, the corresponding density variation as $n(t) = n_0\,[x(t)/h_0]^{-\delta}$, and the outline of both bands of the type II burst as $f(t) = [1,2] \times 0.9 \times 10^4[n(t)]^{1/2}$. By adjusting δ and t_0 in sequential attempts, we endeavor to approach a best outline of the bands. Uncertainties provided by the routine typically do not exceed 1 min for t_0 and 0.2 for δ. The spectrum can be coordinated with measured heights by adjusting n_0, as usually done.

To fit presumable traces of shocks in coronagraph images (in this paper we use measurements from the CME Catalog), we employ two ways. The first way is a manual fit similar to the routine outlined in the preceding paragraph. Input parameters are starting estimates of δ and t_0, the heliocentric distances of the wave origin r_0 and the wave front r_1 measured at a time t_1. The initial approximation of the height–time plot is $r(t) = (r_1 - r_0)[(t-t_0)/(t_1-t_0)]^{2/(5-\delta)} + r_0$. Then sequential attempts are made to approach a best fit. The second way employs a log–log height–time plot, which is a straight line for a power law. We use a second-order fit and iteratively vary δ and t_0 to minimize the second-order term. One should be aware of the fact that the major but unknown uncertainties can be due to identification of the wave front in coronagraph images.

3. Observations

3.1. Event 1: 13 July 2004

This event (Figure 2) was associated with an eruptive M6.7 flare (00:09 – 00:23, *all times hereafter are UT*) in active region 10646 (N13 W46) and a CME observed with SOHO/LASCO (Brueckner *et al.*, 1995) after 00:54. Two parts of the CME (Figure 2b, c) are listed in the CME Catalog as two CMEs measured at position angles (PA) of 294° and 213°. A type II burst was recorded in three observatories. The three estimates of the shock speed progressively decreased in time.

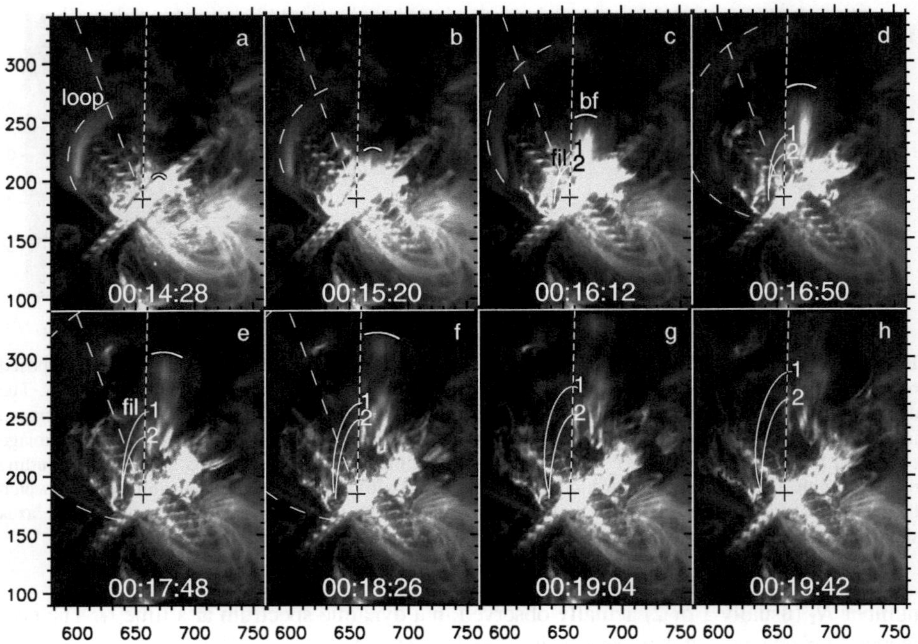

Figure 3 The eruptive system in the 13 July 2004 event (TRACE, 173 Å). The cross marks the initial position of the filament. Oval arcs outline the eruptive loop (dashed), filaments (solid), and bright feature (short arc). The broken lines denote the expansion directions of the loop (dashed) and the filament (dotted). Axes show arc seconds from the solar disk center.

Grechnev *et al.* (2008) revealed signatures of a probable blast wave as an Hα Moreton wave (gray in Figure 2a) and an EUV wave (white). Both disturbances were kinematically close to each other and to a kinematical curve expected for a lower trail of a decelerating coronal blast wave. Its exciter was not discussed. The authors proposed that the decreasing estimates of the shock speed reflected deceleration of a single shock wave, but they did not consider the type II burst.

Pohjolainen, Hori, and Sakurai (2008) [afterwards PHS] analyzed the type II burst in this event, but could not reconcile the overall drift with propagation of a single shock wave. The authors proposed that two shock waves were excited, one by a flare blast, and the second one by an expanding loop, a part of a CME.

We carried out an additional analysis of this event, measured kinematics of an eruptive system in order to find out a probable origin of a shock wave(s), and to reconcile its (their) propagation with the EUV/Moreton waves and the CME.

3.1.1. Eruptive System

Figure 3 shows the eruptive system: a leading bright feature 'bf', two filament segments '1' and '2', and several eruptive loops, of which one ('loop') was conspicuous and thus will be discussed henceforth. Long exposure times (33 – 46 s) caused a blurring of fast features, *e.g.*, a jet-like appearance of the bright feature.

To measure expansion of a feature, we outline it with an oval arc that allows us to trace it, even if its leading edge is difficult to detect. The obtained distance–time plot is used as a starting estimate. Then we choose a regular function to match the distance–time plot and

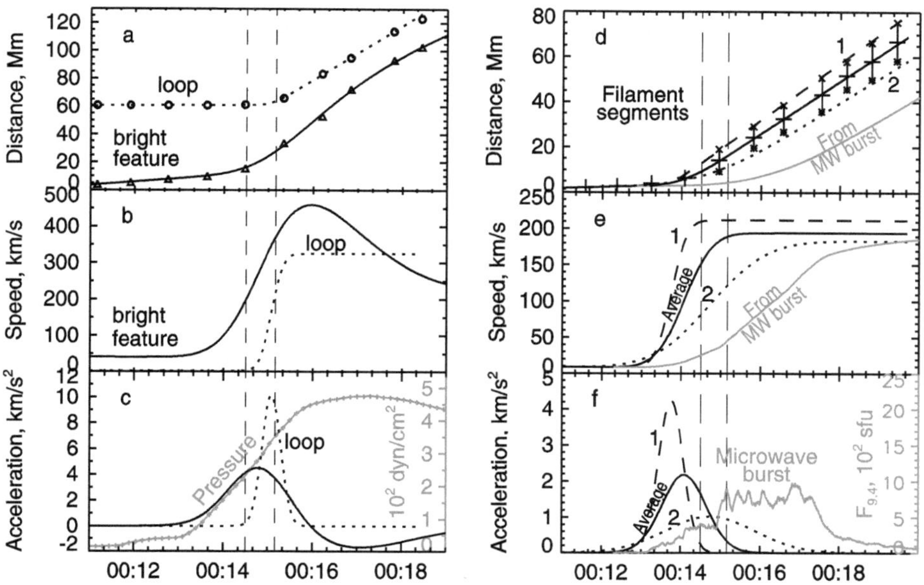

Figure 4 Kinematical plots of the eruptions in the 13 July 2004 event: the loop and bright feature (left) and the filament segments 1 and 2 (right; the solid curves correspond to averages between 1 and 2). Symbols mark the measured plane-of-sky distances, and curves represent their fit. The gray curve in panel (c) displays the plasma pressure computed from GOES SXR fluxes. The gray curves in panels (d – f) show kinematical plots calculated by assuming the correspondence of the acceleration plot to the microwave one. The vertical dashed lines delimit the start time of the wave estimated by Grechnev *et al.* (2008).

estimate its parameters. Using the analytic fit, we calculate expected distance–time points, compare them with observations, and improve the fit. All kinematical plots are calculated by means of integration or differentiation of the analytic fit rather than measurements. Our ultimate criterion is to reproduce the motion of an analyzed feature. Observational limitations do not allow us to reveal a detailed time profile of acceleration, and we describe it instead with a smooth bell-like function. We use a Gaussian time profile (see Wang, Zhang, and Shen, 2009). Then acceleration a is

$$a = (v_1 - v_0)\exp\{-[(t - t_0)/\tau_{\mathrm{acc}}]^2/2\}/(\sqrt{2\pi}\,\tau_{\mathrm{acc}}). \tag{3}$$

Here $\tau_{\mathrm{acc}}\sqrt{8\ln 2}$ is a full width at half-maximum of the acceleration time profile, which is centered at the t_0 time; v_0 and v_1 are velocities at the onset and end of the acceleration stage. In cases of a more complex kinematics, we use a combination of Gaussians and adjust their parameters manually.

Uncertainties are mainly determined by difficulties to identify and trace a feature in question. For this purpose we use both non-subtracted and difference images processed in various ways and improve results in sequential attempts. The worst traceable feature in this event is the eruptive filament, which appears as a semitransparent dark feature, whose segments 1 and 2 are faintly visible in Figure 3. We therefore present the measurements of the filament in more detail.

The results of plane-of-sky measurements are shown in Figure 4 for the loop and bright feature (left) and for the filament segments (right). The dashed curve in Figure 4d was calculated from the dashed acceleration profile in Figure 4f to match an upper envelope of the measured positions of segment 1. The dashed curve in Figure 4e is a corresponding velocity profile. The dotted curves in Figure 4d – f correspond to a lower envelope of the positions of segment 2. The black solid curves correspond to an average height–time plot. The horizontal bars show the exposure durations. We consider the maximum acceleration of the middle part of the filament to reach about 2 $\mathrm{km\,s^{-2}}$ (-60%, $+120\%$), and its probable speed and positions to be close to the solid black curves, being within the dotted and dashed boundaries. The bright feature and eruptive loop are visible much better, their leading edges are well defined, and therefore uncertainties of their positions are significantly less. The estimated height–time profiles match the measurements. The main uncertainty here is due to temporal undersampling, which is only crucial for the estimate of acceleration of the loop.

When the measurements began, the bright feature already rose that probably corresponded to the initiation phase, which started at about 00:07 according to soft X-ray (SXR) GOES data. A strong acceleration started at about 00:13, reached ≈ 4 $\mathrm{km\,s^{-2}}$, and then changed to deceleration. The loop was static by 00:14:30; after 00:15:10 its speed sharply changed to $V_{\mathrm{loop\,max}} \approx 320$ $\mathrm{km\,s^{-1}}$ and did not increase afterwards. The last measured point hints at a possible later deceleration. The transition from the initial zero speed to a final one occurred between two samples. Hence, the maximum acceleration of the loop could well exceed 10 $\mathrm{km\,s^{-2}}$. All components of the corresponding CME decelerated (Grechnev *et al.*, 2008), so that the eruption resembled an explosion with a strong impulsive acceleration followed by continuous decreasing deceleration.

Figure 4c also shows the plasma pressure computed from SXR GOES fluxes and a source size of 15 Mm found from RHESSI (Lin *et al.*, 2002) images. The pressure gradually rose while the bright feature suddenly started to decelerate. Thus, the flare pressure was unlikely a driver of either the eruption or the wave, whose estimated start time is delimited with vertical dashed lines.

The filament started to rise nearly simultaneously with the bright feature. However, both the acceleration and speed of the bright feature were higher, and it surpassed the filament (see Figure 3). The nature of this feature is difficult to identify. In some images it resembles an arcade surrounding the filament, but initially it seems to be located below the filament. This feature might also be one more filament which brightened due to heating. An additional possibility is suggested by a scenario proposed by Meshalkina *et al.* (2009): this feature might be a small-scale magnetic rope whose eruption destabilized the filament.

The gray curves in Figure 4d – f show the kinematical plots calculated under the assumption that the acceleration plot corresponded to the 9.4 GHz light curve (Nobeyama Radio Polarimeters, Torii *et al.*, 1979; Nakajima *et al.*, 1985; ftp://solar.nro.nao.ac.jp/pub/norp/xdr/). Microwave bursts are known to be similar in shape to hard X-ray (HXR) ones, while accelerations of eruptions have been found to be temporally close to HXR bursts (see, *e.g.*, Temmer *et al.*, 2008). The kinematical plots calculated from the microwave burst lag behind the actual plots of the filament by about two minutes indicating that, most likely, the flare was caused by the eruption. This fact suggests that the eruptive filament accelerated almost independently of the flare reconnection rate and HXR emission, at least, in this event.

There are two options regarding a relation between the bright feature and the loop. One possibility is that $\approx 1.5-2$ minutes after the start of the acceleration of the bright feature, the loop suddenly and independently underwent much higher impulsive acceleration. Alternatively, the loop was expelled by a shock front that appears to be more probable. For the latter case, the strength of the shock can be estimated. The Mach number is $M = V_{\mathrm{sh}}/V_{\mathrm{fast}}$, where V_{sh} is the shock speed, and V_{fast} is the fast-mode speed. The shock speed V_{sh} at the onset of the loop motion can be roughly estimated from a PL fit (Grechnev *et al.*, 2008) to be about 1000 $\mathrm{km\,s^{-1}}$, but it is rather uncertain because of insufficient temporal coverage by

TRACE images and their long exposure times. The fast-mode speed can be estimated from an expression $V_{\mathrm{sh}} \approx V_{\mathrm{fast}} + \kappa U_{\mathrm{sh}}/2$, where κ is a coefficient determining the steepening rate of the wave front. This coefficient, $1/2 \leq \kappa \leq 3/2$, depends on plasma beta and the propagation direction. We take the speed of the loop as the gas speed behind the shock front, $U_{\mathrm{sh}} = V_{\mathrm{loop\,max}} \approx 320\ \mathrm{km\,s^{-1}}$ (see Figure 4b). With these quantities and $\kappa \approx 3/2$ for the wave propagation perpendicular to the magnetic field in low-beta plasma, the Mach number is $M \approx 1.3$. One might suppose that the steepening time was about the interval between the peaks of the solid and dotted curves in Figure 4c corresponding to the acceleration profiles of the bright feature and the loop, *i.e.*, about 0.5 min.

We computed coronal magnetic fields from a SOHO/MDI magnetogram using an extrapolation package of Rudenko and Grechnev (1999) based on a potential approximation (Rudenko, 2001). The result showed that the eruptive loops were strongly tilted to the photosphere, and their tops were at a height of about 30 Mm. With a height of the pre-eruptive filament of > 10 Mm, the shock front hit the loop almost horizontally, and the estimated Mach number is related to the horizontal direction, while in the vertical direction the shock was probably stronger.

Thus, the loop itself was most likely passive, and its motion was driven by the shock wave. All observed products of the eruption monotonically decelerated starting from TRACE observations and up to LASCO/C3 ones. The loop therefore is unlikely to have excited the second shock wave, as PHS hypothesized.

3.1.2. Moreton/EUV Wave, Type II Burst, and CME

Grechnev *et al.* (2008) could not find out from kinematics if the leading edge of a coronal transient observed by LASCO (see Figure 2b) was a mass ejection or a trace of a wave. Comparison with a non-subtracted image of the corona before the CME in Figure 2c suggests its spiky leading fringe to be coronal rays deflected by a wave (pronounced for feature F1 at PA = 294°). Vourlidas *et al.* (2003) and Magdalenić *et al.* (2008) interpreted such phenomena as signatures of shocks. Assuming that the CME leading edge was due to the wave, we outlined the measurements from the CME Catalog with a shock-PL fit. The calculated curves with $\delta \approx 2.6$ corresponding to the Saito model agree with the measurements (Figure 5). The decreasing speeds computed from these height–time plots at distances corresponding to the LASCO/C2 and C3 fields of view agree with the linear-fit speeds estimated in the CME Catalog, while within the interval when the type II burst was observed the speeds are higher by a factor of 2 – 2.5.

Figure 6c shows a type II burst recorded by the HiRAS radio spectrograph and our shock-PL fit of both the fundamental and second-harmonic emission (solid lines). We use the density falloff index $\delta = 2.1$, which is close to the Newkirk model expected for a streamer. The interpretation of features QF1 and QF2 is open to question; their presence hinted PHS at two different shock waves. The dotted lines approximately reproduce the outline of PHS following the logic suggested by their Figures 4 and 6. They correspond to fixed velocities of the type II exciters. However, a flare blast wave proposed by the authors is expected to decelerate: our outline corresponds exactly to a freely propagating blast wave.

The solid shock-PL fit outlines the whole slowly drifting structure from the decimetric range up to the lowest frequency. A question remains about features QF1 and QF2. The former feature with an uncertain harmonic structure does not seem to favor the dotted outline relative to the solid one. The weaker QF2 feature, which PHS considered as the onset of the second type II burst, indeed seems to have a harmonic structure. Its shape in the higher-resolution spectrum recorded at the Learmonth station (US Air Force RSTN) resembling "(" outlined with a black arc suggests an encounter of a shock wave with a dense structure (see

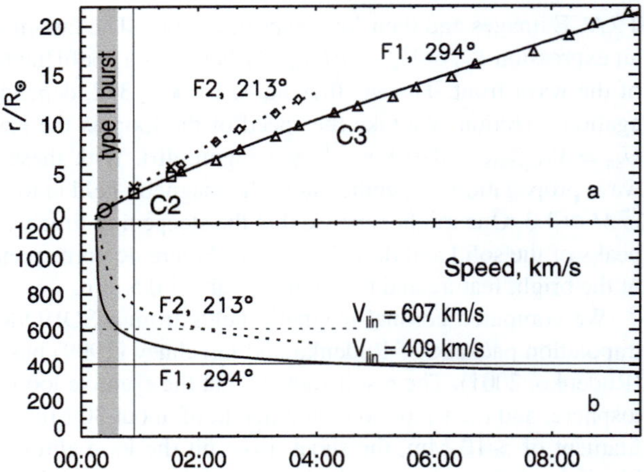

Figure 5 (a) Measurements from the CME Catalog for the F1 and F2 CME components fitted with the shock-PL model, $\delta \approx 2.6$. The circle denotes the position of the off-limb EUV wave at 294°. (b) Speed–time plots corresponding to each fit. Linear-fit speeds are also specified. Shading denotes the interval in which the type II burst was observed.

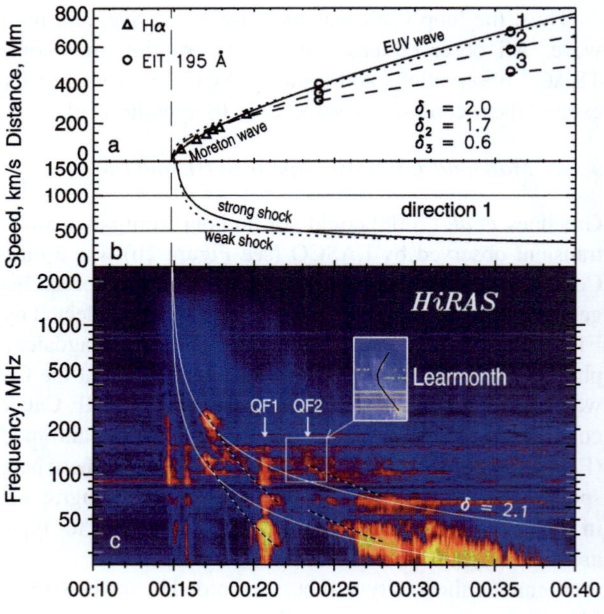

Figure 6 The 13 July 2004 event. Height–time (a) and velocity–time (b) plots of the Moreton and EUV waves, and (c) the fit of the type II burst ($\delta = 2.1$). QF1 and QF2 are features open to question. A reverse drift of QF2 is detectable in the inset which shows a portion of the dynamic spectrum recorded in Learmonth with a higher spectral resolution.

Section 2.2). These facts support association of the type II burst with a single decelerating shock wave. The drifting continuum, which PHS found to start at 00:13 (confirmed by the acceleration profile in Figure 4f), might be due to emission from outside of a pre-shock region expanding toward a decreasing density or, alternatively, from inside of the expanding region with a progressively depleting density, as PHS proposed. Compression of the environment in the pre-shock interval from 00:13 to about 00:14:50 might have produced an excessive plasma density, which we describe with a radial power-law falloff.

For comparison we show in Figure 6a the distance–time measurements of the Moreton wave (triangles) and the EUV wave (open circles) from Grechnev *et al.* (2008) along with shock-PL plots for the three directions 1, 2, and 3 denoted in Figure 2a, and the dotted plot of a spherical weak shock propagating in uniform plasma calculated by using expressions of Uralova and Uralov (1994). Figure 6b presents the speeds for the strong (solid) and weak (dotted) shock approximations along direction 1. Both approximations are close to each other far from the eruption center ($r \gtrsim R_1 \approx 200$ Mm) being somewhat different at shorter distances. The shock wave propagating along the surface probably became weak at $r \gtrsim R_1$, when it left the active region and entered quiet-Sun areas where the coronal density and the fast-mode speed were nearly constant, *i.e.*, $\delta \rightarrow 0$. Closer to the eruption site ($r < R_1$), the plasma density presumably had a power-law falloff, and the shock was not weak. These conditions seem to favor the self-similar shock approximation, in which *i*) the shock propagation speed is proportional to its intensity and does not depend on the fast-mode speed, and *ii*) the wavelength is equal to the distance passed by the wave, *i.e.*, its duration increases. The density falloff of $\delta < 3$ within an active region and at its periphery corresponds to deceleration and damping of such a shock. The limit of a strong shock is a convenient idealization to describe the formation stage of a single shock wave, which propagates far from its source. A real forming shock wave does not seem to be so strong that the decrease of the fast-mode speed could be neglected when the shock leaves the active region. This issue is beyond our scope. We only note that formation of the shock discontinuity in a disturbance produced by an impulsive piston presumably brings to an end (and its intensity reaches maximum) in a region where the falloff of the fast-mode speed is steepest.

Note that deceleration of the EUV wave sweeping over the quiet solar area was stronger toward the equator, as comparison of the three fronts in Figure 2a for the 1 – 3 directions shows. This is expected for a strong shock, whose deceleration is determined by the density distribution, which is maximal at the equator (see the Saito model). This is also expected for a weak shock, whose propagation is governed by the Alfvén velocity decreasing toward the equator due to both the density distribution and the dipole magnetic field of the Sun.

Our analysis of Event 1 has revealed a probable excitation of a single wave by an impulsively accelerated eruption and steepening into a shock within one minute. Then the wave freely propagated like a decelerating blast wave and probably formed the leading edge of the CME. Our results are consistent with the conclusions of PHS about the role of a rapidly expanding eruption, formation of the shock wave at a very low altitude, and their estimates of the shock speed.

3.2. Event 2: 1 June 2002

Meshalkina *et al.* (2009) revealed a possible coronal shock wave presumably excited by a collision of an erupting magnetic rope with a magnetic obstacle in the 1 June 2002 event. An M1.5 flare (S19 E29) started at 03:50 and lasted only 11 min. SOHO/EIT carried out the 'High cadence 195' program, and LASCO did not observe at that time. Figure 7 shows an off-limb EUV wave in this event.

Figure 8a – f presents the eruption with extreme outlines (dotted and dashed) of its leading edge found from two sets of images processed in different ways. Figure 8g – i presents plane-of-sky measurements of kinematics using the same technique as for Event 1. The eruption accelerated up to $\approx 7\ \mathrm{km\,s^{-2}}$ and then decelerated. The deceleration might be overestimated, because the eruption started to disintegrate and become transparent. Similarly to Event 1, acceleration occurred during the rise of the HXR burst recorded with RHESSI, while the plasma pressure gradually increased during the whole interval considered.

Figure 9 shows kinematics of the EUV wave (a, b) and the type II burst (c) similarly to Figure 6. To reveal the harmonic structure of the burst, we use again the record made in Learmonth (the inset). The burst consisted of two pairs of emission bands with frequency ratios in pairs of 2.0 and ≈ 1.5 between the pairs. The two pairs of bands resemble band-splitting usually interpreted as plasma radiation from the regions upstream and downstream of a shock. However, the relative split is atypically large for the metric range (Vršnak *et*

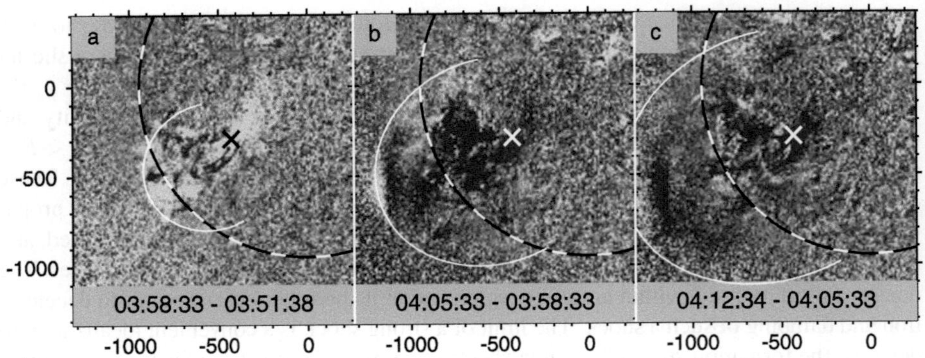

Figure 7 An EUV wave (outlined with ovals) on 1 June 2002 in EIT 195 Å running-difference images. The dashed circles denote the solar limb. The slanted cross marks the flare site.

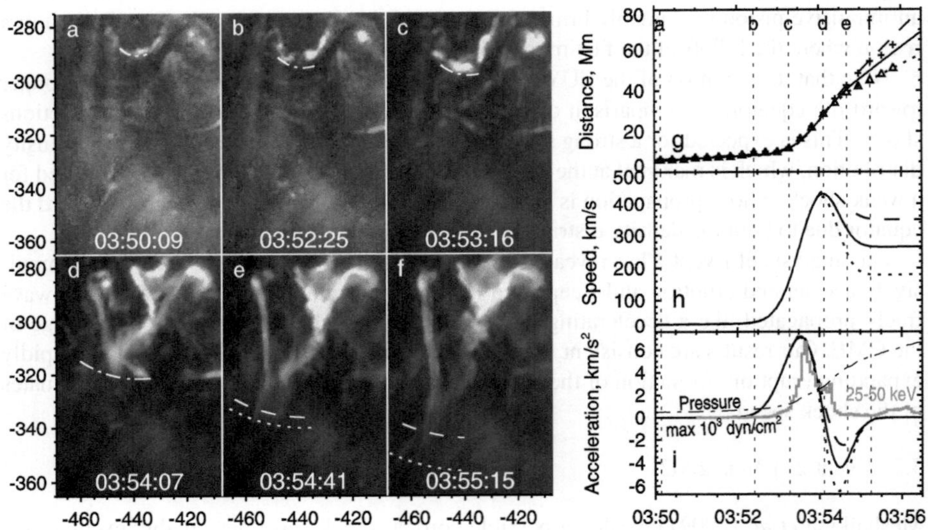

Figure 8 The eruption in TRACE 195 Å images (a – f) and its kinematics (g – i). The dotted and dashed lines correspond to extreme outlines of the eruption. Panel (i) also shows a 25 – 50 keV RHESSI time profile (gray) and the plasma pressure calculated from GOES data (dash-dotted). The vertical dotted lines mark the observation times of images a – f.

al., 2001). Alternatively, this situation suggests propagation of the shock front along two streamers located close to each other. We outline the burst structure with two harmonic pairs of power-laws 1 and 2 with 'f' indicating the fundamental emission and 'h' the second harmonic (twice the fundamental frequency). The difference in δ (2.4 and 2.6) might be due to differences of density falloffs in the streamers as well as different angles between the shock front and the axes of the streamers. The estimated wave onset time is 03:53:40, close to the acceleration peak time (Figure 8i).

Figure 9a shows height–time plots corresponding to the drift of the type II burst along with the EUV wave expansion. The type II emission was observed when its source was presumably located at heights from 190 Mm (heliocentric distance of $1.27 R_\odot$) to 500 – 600 Mm [$(1.7 - 1.9) R_\odot$], *i.e.*, lower than usually assumed. Moreover, the outline of the shock front matches the fast-drifting decimetric continuum suggesting its relation to the shock

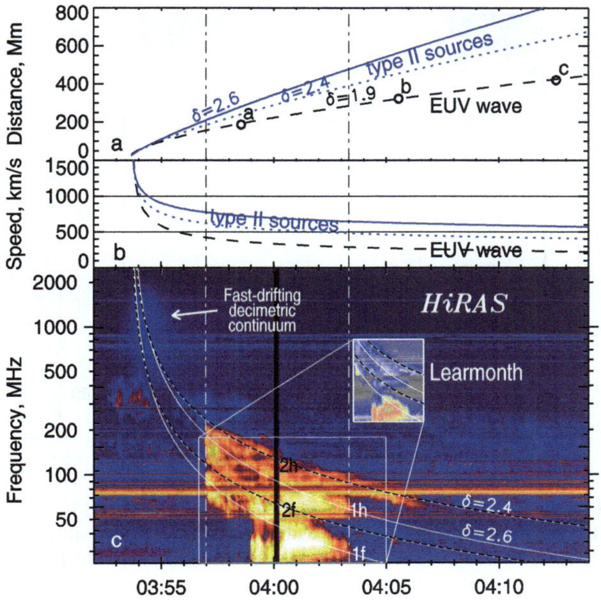

Figure 9 The 1 June 2002 event. Height–time (a) and velocity–time (b) plane-of-sky plots of the EUV wave (black) and the fit of the type II burst source (blue). Open circles in panel (a) represent the distances of the foremost wave fronts from the eruption center, and labels 'a' – 'c' correspond to the images in Figure 7. Panel (c) shows the dynamic spectrum with a shock-PL fit. The vertical broken lines delimit the interval of the type II burst.

wave, which presumably originated still lower. The type II burst probably started when the shock front reached the streamer and ceased due to deceleration and damping of the wave.

3.3. Event 3: 19 May 2007

This event associated with a B9.5 flare at 12:48 – 13:19 (N07 W06) and a fast CME has been well studied due to efforts of several researches mostly from observations made with EUV Imager (EUVI) of SECCHI complex (Howard *et al.*, 2008) on STEREO (Kaiser *et al.*, 2008). Nevertheless, some questions remain.

Long *et al.* (2008) measured kinematics of the EUV wave, found distinct deceleration, and stated that the low velocities of "EIT waves" could be due to their temporal undersampling. The observations were found to be consistent with an impulsively generated fast-mode magnetosonic wave or propagating MHD shock. However, they revealed an initial acceleration of the disturbance from a nearly zero speed. This mismatches the expected behavior of an MHD wave.

Veronig, Temmer, and Vršnak (2008) also found deceleration of this disturbance indicative of a freely propagating MHD wave and revealed a wave reflection at a coronal hole. They assumed that the wave was initiated by the CME, because the associated flare was very weak and occurred too late to account for the wave initiation. They also revealed two eruptions in close succession.

Gopalswamy *et al.* (2009) measured propagation of reflected wave fronts and considered the reflections as an argument in favor of a wave nature of EUV transients (see also Schmidt and Ofman, 2010). However, Attrill (2010) proposed that the reflections resulted from a misinterpretation of the running-difference data and suggested instead that two EUV wave fronts developed during the event. Indeed, running differences reliably show only the outer boundaries of expanding disturbances, while the inner picture reflects all the changes occurring between two images subjected to subtraction (Chertok and Grechnev, 2005).

To see whether or not the reflections actually occurred, we use a movie composed of non-subtracted 171 Å images of STEREO-A/EUVI (euvi_ahead_171.gif in the electronic version

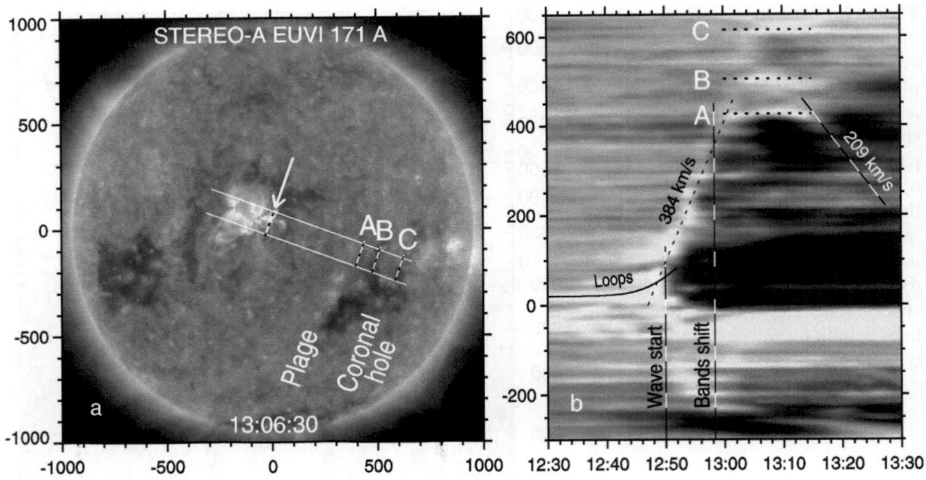

Figure 10 EUV wave observed with STEREO-A/EUVI at 171 Å on 19 May 2007. A set of spatial profiles (b) computed from ratio images within the strip contoured in panel (a). The arrow points at the origin of measurements. The A, B, and C broken bars in both panels mark the reflection positions. The slanted broken lines in panel (b) outline the steepest slopes.

of the paper). A backward motion suggesting a reflection is visible after 13:11 northeast from a plage region denoted in Figure 10a. We repeated the measurements of Gopalswamy *et al.* (2009), but without any subtraction. We only consider the first probable reflection in the direction exactly backwards, because it is difficult to reveal wave fronts in non-subtracted images of this complex event. Figure 10 corresponds to Figures 1 and 2 from their paper. To enhance the sensitivity, we use in panel (b) spatial profiles computed as the sums over the width of each slice extracted from an image as panel (a) shows. Each image was normalized to a pre-event image at 12:16:30 (fixed-base ratios).

Figure 10b shows that the plage region inflected after arrival of the wave front and then returned back. A backwards motion from plage A is faintly visible after 13:14. Region B exhibited a weaker bend. It is not clear from which region of A and B was the wave reflected. The latter seems to be preferable, because continuations of both the direct and reflected slanted traces intersect farther from region A, while a prolonged standing of the wave at region A is doubtful. The slanted broken lines represent the speeds found by Gopalswamy *et al.* (2009) and agree with the slopes in Figure 10b. Thus, the results of the authors were correct, at least, for the first reflection. The wave reflected backwards was considerably slower than the incident wave. This fact supports the shock-wave nature of the disturbance. Indeed, if an incident shock wave propagating with a velocity $V_{\mathrm{inc\,sh}}$ encounters a semitransparent 'wall' like a coronal hole, then the shock reflected backwards is slower: $V_{\mathrm{back\,sh}} \approx V_{\mathrm{inc\,sh}} - V_{\mathrm{gas}}$, where V_{gas} is a velocity of the gas trailing the incident shock front. V_{gas} can be up to the sound speed.

What does the accelerating part prior to 12:50 display? Figure 11 shows EUVI 171 Å images with subtraction and without it. The outermost boundary of the expanding bright feature coincides with the edge of coronal loops visible in the earlier non-subtracted images. Then eruptive loops rapidly lose brightness due to expansion and become invisible in non-subtracted images. We conclude that the accelerating part measured by Long *et al.* (2008) was related to the expanding loops, while the decelerating part was related to the wave. It is difficult to distinguish an appearing wave, which brightens, from a piston (loops), which

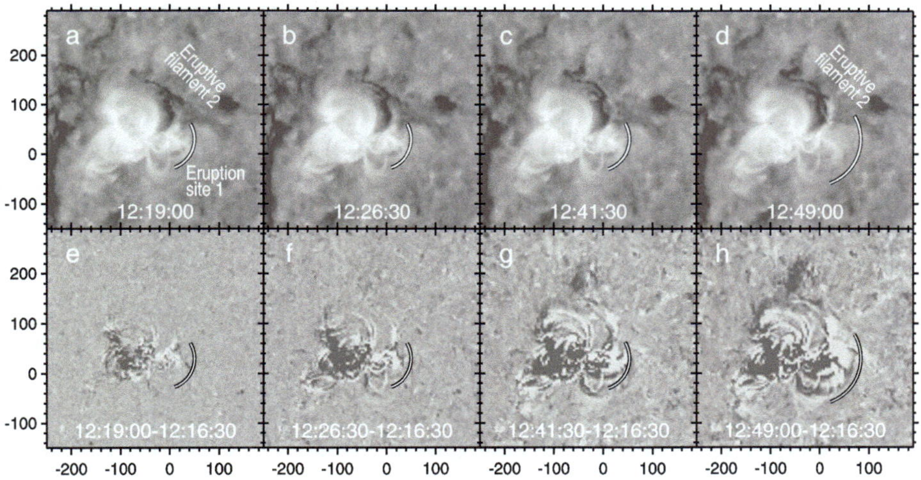

Figure 11 The first eruption on 19 May 2007 observed with STEREO-A/EUVI at 171 Å. Top: non-subtracted images, bottom: fixed-base differences. The arcs outline the foremost edge of the eruptive loop system according to the fit shown with the solid line in Figure 10b.

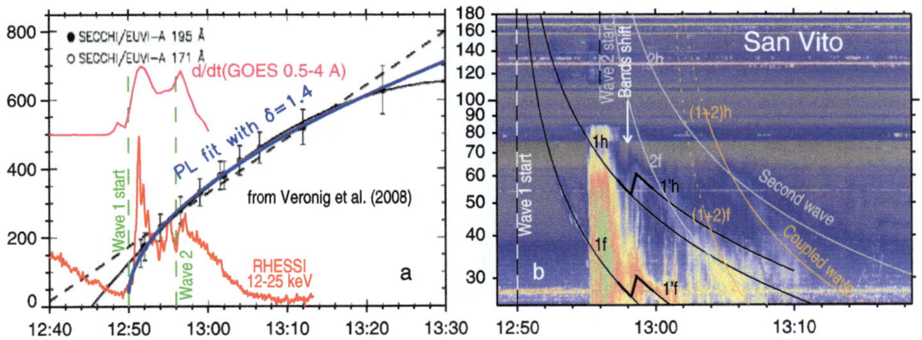

Figure 12 The 19 May 2007 event. (a) Propagation of the EUV wave measured by Veronig, Temmer, and Vršnak (2008, black) along with an HXR time profile (red), a derivative of the SXR flux (pink), and a shock-PL fit (blue). (b) Type II burst with shock-PL-fitted four pairs of bands. Paired bands are shown with the same colors. See details in the text.

becomes transparent. For this reason the acceleration of the loops is uncertain within 120–270 m s^{-2}; the outline in Figures 10 and 11 corresponds to 260 m s^{-2}.

Figure 12a shows the measurements of the wave presented by Veronig, Temmer, and Vršnak (2008) and their shock-PL fit (thick blue curve). The wave start (12:50) corresponds to the early rise phase of the HXR burst (red, also from their paper). The power-law fit corresponds to the measurements of the authors better than their quadratic fit and the linear one. The eruption in this event also accelerated before the appearance of manifestations of flare reconnection.

The suggestion of Attrill (2010) about the second wave appears to be correct. The second eruption was probably triggered by the first one. Filament 2 activated at about 12:47 and erupted at 12:55 – 12:57 according to TRACE 173 Å images. The HXR time profile was complex, but two distinct episodes are detectable in the derivative of the SXR flux recorded

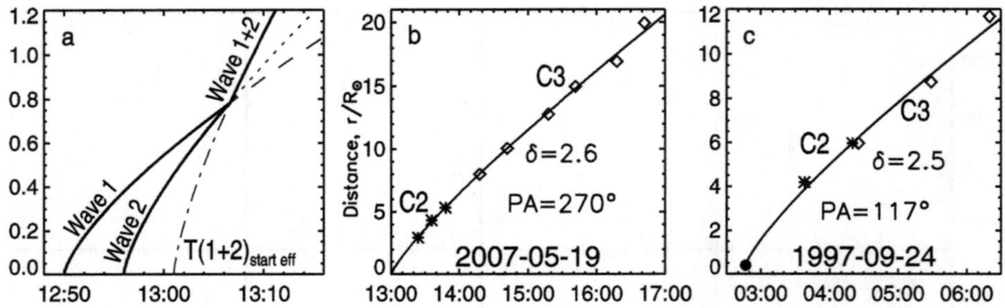

Figure 13 Coalescence of two shock waves (a) and height–time plots of CMEs observed on 19 May 2007 (b) and 24 September 1997 (c). Symbols present data from the CME Catalog, lines show their shock-PL fit. The filled circle in panel (c) denotes the origin of the wave.

with GOES (pink in Figure 12a). The onset times of the two waves were about 12:50 and 12:56.

A shock wave trailing a preceding one must reach the leading front due to properties of shock waves; the two shock waves coalesce to produce a single shock front (and a weak backwards disturbance, which we are not interested in). Its speed is less than the sum of the initial fronts' speeds; however, the resulting shock is stronger and faster than either of the initial ones. So the slope of its distance–time plot is steeper than the initial waves had, and its virtual onset time is later than for either of the initial waves, as Figure 13a outlines.

Let us try to understand a complex dynamic spectrum in Figure 12b recorded in San Vito (USAF RSTN). A harmonic pair of weak bands 1f, 1h is sometimes detectable after 12:52. A strong type III burst at 12:55 – 12:58 probably associated with the second eruption partially blocked the weaker bands 1f, 1h. Two stronger type II bands 1'f, 1'h appeared at 12:58. Their appearance looks like an inverse-N-like shift of the initial bands to higher frequencies (thicker outline in Figure 12b). The initial bands 1f, 1h are still detectable sometimes. The outline of both these pairs of bands has the same start time of the first wave, 12:50. The appearance of the second pair might be due to the entrance of a part of the shock front into a dense region located rather high above plage A (see Figure 10a, b). The surface EUV wave reached plage A slightly later, as expected for a convex front tilted toward the solar surface (see Section 2.2).

The second shock front was probably manifest in weak bands 2f, 2h (pale, the start time is 12:56). The bands denoted $(1 + 2)$f and $(1 + 2)$h (orange outline) probably reveal the resulting shock with a virtual start time of 13:01. All of these bands overlap with others increasing the total emission at the intersections. We remind that various bands were most likely emitted from spatially different sites.

We have not revealed manifestations of the two merging shock waves in images or in the kinematical plot. This result could be expected due to different propagation conditions for a shock wave upwards and along the solar surface. The near-surface portion of a shock front decelerates stronger and experiences significant damping, thus becoming weak and approaching a linear disturbance. If the trailing shock wave succeeded to catch up or intersect with the leading one, the effect is expected to approach an interaction of linear disturbances, when two waves pass through each other experiencing interference, while the scheme in Figure 13a shows an essentially nonlinear effect.

The CME was fast (958 $\mathrm{km\,s^{-1}}$) and decelerated. We assume that its leading edge was a trace of a wave and apply a shock-PL fit to the measurements in the CME Catalog at PA = 270° (Figure 13b). The onset corresponds to the virtual start time of the coupled shock wave, and the density falloff corresponds to the Saito model. A trailing, poorly observed, CME at PA = 310° and probably related to the same event was considerably slower (294 $\mathrm{km\,s^{-1}}$). Its deceleration might be due to the influence of the wave running ahead and problems of measurements.

Our analysis does not pretend to be perfect, but it shows that even such a complex dynamic spectrum can be reconciled with EUV observations and CME expansion under assumption of the shock-wave nature of related disturbances. Oversimplified considerations of such a complex event can be misleading. In particular, Yang and Chen (2010) concluded that the EUV wave in this event ran slower in regions of stronger magnetic field that seemed to be a challenge for the wave hypothesis. However, *i*) the authors considered the radial component of the magnetic field only, whereas the Alfvén speed depends on its magnitude. *ii*) By taking the range of the magnetic field strengths ≤ 0.6 G, for which Yang and Chen obtained anticorrelation with the EUV wave speed, and a density of $\gtrsim 2 \times 10^8\ \mathrm{cm^{-3}}$ from the Saito model, one obtains $\beta = (2/\gamma)C_\mathrm{s}^2/V_\mathrm{A}^2 \gtrsim 4$, *i.e.*, the wave must be almost insensitive to the magnetic field. *iii*) The fronts in their Figure 4 stretch west-southwestward and become sharply pointed at 12:59, whereas Long *et al.* (2008), Veronig, Temmer, and Vršnak (2008), and Attrill (2010) showed the front to be blunt in this direction at that time. *iv*) Usage of the Huygens plotting to find trajectories of the wave front resulted in a strange picture of intermittently condensed and rarefied ray trajectories in their Figure 4. Thus, the results of Yang and Chen (2010) do not offer problems for the shock-wave hypothesis.

3.4. Event 4: 24 September 1997

This event was associated with a short M5.9 flare (02:43 – 02:52, S31 E19). An Hα Moreton wave and EUV wave in this event were first analyzed by Thompson *et al.* (2000). Warmuth *et al.* (2004a, 2004b) found kinematical closeness of both wave fronts to each other and their deceleration. The first EUV wave front (Figure 14a) was sharp and bright suggesting that the main EUV-emitting layer was low. A difference ratio image (Figure 14b) reveals weak wave manifestations south, southwest, and slightly west from drawn by outline of Warmuth *et al.* (2004a). The third front in Figure 14c is close to their outline. Deceleration of the EUV wave was therefore even slightly stronger than the authors estimated.

White and Thompson (2005) analyzed wave signatures in microwave images at 17 GHz but did not reveal any deceleration. They also found that *i*) the speed of the microwave disturbance was 830 $\mathrm{km\,s^{-1}}$ against $\approx 500\ \mathrm{km\,s^{-1}}$ estimated for the Moreton wave; *ii*) the brightness temperature at 17 GHz was about five times higher than an estimate from EIT data, and the discrepancy could be reduced if the kinetic temperature at 17 GHz would be different (preferentially higher) from the characteristic temperature of the 195 Å channel. The authors also concluded that the timing of images should be corrected by ≈ 100 s for EIT and by ~ 180 s for Hα to reconcile all observations. These facts indicate that the layers emitting microwaves and EUV were not identical. The higher speed, lesser deceleration, and higher brightness temperature (*i.e.*, column emission measure) observed at 17 GHz with respect to EUV hint at a possibly higher location of the microwave-emitting layer. Figure 15 shows our suggestion in panel (c); panel (a) presents the distance–time plots from both papers. The corrected times of EIT images (+ 99 s) and Hα ones (+ 170 s) are indicated at the corresponding data points.

To reconcile kinematics of the microwave and EUV/Hα wave fronts, we shift the White and Thompson data by 40 Mm and fit both data sets with the same onset time of 02:46:50 but different power-law exponents. Figure 15c explains the idea: the lower part of the front propagating in high-density regions decelerated stronger ($\delta \approx 2.3$ from Warmuth *et al.* measurements), while microwaves were dominated by long cross sections of the wave front

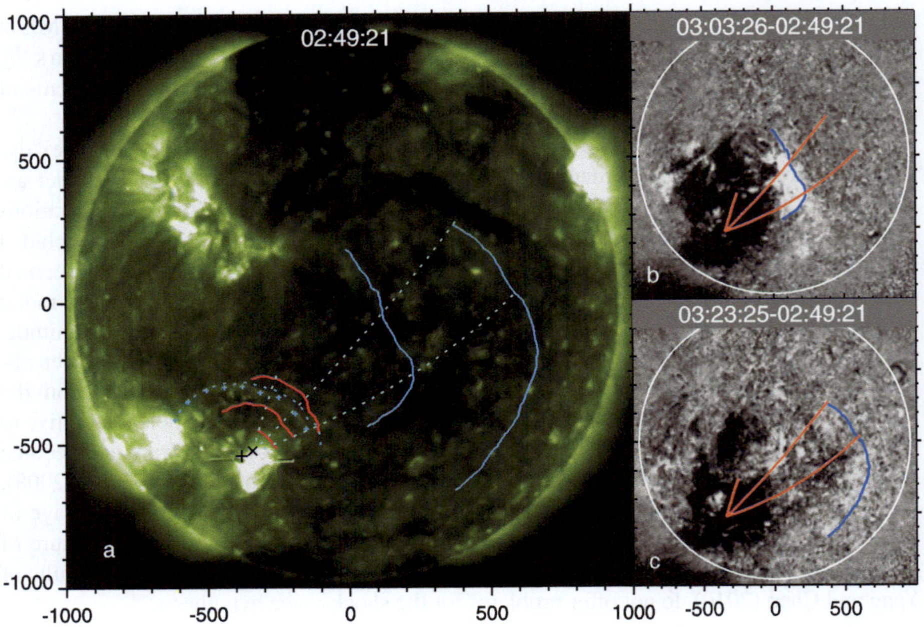

Figure 14 EUV wave (blue) and Moreton wave in the 24 September 1997 event. (a) Non-subtracted EIT 195 Å image with the first front (blue crosses). Blue dotted line outlines its foremost edge. Red fronts outline the Moreton wave. White broken lines trace the measurement great circles. Black crosses denote the origins of measurements. (b, c) Difference ratio images with blue outlines of EUV wave fronts and red measurement great circles. The outlines correspond to Warmuth *et al.* (2004a) (courtesy A. Warmuth).

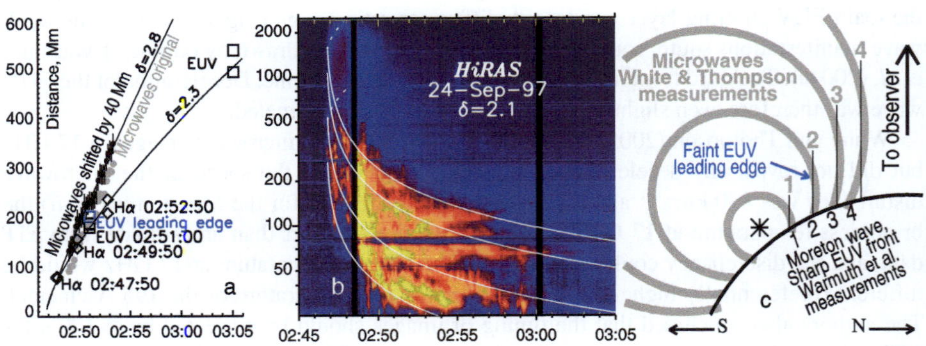

Figure 15 (a) Propagation of the EUV wave (squares) and Moreton wave (diamonds) measured by Warmuth *et al.* (2004a). Filled gray circles show the White and Thompson (2005) data, black open circles with error bars show them shifted by 40 Mm. Both data sets are outlined with shock-PL fit. (b) The HiRAS dynamic spectrum outlined with shock-PL fit and a negatively drifting continuum (dotted). (c) Presumable relation between the layers emitting microwaves and EUV/Hα. Thin vertical lines show cross sections of the wave fronts of the largest column emission measure contributing at 17 GHz. The star is the origin of the wave.

(bars 1 – 4) running in lower-density regions, $\delta \approx 2.8$ (*cf.* Paper III). The large-height EUV wave's leading edge detectable close to the eruption center diminished at large distances, where the EUV wave was almost exclusively due to response from low structures. The wave presumably appeared at a considerable height (the star). With a difference between the origins of measurements (black crosses in Figure 14a) of ≈ 32 Mm, our shift of 40 Mm, and spherical wave fronts, we get a height of 117 Mm. It seems to be overestimated; an estimate of about ≈ 75 Mm from the dynamic spectrum appears to be more plausible. Thus, the wave could be strongly anisotropic starting from its appearance, or the wave exciter was large, or both.

The complex dynamic spectrum in Figure 15b has a questionable harmonic structure, which is beyond our scope; we are only interested in the drift of the burst envelope. We have outlined presumable fundamental and second-harmonic bands. A possible higher-frequency band and the highest-frequency envelope of the burst are formally outlined as the fourth and sixth harmonics (this does not mean reality of emissions at such high harmonics). The outline corresponds to the initial height of 75 Mm, the same wave start time of 02:46:50, and $\delta \approx 2.1$ typical of streamers. A negatively drifting continuum (broken outline) at the initial stage indicates propagation of the shock front toward the chromosphere.

A poorly observed CME centered at 137° with a speed of 531 $\mathrm{km\,s^{-1}}$ was injected into a preceding CME. The CME Catalog estimates its acceleration to be positive with a remark about uncertainty. By adding the known origin of the wave (the filled circle in Figure 13c), we get a shock-PL fit of the measurements from the CME Catalog with an exponent $\delta \approx 2.5$ close to the Saito model.

Our considerations confirm the results of both Warmuth *et al.* (2004a) and White and Thompson (2005), reconcile them with each other and with the type II burst as well as the CME. Deceleration of the front portion detectable at 17 GHz was much less than its lowest part visible in EUV and Hα had. It was not possible to reveal deceleration from microwave observations, which allowed detection of the wave within an interval as short as 4.5 min.

4. Discussion

TRACE observations of abrupt eruptions in events 1 and 2 have revealed plane-of-sky accelerations of magnetic rope structures of $4-7$ $\mathrm{km\,s^{-2}}$, *i.e.*, $(15-25)$-fold gravity acceleration. Then the eruptions in both events decelerated. Coronal waves appeared in events $1-3$ approximately at the peak of acceleration. The onsets of the waves in all four events corresponded to the rise phases of HXR or microwave bursts (in event 4 according to White and Thompson, 2005). The wave in event 1 steepened into a shock within one minute and reached a Mach number of about 1.3 in a horizontal direction, while the upwards shock could be stronger. In the next section we consider which of known shock formation scenarios (see, *e.g.*, Vršnak and Cliver, 2008) appear to match the observations.

4.1. Comparison of Observations with Shock Formation Scenarios

A scenario of the shock formation by a flare pressure pulse is based on an idea that the β ratio of the plasma pressure to the magnetic pressure sharply changes from $\beta \ll 1$ to $\beta > 1$. Such a change of β is believed to be possible in a flare loop and considered as a cause of an omnidirectional disturbance. Dramatic changes of the volume of a loop or its abrupt motion are necessary to get a significant intensity of a wave excited in this way. Objections against this scenario do exist.

i) As Grechnev *et al.* (2006c) showed, the effect of a high β (even $\beta > 1$) in a flare loop is not dramatic, only results in an increase of all its dimensions by $\sqrt[4]{1+\beta}$.
ii) The idea that a situation of $\beta \rightarrow 1$ can cause instability of a loop is not confirmed by observations. Ichimoto *et al.* (1993) and Grechnev *et al.* (2006a) showed that the $\beta < 1$ condition was not satisfied in long-lived hot coronal loops.

iii) The flare pressures in events 1 and 2 rose gradually, without any signatures of the wave appearance.
iv) RHESSI showed the SXR-emitting regions in events 1 and 2 to be fixed when the pressure increased.

Veronig, Temmer, and Vršnak (2008) concluded that the wave ignition in event 3 by the flare was unlikely. Thus, the theoretical considerations supported by the observational facts make the ignition of waves by flares doubtful in agreement with the conclusion of Cliver *et al.* (2004) (note that the authors implied the shocks to be driven by outer surfaces of CMEs, whereas we consider the shocks to appear inside CMEs).

This conclusion is seemingly opposed by the results of Magdalenić *et al.* (2008, 2010) who presented five events, in which shocks appeared during the flare impulsive phase, whereas related CMEs were slow. However, the authors did not consider a rare type of CME kinematics with very strong initial acceleration followed by deceleration. An example is our event 1 (Section 3.1, Figure 4). The acceleration stage in this event lasted three minutes. The fastest feature accelerated up to 470 km s^{-1} and continuously decelerated later on. From an overall height–time plot, including measurements from LASCO images, Grechnev *et al.* (2008) estimated the speed of this feature to approach 100 km s^{-1} far from the Sun. Nevertheless, the shock in this event was most likely excited just by this feature as an impulsive piston rather than by the flare. Thus, a slower motion of the main body of a CME well behind a shock front does not guarantee that the CME or its components were not implicated in excitation of the shock.

The maximum plane-of-sky speeds of the eruptions in events 1 – 3 appear to be well below the Alfvén speed expected at moderate heights (< 100 Mm) above active regions (see, *e.g.*, Mann *et al.*, 2003), where the waves appeared. It is possible that the angles between the velocity vectors and the line of sight significantly differed from 90°, so that the real velocities could be higher, but the corresponding corrections are insufficient to increase the velocities above the Alfvén speed. The time profiles of the velocities estimated for the eruptions and waves were quite different. Thus, the bow-shock scenario is also unlikely.

4.2. Impulsive Piston Scenario

In the simplest scheme, a piston moving with a speed U has a box-like acceleration profile with a value a during the acceleration phase and zero before and afterwards. An important condition is that plasma cannot flow around the piston (this occurs, *e.g.*, in 3D expansion of an arcade). In a flat geometry and homogeneous medium the plasma flow ahead of the piston corresponds to a simple wave. The discontinuity appears at $t^0_{\mathrm{sh}} = V_{\mathrm{fast}}/(\kappa a)$ at a distance $r^0_{\mathrm{sh}} = V_{\mathrm{fast}} t^0_{\mathrm{sh}}$ with $1/2 \le \kappa \le 3/2$ that is similar to a solution of an analogous gas-dynamic problem (Landau and Lifshitz, 1987). Then the speed jump in the discontinuity increases up to the piston's maximum speed $U_{\max}$. The condition $U > V_{\mathrm{fast}}$ essential for bow-shock formation is unnecessary in the impulsive piston scenario.

Accelerations and their durations before the wave onsets were (Section 3): 4 km s^{-2} and 90 s in event 1; 7 km s^{-2} and 70 s in event 2; 0.12 – 0.27 km s^{-2} and 800 s in event 3. The stronger acceleration, the faster a shock appeared.

The shock waves in the four events were most likely excited by eruptive structures as impulsive pistons, which one might call the forming CMEs. Then the waves rapidly steepened into shocks, detached from the pistons, and freely propagated afterwards like blast waves. The shock excitation mechanism implies a source height to be nonzero, but rather low, probably < 100 Mm, as suggested by all dynamic spectra and implied by event 4. Magdalenić *et al.* (2008, 2010) found the heights of sources of metric type II bursts to be between 70 and 280 Mm; however, the apparent heights of limb sources in the metric range could be noticeably reduced due to refraction (see, *e.g.*, Zheleznyakov, 1970). The shock character of the waves is supported by the correspondence of their kinematics to the expected propagation of shock waves as well as the drift rates of type II bursts and drifting continua. The shock-wave nature of EUV waves is supported with probable reflections and coupling of two shock waves in event 3. Expansion of leading edges of CMEs produced in three events corresponded to propagation of the lower skirts of the shock fronts observed as Moreton/EUV waves. Thus, the wave excitation by an impulsive piston appears to match all the considered observations, basically corresponding to a picture proposed by Uchida (1974).

So far we did not relate a piston with a particular structure or its surface. Presumable pistons could be either an eruptive filament (EF) or a CME frontal structure (FS). Both expand as an entire ensemble in a completely formed CME; only its outer sheath can be a piston. This sheath is believed to be both the surface of contact discontinuity and the outer FS surface. Expansion of such an FS-piston determines propagation of an interplanetary piston-driven wave and the drag force affecting a CME. The situation is different during the early CME formation inside an active region, when EF moves faster than it would be adequate for a self-similar expansion of the whole CME. The EF acts here as an impulsive piston. The piston excites a wave inside a future CME. The wave freely propagates outwards as a shock wave. In the four events we revealed just this excitation scenario of waves, which resembled blast ones. Propagating upwards, such a wave inevitably would pass through the FS and appear ahead of it.

4.3. CME Components and Waves

4.3.1. Particularities of Expansion

Expansion of magnetoplasma constituents of the CME is different from kinematics of a wave traced, *e.g.*, with a leading edge of a plasma flow driven by a shock. The CME expansion is known to be about self-similar at moderate distances from the Sun. *The self-similar approach does not apply to early stages of expansion, when the structure and shape of a CME have not yet been established.* When an instability driving an eruption is over and drag from the solar wind is not yet significant, self-similar CME kinematics can be obtained from considerations of forces affecting a CME (Low, 1982; Uralov, Grechnev, and Hudson, 2005).

Expansion of magnetoplasma structures is governed by magnetic forces, plasma pressure, and gravity as long as the effect of the solar wind is small. With the polytropic index $\gamma \approx 4/3$ all the forces integrated over the boundary and volume of a CME scale with distance r by the same factor of r^{-2}. This leads to an expression for the CME velocity $V_{\mathrm{CME}}^2 = V_0^2 + (V_\infty^2 - V_0^2)(1 - R_0/r)$, $V_{\mathrm{CME}} = \mathrm{d}r/\mathrm{d}t$ (Grechnev *et al.*, 2008). Here R_0 is the initial size of self-similar expansion, V_0 the initial velocity at R_0, and V_∞ the asymptotic velocity in infinity. At large distances acceleration $\propto r^{-2} \to 0$ and $V_{\mathrm{CME}} \to V_\infty$. The expression for the CME velocity describes different types of kinematics. The situation $V_\infty \gg V_0 \sim 0$ appears to be typical. Event 1 showed a different behavior, $V_0 > V_\infty$, resembling an explosion with an impulsive acceleration followed by deceleration. A special type $V_0 \approx V_\infty$ (an impulsive acceleration is required to reach V_0) might correspond to some jet-like ejections, where magnetic reconnection destroys a structure of an eruption (Meshalkina *et al.*, 2009; Filippov, Golub, and Koutchmy, 2009; Liu *et al.*, 2011). Event 2 might have belonged to this type.

Since a FS starts to expand practically from a static equilibrium in a typical situation $V_\infty \gg V_0 \sim 0$, the FS-piston usually either accelerates or moves with a nearly constant speed at the self-similar stage. By contrast, shock waves in all considered events decelerated. Hence, a typical FS-piston is expected to eventually approach the wave front. What does such a relation between the speeds of the shock front and piston mean? Do MHD equations allow a decelerating shock wave to run for a long time ahead of a non-decelerating FS-piston? We search answers in a theory developed by Low (1984) in solving a problem of self-similar expansion of a CME preceded by a strong shock wave. Though the problem was solved in a limit of very strong shock propagating in plasma with too steep density falloff $r^{-26/7}$, the solution correctly relates accelerations of the piston and piston-driven shock. Assuming a common linear profile of the plasma velocity in the whole region from the expansion center up to the shock front, it is possible to relate kinematics of the shock front and the contact discontinuity, *i.e.*, the FS-piston, without a complete solution of the problem. By fitting the motion of the piston with a function $r_{\mathrm{pist}} = bt^m$, we express the sign of the shock acceleration a_{sh} *vs.* m: $a_{\mathrm{sh}} \propto (\alpha m - 1)t^{\alpha m - 2}$ with $\alpha = (\gamma + 1)/2$. Thus, the conditions $a_{\mathrm{sh}} < 0$ and $a_{\mathrm{pist}} = \mathrm{d}^2 r_{\mathrm{pist}}/\mathrm{d}t^2 > 0$ are incompatible. Such an FS-piston and the shock front expand in different ways, which cannot be coordinated with each other, unlike a bow shock. An impulsively excited freely propagating shock wave must eventually change to a piston-driven mode. Presumably this typically occurs at large distances, probably beyond the LASCO/C3 field of view. The transformation of a blast shock wave into a piston-driven one marks switch-on the aerodynamic drag and termination of the self-similar expansion regime. The drag force becomes significant, which means establishment of a continuous energy transport from the FS-piston to the shock wave. By contrast, a blast-like wave excited by an EF and running ahead of FS, which does not yet act as a piston, facilitates expansion of a CME into the solar wind. The shock wave forwards a part of its energy to the FS-piston, and the drag force is absent.

Most likely, real shock waves are neither purely blast waves nor purely piston ones. A shock front is sensitive to any events occurring behind it, *e.g.*, changes of the FS-piston speed, because the fast-mode speed behind the shock front is higher than its phase speed. To produce one more shock wave, an FS-piston has to repeat the maneuver, which produced the first shock. This is improbable when an instability driving an eruption has completed and a CME left the Sun.

4.3.2. Distinguishing Between Shock Signatures and CME Components

Sheeley, Hakala, and Wang (2000) and Vourlidas *et al.* (2003) considered distortions of coronal streamers as a morphological suggestion of presumable shocks. Indeed, moving wave-like kinks or deflections of coronal rays resemble an expected effect of a propagating shock wave. Some fast "CMEs" in difference images might be actually combinations of coronal rays deflected by shock waves. However, Filippov and Srivastava (2010) demonstrated that deflection of coronal rays could be due to expansion of a CME in magnetized corona without a shock. Irrespective of the particular type of possible shock, a more reliable morphological suggestion might be a spike-like leading feature due to deviation of a coronal ray by a wave. Magdalenić *et al.* (2008) showed such a situation in their Figure 4, where deflected coronal rays were visible well ahead of trailing CME structures.

We remind the reader that the conic bow-shock shape is not expected, at least, for wide super-Alfvénic CMEs. The shock front must cling to its foremost edge and closest flanks, while far flanks and a rear part can be constituted by a freely propagating shock front, so that the shape of the whole front would resemble an egg.

An attractive way to detect a shock front is to search for discontinuity in the density distribution shown by coronagraph images (Vourlidas *et al.*, 2003). However, this is model dependent. Besides modeling the coronal density, one has to distinguish between the shock discontinuity and the contact surface separating the CME and environment. The three-dimensional contact surface should be also modeled. Thus, identifying a shock front in coronagraph images does not seem to be a simple task. Combinations of different indications seem to be useful. A non-structured faint density enhancement forming the envelope of a transient could be among them (Vourlidas *et al.*, 2003). Fast decelerating halo CMEs with such edges might be shock candidates. An effective shock indication might be such a halo edge crossing a distorted streamer. An important complement of morphological suggestions is kinematics of an expected shock wave and its correspondence to the drift of a type II burst. We also address in Paper III changes in shape of a shock front occurring in its propagation in the corona.

4.4. Presumable Scenario

Our observations and considerations suggest the following presumable scenario of a flare-related eruptive event. An eruption occurs due to a rapid development of an instability in a magnetic structure. An abruptly accelerating eruption destroys a pre-existing magnetic configuration, thus causing a flare, and produces an MHD disturbance as an impulsive piston. The disturbance appears at a height of ~ 50 Mm during the rise phase of an HXR/microwave burst, leaves the piston, rapidly steepens into a shock, and then freely propagates like a blast wave. Its lower trail might be observed as a decelerating Moreton wave as well as an EUV wave, and the wave dome is sometimes observed to expand above the limb.

The motion of the shock front shows up in radio spectra as a drifting continuum and, when the shock front reaches the current sheet of a coronal streamer, as a type II burst. For the fundamental emission this usually occurs at ~ 100 MHz ($r \sim 1.5 R_\odot$). Metric type II bursts are expected to cease due to damping of shock waves that typically occurs above 20 MHz ($r < 3 R_\odot$). Revival of a shock is possible at a few $R_\odot$ due to decreasing Alfvén speed, and decametric/hectometric type II emission can appear. The complex piston-blast-piston transformations of shock waves traveling in the corona with significantly varying parameters and possible coupling of multiple shocks imply well-known disagreement between metric and interplanetary type II events (see, *e.g.*, Cane and Erickson, 2005).

Expanding shock fronts can form envelopes of CMEs. Measurements in the CME Catalog referring to a fastest feature might be related to shock waves for fast decelerating CMEs, especially halos. As a shock wave decelerates, a trailing mass must eventually approach its front. The shock becomes a piston-driven one presumably at distances $r > 20 R_\odot$. The aerodynamic drag becomes important. This picture is consistent with results of several cited papers. The story of shock waves associated with flare-related CMEs appears to be more complex than often assumed, in fact combining different scenarios.

5. Concluding Remarks

Our seemingly simplified approach has resulted in surprisingly fine reconciliation of EUV waves, Moreton waves, metric type II bursts, and leading edges of CMEs. The first consequence is that independent of the quality of our approximation, all these phenomena are really manifestations of a common agent, *i.e.*, a traveling coronal shock wave excited by an eruption. Second, our approach indeed provides a promising instrument for analyses of slow-drifting bursts and their comparison with other eruption-related phenomena. The power-law approximation turns out to work well beyond conditions, for which it was derived. An results clarify relations between flares, traveling coronal shocks, CMEs, associated wave-like manifestations, type II bursts, and provide a common quantitative description for some of these phenomena. An important by-product is an indication of the leading role of eruptions with respect to flares, *i.e.*, the acceleration of an eruption occurs almost independently of the flare reconnection rate.

Our approach and analysis needs elaboration and continuation. There remains a number of issues to be addressed. Low heights, at which type II emission sometimes appears, indicate that our consideration of its generation in the current sheet of a coronal streamer needs elaboration. Data sets similar to those analyzed in our paper should be compared with imaging observations in the metric range. The analysis of decimetric to metric drifting bursts should be extended to longer radio waves in conjunction with coronagraphic observations. Despite the success of our self-similar shock approximation, a more realistic weak shock approximation should be considered. The latter issue is a subject of our paper II.

Acknowledgements We thank A. Warmuth for data, which he made available to us, and M. Eselevich, E. Ivanov, E. Schmahl, M. Temmer, V. Eselevich, A. Altyntsev, G. Rudenko, L. Kashapova, V. Fainshtein, N. Prestage, S. Pohjolainen, S. White, A. Zhukov, and J. Magdalenić for fruitful discussions and cooperation. We gratefully remember Mukul Kundu who inspired a significant part of our study. We thank an anonymous referee for valuable remarks.

We thank the teams operating all instruments whose data are used in our study for their efforts and open data policies: the ESA & NASA EIT, LASCO, and MDI instruments on SOHO; TRACE and STEREO/SECCHI telescopes; the Mauna Loa Solar Observatory; the NOAA/SEC GOES satellites; the NICT HIRAS (Japan), the IPS Radio and Space Services Learmonth Observatory (Australia), and the USAF RSTN radio telescopes. We appreciatively use the CME catalog generated and maintained at the CDAW Data Center by NASA and the Catholic University of America in cooperation with the Naval Research Laboratory. This research was supported by the Russian Foundation of Basic Research under grant 09-02-00115.

References

Afanasyev, A.N., Uralov, A.M.: 2011, *Solar Phys.* doi:10.1007/s11207-011-9730-9. this issue.
Attrill, G.D.R.: 2010, *Astrophys. J.* **718**, 494.
Biesecker, D.A., Myers, D.C., Thompson, B.J., Hammer, D.M., Vourlidas, A.: 2002, *Astrophys. J.* **569**, 1009.
Brueckner, G.E., Howard, R.A., Koomen, M.J., Korendyke, C.M., Michels, D.J., Moses, J.D., *et al.*: 1995, *Solar Phys.* **162**, 357.
Cane, H.V., Erickson, W.C.: 2005, *Astrophys. J.* **623**, 1180.
Chen, P.F., Fang, C., Shibata, K.: 2005, *Astrophys. J.* **622**, 1202.
Chertok, I.M., Grechnev, V.V.: 2005, *Solar Phys.* **229**, 95.
Chertok, I.M., Grechnev, V.V., Uralov, A.M.: 2009, *Astron. Rep.* **53**, 355.
Cliver, E.W., Nitta, N.V., Thompson, B.J., Zhang, J.: 2004, *Solar Phys.* **225**, 105.
Cohen, O., Attrill, G.D.R., Manchester, W.B., Wills-Davey, M.J.: 2009, *Astrophys. J.* **705**, 587.
Delaboudinière, J.-P., Artzner, G.E., Brunaud, J., Gabriel, A.H., Hochedez, J.F., Millier, F., *et al.*: 1995, *Solar Phys.* **162**, 291.
Delannée, C.: 2000, *Astrophys. J.* **545**, 512.
Filippov, B., Srivastava, A.K.: 2010, *Solar Phys.* **266**, 123.
Filippov, B., Golub, L., Koutchmy, S.: 2009, *Solar Phys.* **254**, 259.
Gallagher, P.T., Long, D.M.: 2010, *Space Sci. Rev.* doi:10.1007/s11214-010-9710-7.
Gopalswamy, N., Yashiro, S., Temmer, M., Davila, J., Thompson, W.T., Jones, S., McAteer, R.T.J., Wuelser, J.-P., Freeland, S., Howard, R.A.: 2009, *Astrophys. J. Lett.* **691**, 123.
Grechnev, V.V., Kuzin, S.V., Urnov, A.M., Zhitnik, I.A., Uralov, A.M., Bogachev, S.A., Livshits, M.A., Bugaenko, O.I., *et al.*: 2006a, *Solar Syst. Res.* **40**, 286.
Grechnev, V.V., Uralov, A.M., Zandanov, V.G., Baranov, N.Y., Shibasaki, K.: 2006b, *Publ. Astron. Soc. Japan* **58**, 69.
Grechnev, V.V., Uralov, A.M., Zandanov, V.G., Rudenko, G.V., Borovik, V.N., Grigorieva, I.Y., Slemzin, V.A., Bogachev, S.A., *et al.*: 2006c, *Publ. Astron. Soc. Japan* **58**, 55.
Grechnev, V.V., Uralov, A.M., Slemzin, V.A., Chertok, I.M., Kuzmenko, I.V., Shibasaki, K.: 2008, *Solar Phys.* **253**, 263.

Grechnev, V.V., Afanasyev, A.N., Uralov, A.M., Chertok, I.M., Eselevich, M.V., Eselevich, V.G., Rudenko, G.V., Kubo, Y.: 2011, *Solar Phys.* doi:10.1007/s11207-011-9781-y this issue.
Handy, B.N., Acton, L.W., Kankelborg, C.C., Wolfson, C.J., Akin, D.J., Bruner, M.E., Caravalho, R., Catura, R.C., *et al.*: 1999, *Solar Phys.* **187**, 229.
Howard, R.A., Moses, J.D., Vourlidas, A., Newmark, J.S., Socker, D.G., Plunkett, S.P., Korendyke, C.M., Cook, J.W., *et al.*: 2008, *Space Sci. Rev.* **136**, 67.
Hudson, H.S., Warmuth, A.: 2004, *Astrophys. J. Lett.* **614**, 85.
Hudson, H.S., Khan, J.I., Lemen, J.R., Nitta, N.V., Uchida, Y.: 2003, *Solar Phys.* **212**, 121.
Ichimoto, K., Sakurai, T., Flare Telescope, Norikura Teams: 1993, In: *Proc. 2nd Japan–China Seminar on Sol. Phys.* 151.
Kaiser, M.L., Kucera, T.A., Davila, J.M., St. Cyr, O.C., Guhathakurta, M., Christian, E.: 2008, *Space Sci. Rev.* **136**, 5.
Khan, J.I., Aurass, H.: 2002, *Astron. Astrophys.* **383**, 1018.
Klassen, A., Aurass, H., Mann, G., Thompson, B.J.: 2000, *Astron. Astrophys. Suppl. Ser.* **141**, 357.
Landau, L.D., Lifshitz, E.M.: 1987, *Fluid Mechanics*, 2nd edn., Pergamon, Oxford.
Lin, R.P., Dennis, B.R., Hurford, G.J., Smith, D.M., Zehnder, A., Harvey, P.R., *et al.*: 2002, *Solar Phys.* **210**, 3.
Liu, W., Berger, T.E., Title, A.M., Tarbell, T.D., Low, B.C.: 2011, Astrophys. J. **728**, article id. 103.
Long, D.M., Gallagher, P.T., McAteer, R.T.J., Bloomfield, D.S.: 2008, *Astrophys. J. Lett.* **680**, L81.
Low, B.C.: 1982, *Astrophys. J.* **254**, 796.
Low, B.C.: 1984, *Astrophys. J.* **281**, 381.
Magdalenić, J., Vršnak, B., Pohjolainen, S., Temmer, M., Aurass, H., Lehtinen, N.J.: 2008, *Solar Phys.* **253**, 305.
Magdalenić, J., Marqué, C., Zhukov, A.N., Vršnak, B., Žic, T.: 2010, *Astrophys. J.* **718**, 266.
Mancuso, S., Abbo, L.: 2004, *Astron. Astrophys.* **415**, L17.
Mancuso, S., Raymond, J.C.: 2004, *Astron. Astrophys.* **413**, 363.
Mann, G., Klassen, A., Aurass, H., Classen, H.-T.: 2003, *Astron. Astrophys.* **400**, 329.
Meshalkina, N.S., Uralov, A.M., Grechnev, V.V., Altyntsev, A.T., Kashapova, L.K.: 2009, *Publ. Astron. Soc. Japan* **61**, 791.
Moreton, G.E.: 1960, *Astron. J.* **65**, 494.
Nakajima, H., Sekiguchi, H., Sawa, M., Kai, K., Kawashima, S.: 1985, *Publ. Astron. Soc. Japan* **37**, 163.
Newkirk, G. Jr.: 1961, *Astrophys. J.* **133**, 983.
Pohjolainen, S., Hori, K., Sakurai, T.: 2008, *Solar Phys.* **253**, 291.
Pomoell, J., Vainio, R., Kissmann, R.: 2008, *Solar Phys.* **253**, 249.
Reiner, M.J., Vourlidas, A., St. Cyr, O.C., Burkepile, J.T., Howard, R.A., Kaiser, M.L., Prestage, N.P., Bougeret, J.-L.: 2003, *Astrophys. J.* **590**, 533.
Rudenko, G.V.: 2001, *Solar Phys.* **198**, 5.
Rudenko, G.V., Grechnev, V.V.: 1999, In: *Astronomical Data Analysis Software and Systems VIII* **CS-172**, Astron. Soc. Pac., San Francisco, 421.
Saito, K.: 1970, *Ann. Tokyo Astron. Obs.* **12**, 53.
Schmidt, J.M., Ofman, L.: 2010, *Astrophys. J.* **713**, 1008.
Sheeley, N.R. Jr., Hakala, W.N., Wang, Y.-M.: 2000, *J. Geophys. Res.* **105**(A3), 5081.
Temmer, M., Veronig, A.M., Vršnak, B., Rybák, J., Gömöry, J., Stoiser, S., Maričić, D.: 2008, *Astrophys. J. Lett.* **673**, 95.
Thompson, B.J., Plunkett, S.P., Gurman, J.B., Newmark, J.S., St. Cyr, O.C., Michels, D.J.: 1998, *Geophys. Res. Lett.* **25**, 2465.
Thompson, B.J., Gurman, J.B., Neupert, W.M., Newmark, J.S., Delaboudinière, J.-P., St. Cyr, O.C., Stezelberger, S., Dere, K.P.: 1999, *Astrophys. J. Lett.* **517**, L151.
Thompson, B.J., Reynolds, B., Aurass, H., Gopalswamy, N., Gurman, J.B., Hudson, H.S., Martin, S.F., St. Cyr, O.C.: 2000, *Solar Phys.* **193**, 161.
Torii, C., Tsukiji, Y., Kobayashi, S., Yoshimi, N., Tanaka, H., Enome, S.: 1979, *Proc. Res. Inst. Atmos. Nagoya Univ.* **26**, 129.
Uchida, Y.: 1968, *Solar Phys.* **4**, 30.
Uchida, Y.: 1974, *Solar Phys.* **39**, 431.
Uralov, A.M., Grechnev, V.V., Hudson, H.S.: 2005, *J. Geophys. Res.* **110**, A05104.
Uralova, S.V., Uralov, A.M.: 1994, *Solar Phys.* **152**, 457.
Veronig, A.M., Temmer, M., Vršnak, B.: 2008, *Astrophys. J. Lett.* **681**, L113.
Veronig, A.M., Muhr, N., Kienreich, I.W., Temmer, M., Vršnak, B.: 2010, *Astrophys. J. Lett.* **716**, L57.
Vourlidas, A., Wu, S.T., Wang, A.H., Subramanian, P., Howard, R.A.: 2003, *Astrophys. J.* **598**, 1392.
Vršnak, B., Cliver, E.W.: 2008, *Solar Phys.* **253**, 215.
Vršnak, B., Aurass, H., Magdalenić, J., Gopalswamy, N.: 2001, *Astron. Astrophys.* **377**, 321.

Wang, Y., Zhang, J., Shen, C.: 2009, *J. Geophys. Res.* **114**, 10104.
Warmuth, A.: 2007, In: *Lecture Notes in Physics* **725**, Springer, Berlin, 107.
Warmuth, A.: 2010, *Adv. Space Res.* **45**, 527.
Warmuth, A., Vršnak, B., Aurass, H., Hanslmeier, A.: 2001, *Astrophys. J. Lett.* **560**, L105.
Warmuth, A., Vršnak, B., Magdalenić, J., Hanslmeier, A., Otruba, W.: 2004a, *Astron. Astrophys.* **418**, 1101.
Warmuth, A., Vršnak, B., Magdalenić, J., Hanslmeier, A., Otruba, W.: 2004b, *Astron. Astrophys.* **418**, 1117.
Wills-Davey, M.J., Attrill, G.D.R.: 2009, *Space Sci. Rev.* **149**, 325.
Wills-Davey, M.J., DeForest, C.E., Stenflo, J.O.: 2007, *Astrophys. J.* **664**, 556.
White, S.M., Thompson, B.J.: 2005, *Astrophys. J. Lett.* **620**, L63.
Yang, H.Q., Chen, P.F.: 2010, *Solar Phys.* **266**, 59.
Yashiro, S., Gopalswamy, N., Michalek, G., St. Cyr, O.C., Plunkett, S.P., Rich, N.B., Howard, R.A.: 2004, *J. Geophys. Res.* **109**, A07105.
Zheleznyakov, V.V.: 1970, *Radio Emission of the Sun and the Planets*, Pergamon, Oxford.
Zhukov, A.N., Auchère, F.: 2004, *Astron. Astrophys.* **427**, 705.
Zhukov, A.N., Rodriguez, L., de Patoul, J.: 2009, *Solar Phys.* **259**, 73.

Solar Phys (2011) 273:461–477
DOI 10.1007/s11207-011-9781-y

Coronal Shock Waves, EUV Waves, and Their Relation to CMEs. III. Shock-Associated CME/EUV Wave in an Event with a Two-Component EUV Transient

V.V. Grechnev · A.N. Afanasyev · A.M. Uralov · I.M. Chertok · M.V. Eselevich · V.G. Eselevich · G.V. Rudenko · Y. Kubo

Received: 28 November 2010 / Accepted: 19 April 2011 / Published online: 2 June 2011

Abstract On 17 January 2010, STEREO-B observed in extreme ultraviolet (EUV) and white light a large-scale dome-shaped expanding coronal transient with perfectly connected off-limb and on-disk signatures. Veronig *et al.* (*Astrophys. J. Lett.* **716**, L57, 2010) concluded that the dome was formed by a weak shock wave. We have revealed two EUV components, one of which corresponded to this transient. All of its properties found from EUV, white light, and a metric type II burst match expectations for a freely expanding coronal shock wave, including correspondence with the fast-mode speed distribution, while the transient sweeping over the solar surface had a speed typical of EUV waves. The shock wave was presumably excited by an abrupt filament eruption. Both a weak shock approximation and a power-law fit match kinematics of the transient near the Sun. Moreover, the power-law fit matches the expansion of the CME leading edge up to 24 solar radii. The second, quasi-stationary EUV component near the dimming was presumably associated with a stretched CME structure; no indications of opening magnetic fields have been detected far from the eruption region.

Energy Storage and Release through the Solar Activity Cycle – Models Meet Radio Observations
Guest Editors: Christophe Marqué and Alexander Nindos

Electronic supplementary material The online version of this article (doi:10.1007/s11207-011-9781-y) contains supplementary material, which is available to authorized users.

V.V. Grechnev (✉) · A.N. Afanasyev · A.M. Uralov · M.V. Eselevich · V.G. Eselevich · G.V. Rudenko
Institute of Solar-Terrestrial Physics SB RAS, Lermontov St. 126A, Irkutsk 664033, Russia
e-mail: grechnev@iszf.irk.ru

I.M. Chertok
Pushkov Institute of Terrestrial Magnetism, Ionosphere and Radio Wave Propagation (IZMIRAN), Troitsk, Moscow Region 142190, Russia
e-mail: ichertok@izmiran.ru

Y. Kubo
National Institute of Information and Communications Technology, Tokyo, Japan
e-mail: kubo@nict.go.jp

Keywords Coronal mass ejections, low coronal signatures · Coronal mass ejections, initiation and propagation · Radio bursts, type II · Waves, shock

1. Introduction

Large-scale wave-like transients called EUV waves or "EIT waves" are observed in extreme ultraviolet (EUV) and soft X-rays in association with coronal mass ejections (CMEs) and flares (Thompson *et al.*, 1998, 1999). Efforts of researchers to understand the nature of EUV waves meet difficulties. The main observational material acquired with SOHO/EIT suffers from insufficient temporal coverage. Bright flare emission hinders detection of faint EUV waves. Properties of their propagation, association with flares and metric type II bursts, *etc.* appear to be diverse and contradictory (see, *e.g.*, Biesecker *et al.*, 2002; Klassen *et al.*, 2000). Deficiency of observations stimulated development of conflicting concepts based mainly on *i*) MHD fast-mode disturbances (Thompson *et al.*, 1999; Warmuth *et al.*, 2001; Khan and Aurass, 2002; Long *et al.*, 2008; Gopalswamy *et al.*, 2009) or *ii*) plasma compression in bases of stretching loops (Delannée and Aulanier, 1999; Chen *et al.*, 2002; Chen, Fang, and Shibata, 2005; Attrill *et al.*, 2007) both caused by a CME eruption (see also Zhukov and Auchère, 2004; reviews by Wills-Davey and Attrill, 2009; Gallagher and Long, 2010). The former set of hypotheses *i*) implies CME-related opening or reconnection of magnetic fields in the vicinity of an eruption site; the latter one *ii*) assumes it to be global to describe both standing 'EUV waves' and those propagating over large distances.

The launch in 2006 of the twin-spacecraft *Solar-Terrestrial Relations Observatory* (STEREO; Kaiser *et al.*, 2008) carrying the Sun Earth Connection Coronal and Heliospheric Investigation instrument suites (SECCHI; Howard *et al.*, 2008) significantly enhanced opportunities to study EUV waves, including their temporal coverage, with the Extreme Ultraviolet Imagers (EUVI). However, this has not lead to consensus about their nature. Several studies argued the shock-wave nature of observed EUV waves (see, *e.g.*, Kienreich, Temmer, and Veronig, 2009; Patsourakos and Vourlidas, 2009; Patsourakos *et al.*, 2009). Conversely, Zhukov, Rodriguez, and de Patoul (2009) presented an EUV wave incompatible with the fast-mode wave interpretation. Disappointing was the study of the 19 May 2007 event, which was considered both in favor of the shock-wave hypothesis (Long *et al.*, 2008; Veronig, Temmer, and Vršnak, 2008; Gopalswamy *et al.*, 2009) and against it (Attrill, 2010; Yang and Chen, 2010) [our analysis in Paper I (Grechnev *et al.*, 2011) supports the shock-wave interpretation]. A recent analysis of an EUV wave observed in still more detail with *Solar Dynamics Observatory* provides more questions than answers (Liu *et al.*, 2010).

The diversity of EUV waves implies their probable relation to different phenomena (Zhukov and Auchère, 2004; Grechnev *et al.*, 2008; Cohen *et al.*, 2009). Our companion Papers I – II consider EUV waves associated with coronal shock waves. Paper I shows how to reconcile shock-related EUV waves, type II bursts, and corresponding CMEs. We propose a simple quantitative description for all of these phenomena based on an approach of a self-similar shock wave. The large length of such a wave is comparable with the curvature radius of the wave front. Its deceleration is determined by the increasing mass inside the volume limited by the shock front. The self-similar approach describes propagation of strong shock waves. Our experience has revealed a convenient way to fit the kinematics of real coronal waves with direction-dependent power-law (PL) functions (abbreviated henceforth 'shock-PL fit'). Afanasyev and Uralov (2011; Paper II) have considered the opposite limit of a weak shock calculated analytically in terms of the WKB (Wentzel–Kramers–Brillouin) approach taking account of nonlinear effects.

An eruptive event on 17 January 2010 produced a CME and wave, whose expanding three-dimensional dome with its lower skirt sweeping over the solar surface was observed in unprecedented detail by EUVI and coronagraphs on the STEREO-B spacecraft. EIT (Delaboudinière *et al.*, 1995) and LASCO (Brueckner *et al.*, 1995) instruments on SOHO also observed this transient. A weak type II burst was recorded by HiRAS (NICT, Japan) and Learmonth (US Air Force RSTN) radio spectrographs. Veronig *et al.* (2010) analyzed this backside event and concluded that the coronal transient observed both in EUV and white light was a dome of a 'weakly shocked fast-mode MHD wave'. The authors found that the lateral expansion of the wave dome far from the eruption site was free, while, in their opinion, its upward expansion was driven by the CME all the time.

We analyzed this event independently and also inferred the shock-wave nature of this coronal transient. However, the scopes and some conclusions of our and Veronig *et al.* (2010) studies do not coincide. Unlike the authors, we find deceleration of both near-surface and off-limb traces of the wave. Our analysis in the present Paper III shows that the shock wave was most likely excited by the impulsive-piston mechanism and freely propagated omnidirectionally afterwards as considered in Paper I. We study the shock-wave propagation both on-disk and off-limb in comparison with the fast-mode speed (V_{fast}) distribution and the drift rate of the type II burst. We explain the differences between our results and those of Veronig *et al.* (2010), and compare the results, which the self-similar shock approximation and modeling of a weak shock provide being applied to this event, including the wave propagation at larger distances from the Sun. We have revealed one more EUV transient, which adjoined the dimming and was quasi-stationary. This fact confirms that different kinds of "EIT waves" do exist. We analyze the observations in Section 2, compare the revealed properties of the wave with modeling results in Section 3, and summarize the outcome in Section 4.

2. Analysis of Observations

The two STEREO spacecraft were located 69.6° behind the Earth and 64.3° ahead of it. The eruption site shown by the flare was seen from STEREO-B at S25 E59 (heliolatitude $B0 = 3.74°$) and located for observers on the Earth $\approx 37°$ behind the east limb with a projected position onto the visible solar surface of about S32 E55 ($B0 = -4.75°$). The projected positions onto the visible solar surface observed from the Earth (and SOHO) and STEREO-B were close to each other. The radial CME extent and velocity were smaller by a factor of 1.13 for observers on the Earth (SOHO) with respect to observations on STEREO-B. The STEREO-A/COR1 coronagraph registered a wide transient around a position angle of PA $\approx 225°$ (we do not consider STEREO-A or EIT data).

2.1. Eruption and a Probable Shock Wave

Figure 1a–d and the movie euvi_195.mpg in the electronic version of our paper show the onset of the event observed in EUVI 195 Å fixed-base ratio images. A dome-like EUV wave expanded above the limb and propagated along the solar surface. The boundary of the surface front passed into the off-limb dome suggesting their common nature. The front was followed by extended brightenings indicating a large length of the disturbance. Eruption and untwisting of a magnetic structure (probably a filament) is seen inside the EUV wave dome. The motion of the eruption was three-dimensional. This fact, fading out of the eruption, and difficulties to distinguish it from the wave front make measurements of its kinematics unreliable. Nevertheless, it is possible to see in Figure 1a–d and in the movie that the eruption

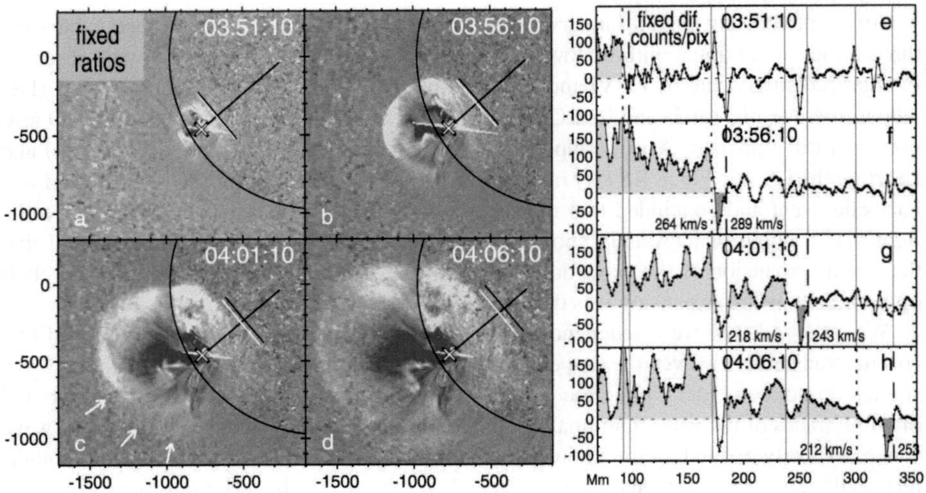

Figure 1 (a – d) EUV wave and eruption in EUVI 195 Å images. White arrows in panel (c) indicate the fastest part of the front. The slanted cross marks the eruption site. The black line going from the eruption site northwest shows the direction where the spatial profiles were computed. The white and black bars across this line mark the presumable fronts suggested by the spatial profiles. In all EUVI images hereafter, the axes show arc seconds from the solar disk center as if viewed from the Earth. (e – h) The spatial profiles of the EUV brightness measured in the direction shown in the left panels. The vertical lines denote presumable fronts. The shading indicates the EUV wave brightening behind the front and a possible negative precursor ahead.

changed shape like a straightening mainspring. Its foremost edge was close to the wave front at 03:56. Thus, just an abrupt eruption of the rope structure could have played a role of an impulsive piston, which excited the wave, as was the case in events considered in Paper I.

The wave front is sharpest in Figure 1b – d and Figure 4b just to the left from the eruption site (slanted cross) in the plane of the sky, while the fastest faint part of the front indicated by the white arrows in Figure 1c is closer to the radial direction. The sharper appearance of the front in the leftwards direction could be due to overlap with expanding loops and a stronger shock in this direction. The latter effect is consistent with the predominant upwards motion of the eruption. The faintness of the front in the South Pole's environment is due to the closeness of the polar coronal hole, where the Alfvén velocity V_A is much higher (Figure 4h). One more outcome is free propagation of the wave in the radial direction, where the front moved ahead of possible loops (Figure 1c).

Figure 1e – h shows plane-of-sky spatial profiles computed from the four fixed-difference 195 Å images within sectors of 1° along the directions indicated with the black lines. The profiles show a relief constituted by variations of compact features. The chosen direction crosses features, which seem to have responded to the passage of the wave front. The EUV wave brightening appears in the profiles as an enhancement (light shading) to the left from the front (dotted). All the profiles show compact darkenings (dashed, darker shading) preceding the brightenings. The darkening and brightening regions in Figure 1e seem to be imperfectly resolved, which reduces the depth of the narrow darkening. The dashed and dotted lines in Figure 1e – h correspond to the black and white bars in Figure 1a – d. Comparison of all the panels e – h with each other reveals slightly variable compact features at the four fixed positions, where the front presumably showed up. Hits by a shock front probably disturbed the features, producing the sharp effect suggested by the profiles, but not a gradual elevation. The EUVI pixel size (small circles show the samples) was ≈ 1190 km;

with exposure times of 16 s, a step-like front moving with plane-of-sky speeds shown in Figure 1f–h must be caught in 3–4 pixels. Thus, just such a response to a shock front is expected.

Comparison of the profiles with the quiet Sun's level of about 290 counts pixel^{-1} shows that if this marginal effect was real, then its value could only be due to disturbance of low structures. The fact that the probable response of different solar features matched arrival of the wave front at different times indicates that the observed effect deserves attention and should be checked in other events.

After the passage of the wave front, small features like coronal bright points got disturbed, but did not disappear (see also the euvi_195.mpg movie (electronic supplementary material)). This implies that closed magnetic fields in these configurations did not open. No irreversible changes are seen. No signatures of magnetic reconnection are detectable.

The plane-of-sky velocities of the presumable fronts systematically decreased, despite their propagation from the near-the-limb eruption site toward the solar disk center, which must increase the projected speed. The surface velocities estimated along an appropriate great circle all exceeded 390 km s^{-1} initially and all were less than 290 km s^{-1} finally, which indicates deceleration of the wave.

2.2. Global EUV Wave Fronts

We divided the problem of identifying the wave fronts into two tasks: *i*) identification of global fronts and *ii*) analysis of smaller-scale properties of the EUV wave propagation (next section). We reveal global wave fronts from ratios of running-difference images to preceding ones. To detect weak portions of the fronts, the images were rebinned to 512 × 512 pixels and deeply filtered using smoothing with a boxcar, a median smoothing, and displayed by means of the *sigrange* SolarSoftware routine. The result is shown in Figure 2 (eight of 12 images used with a total coverage of 55 min). We separately outlined the on-disk and off-limb parts of the fronts with red and pink ovals, trying to catch their outermost envelopes over a maximal spatial extent. The distances were measured along the green great circle. The technique used by Veronig *et al.* (2010) was more sensitive. They analyzed spatial profiles computed within some sectors and searched for their foremost edges close to the visually identified fronts. The blue contours approximately reproduce the fronts, which the authors identified.

Figure 2 shows the following facts.

i) The fronts identified by Veronig *et al.* all lead our fronts with increasing separation.

ii) The southern part of the front indicated by the arrows in Figure 2d, e moved considerably faster in the environment of the polar coronal hole, while the fronts themselves were difficult to detect there.

iii) The wave dome expanded non-radially: with the southeastern position of the eruption site, expansion of the dome was pointed almost exactly to the left in the plane of the sky. Also, the projection of the off-limb dome center onto the solar surface increasingly shifted northeast, so that the fronts in later images were not parallel to the earlier ones (*cf., e.g.,* Figure 2d and h).

The lag of the red fronts behind the blue ones is initially small and nearly constant, and then increases. Since the speed of the blue fronts was constant (Veronig *et al.*, 2010), this behavior implies deceleration of the red fronts. Indeed, our distance–time plots in Figures 7b, 8b (red symbols) show deceleration. The plots are well fitted with PL functions expected for a shock

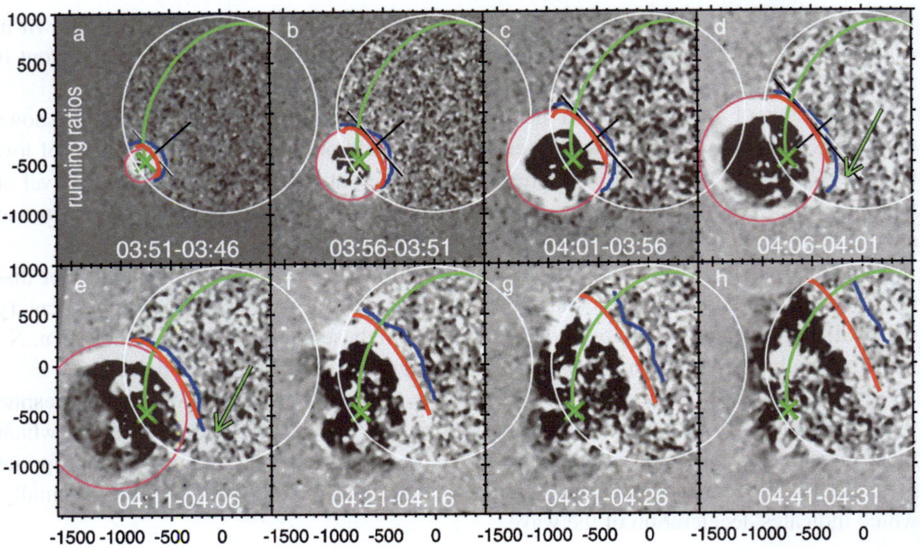

Figure 2 Global wave fronts in EUVI 195 Å images. The red arcs outline the leading edges of the global wave fronts detectable in the figure. The blue lines correspond to the fronts of Veronig *et al.* (2010). The pink ovals outline the off-limb wave dome. The arrows indicate a bend of the fronts into the South Pole's coronal hole. Distances along the solar surface were measured from the eruption cite (the slanted cross) along the green great circle. The white and black bars in panels a – d mark the presumable fronts from Figure 1.

wave (see Paper I):

$$x(t) = x_1\big[(t - t_0)/(t - t_1)\big]^{\alpha}, \quad (1)$$

where t and x are current time and distance, $t_0 = 03{:}47{:}48$ is the wave start time (estimated in sequential attempts to fit the EUV wave and the type II burst), t_1 and x_1 correspond to one of the measured fronts, and the PL exponent $\alpha = 2/(5 - \delta)$ with δ being a density falloff index in this formal approximation. We fitted the kinematics of the wave front with an exponent of $\alpha \approx 0.75$ ($\delta \approx 2.35$) for the surface propagation and $\alpha \approx 0.91$ ($\delta \approx 2.80$) for the off-limb expansion.

The measurements of the velocities along the great circle have the largest uncertainties at the earliest stages of the wave expansion, and for the initial interval of 15 min we also used 171 Å images observed with a higher imaging rate. The velocity corresponding to a power-law distance–time plot has a singularity in the origin t_0 and is not limited from below by V_{fast} at large distances. Hence, the δ parameter is expected to be slightly different for long and short time intervals beginning with t_0 (actually 55 min for 195 Å images and 15 min for 171 Å ones).

Figure 3 presents four of 12 EUVI 171 Å images which we used. Since the shock-PL fit applies, we used its parameters found from the 195 Å data as an initial estimate and endeavored to outline each of the on-disk and off-limb wave portions with ovals calculated from the shock-PL fit according to the observation times at 171 Å. The δ parameter was adjusted to match the fronts. If some parts of the fronts were not detectable, we used their other possible signatures. An extreme example is shown in Figure 3d. Here the reference regions for the off-limb oval were the upper (in the plane of the sky) brightening just above the limb and three faint lowermost compact regions. The on-disk oval was referred to the

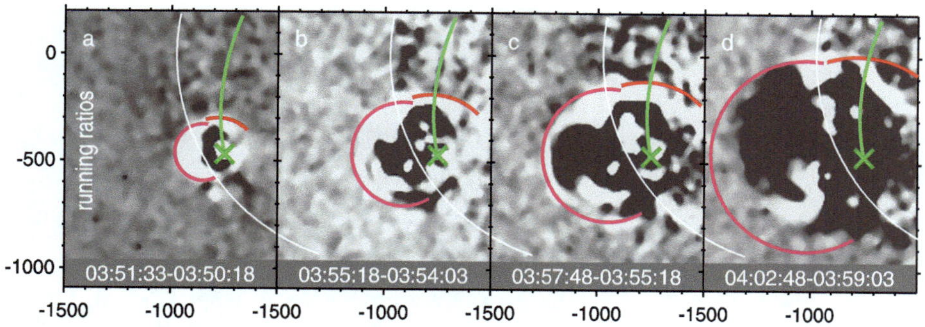

Figure 3 Global wave fronts detectable in EUVI 171 Å images and their outline. The red arcs outline the leading edges of the on-disk wave fronts detectable in the figure. The pink ovals outline the off-limb wave dome. The green arc denotes a great circle along which our measurements were made. The slanted green cross marks the eruption site.

bright feature crossing the limb and a small portion of the front next to the former feature. The results of the measurements shown in Figures 7b and 8b with blue triangles are fitted with $\delta = 2.74$ for the off-limb dome and $\delta = 2.1$ for the surface propagation (blue curve). The difference between δ found from the 195 Å and 171 Å images is detectable in the velocity–time plot in Figure 8c.

The measurements based on outline of entire wave fronts reveal some indistinct effects such as the motion of the wave center and provide tighter uncertainties than, *e.g.*, measurements of a fastest front portion do. Preconditioning with an expected analytic function still tightens the uncertainties. Estimating them is not a simple task. One way is to find the extreme envelopes enclosing possible options of the outline, but it is time consuming. We alternatively assumed extreme uncertainties of the front identification of 1.5 minor ticks (110 Mm) in the latest 195 Å images and twice lesser ones at 03:56. The uncertainty in t_0 estimated from the type II burst was assumed to be ± 30 s. The resulting gray scatter bands in Figure 8b, c for the surface distance and velocity plots *vs.* time correspond to $\delta = 2.35 \pm 0.05$. The assumed uncertainties of the front identification appear to be well overestimated; realistic bands should be considerably narrower.

2.3. EUV Wave Components and Fast-Mode Speed Distribution

Running-difference images are best suited to emphasize outermost fronts, but inner quasi-stationary features do not show up in such images. The EUV wave in this event is well visible in fixed-base ratio EUVI 195 Å images in Figure 4a – g allowing us to see what happened behind the expanding front. The whole large-scale brightening consisting of small patches was wide and complex. The outer propagating front included another, inner EUV transient. After an initial evolution, its on-disk part adjoining the dimming became stationary. Its brightness initially was comparable with the outer front and exceeded it later on. The inner component appears to have consisted of two parts distinguishable in Figure 4f, g and in the movie, with the northern part slowly moving northeast.

An off-limb brightening (HB1) visible up to 290 Mm above the inner component slowly expanded northwards along the limb for about half an hour following the outer front. This brightening could be due to a portion of the outer front sweeping over the backside part of the solar surface.

A high-altitude brightening HB2 (Figure 4f, g) at about 230 Mm appeared when the wave front reached it suggesting that a lower dense part of a coronal streamer highlighted

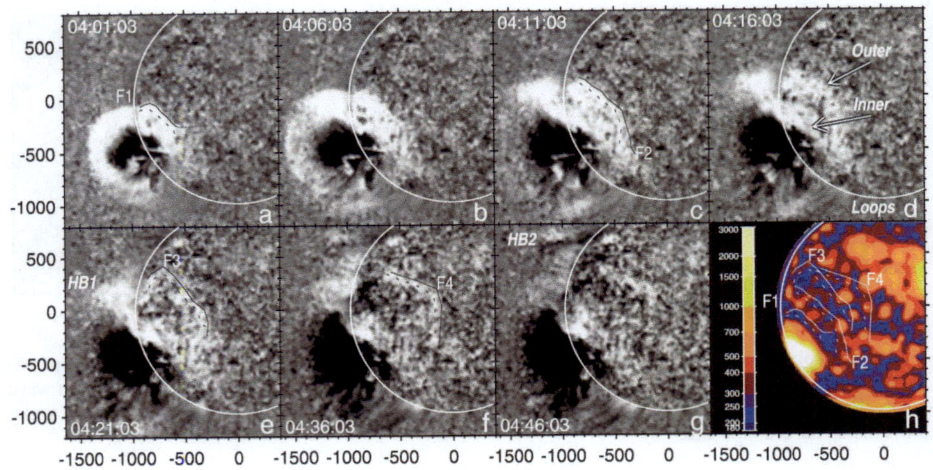

Figure 4 The EUV wave in fixed-base ratio EUVI 195 Å images (a – g) and V_{fast} distribution at 30 Mm (h). The white circles outline the solar limb. The stationary inner component and the propagating outer one are denoted in panel (d). Some portions of the wave fronts are outlined with black contours F1 – F4 at their foremost (solid) and brightest (dotted) parts. The scale bar in panel h quantifies the V_{fast} levels in km s^{-1}.

the front. Long loops connecting the active region with a southwestern area ('Loops' in Figure 4d) also highlighted the wave. Most of the loops outside of the active region did not show any stretching, although the wave front passed through these loops. These facts support a wave nature of the outer EUV wave. There are no manifestations of magnetic field opening aside of the eruption region, where, however, the outer EUV wave was visible. The limited magnetic field opening is also confirmed by the confined coronal dimming region in Figure 4 that was first stated by Veronig *et al.* (2010). Just the stationary inner brightening appears to be related to a stretched CME structure.

The conclusion of Yang and Chen (2010) that '... EIT wave propagates more slowly in the regions of stronger magnetic field' inspired us to compare near-surface EUV wave manifestations with the V_{fast} distribution ($V_{\mathrm{fast}}^2 \approx V_{\mathrm{A}}^2 + C_{\mathrm{s}}^2$ with the sound speed C_{s} is assumed to be 180 km s^{-1} everywhere). We calculate V_{fast} from the *magnitude* $|\mathbf{B}|$ of the magnetic field, which determines the Alfvén speed rather than any magnetic component. The magnetic field was extrapolated to 30 Mm from a SOLIS magnetogram observed at 19:30 on 20 January using potential approximation (Rudenko, 2001). A simplest way to obtain a V_{fast} distribution is to assume a constant temperature and to take densities, *e.g.*, from the Saito model. We attempt to get a somewhat more realistic density distribution using a SOHO/EIT 195 Å image observed on 20 January and an expression $\log n_{\mathrm{e}} = 8.34 + 0.509 \log I_{195}$ obtained by Brosius *et al.* (2002) in a study of a particular region (n_{e} is the electron density, I_{195} is the brightness in the 195 Å EIT channel). This expression cannot be universal, because the EUV brightness, $I_{195} \propto n_{\mathrm{e}}^2 L$, depends on the depth L. However, V_{A} depends on the depth weakly, $\propto L^{1/4}$, and we restricted the density above quiet regions by limiting plasma beta $\beta \leq 0.65$ (see, *e.g.*, Warmuth and Mann, 2005). The resulting V_{fast} distribution is presented in Figure 4h (the highest-speed values above the active region are limited by ≤ 3000 km s^{-1} to reveal low-speed regions throughout the solar disk). This distribution is not accurate for the following reasons. Most likely, the high-speed area in the active region was smaller on 17 January than the three-days later magnetogram shows. The density could be underestimated there, thus somewhat increasing V_{A}. We cannot untangle the height dependence of the density distribution from an EUV image of the solar disk. These inaccuracies are not essential for our results.

The shock formation time can be estimated from this distribution. A disturbance caused by an impulsive filament eruption steepens into a shock presumably in a region of a sharp falloff of V_{fast} due to jam of the wave profile. With a half-width of the high-speed area above the active region (white area in Figure 4h) of about 100 Mm, the shock must form in the lateral direction within one minute [consistent with the estimate of Grechnev *et al.* (2008) for a different event]. With the wave onset time $t_0 \approx$ 03:47:48, this estimate is consistent with the fact that the type II burst was observed as early as 03:51 indicating that the shock already existed in the upwards direction at that time.

The on-disk EUV wave was distinct in an area between the active region and a large northeastern high-speed domain. $V_{\text{fast}} = 290$ km s^{-1} dominated there. The surface EUV wave speed was from 300 – 325 km s^{-1} at 04:01 to 240 – 270 km s^{-1} (see Figure 8c) at its latest observation. Thus, the near-surface portion of the wave front was, most likely, in the weak shock regime in regions of low V_{fast} and propagated almost as a linear fast-mode wave in regions of increased V_{fast}.

Four distinct portions of the EUV wave fronts F1 – F4 are outlined both in EUVI images (Figure 4a, c, e, f) and on the V_{fast} distribution (Figure 4h). The solid lines trace the foremost fronts; the dotted lines trace their brightest parts. Comparison shows that the EUV wave was brightest in regions of lowest V_{fast}. The boundary of the EUV wave corresponded to regions, where V_{fast} increased. Portions of the fronts located in regions of increased V_{fast} stretched and lost brightness. Fronts sharpened, brightened and suspended motion in regions of low V_{fast}. Some suspended front portions are detectable in two or even more images. Practically the same front F4 persists in Figure 4f, g. The euvi_195.mpg movie shows other examples. The southern branch of front F2 is detectable at 04:06 – 04:16. The northern bend of front F3 is visible at 04:21 and 04:26. That is, the small-scale V_{fast} distribution did not determine the overall kinematics of the wave (see Figure 2), but affected the brightness and sharpness of the wave front. Indeed, the Mach number $M = V_{\text{shock}}/V_{\text{fast}}$ increases in regions of reduced V_{fast}, $\Delta M \approx -(M-1)\Delta V_{\text{fast}}/V_{\text{fast}}$, *i.e.*, the plasma compression is stronger.

These facts agree with the perturbation profile evolution revealed by Veronig *et al.* (2010) in averaging over a spherical sector of 60°: the profile initially increased in magnitude and sharpened until 04:01, and thereafter evolved in the reverse manner. Since the shock most likely appeared 10 min before 04:01, the observed steepening was not due to the shock formation. The sharpest and brightest front found by the authors at 04:01 corresponds to our front F1 (Figure 4a, h), which was mostly aligned with a deep valley in the V_{fast} distribution. Veronig *et al.* (2010) estimated the Mach number averaged over the F1 front to be 1.15; we estimate that locally it could reach 1.5. Later on, the shock probably dampened, as the authors concluded. Besides, the depth, homogeneity, and orientation of each subsequent valley relative to the wave front and to the measurement direction were not as favorable as in the first valley. Dispersion of the front over increasing width of the authors' measurement sector also increased.

Thus, kinematics of the shock wave was governed by the large-scale V_{fast} distribution: the wave ran much faster in the polar coronal hole and its environment. Conversely, when the wave expanded enough to exceed compact structures, the effect of the small-scale near-surface distribution of V_{fast} was more pronounced in sharpness and brightness of the EUV wave than in its local speed.

The two EUV wave components in this event remind us of two disturbances in Figure 7 from Chen, Fang, and Shibata (2005). One disturbance is a compressive effect due to opening magnetic fields during the CME lift-off as initially proposed by Delannée and Aulanier

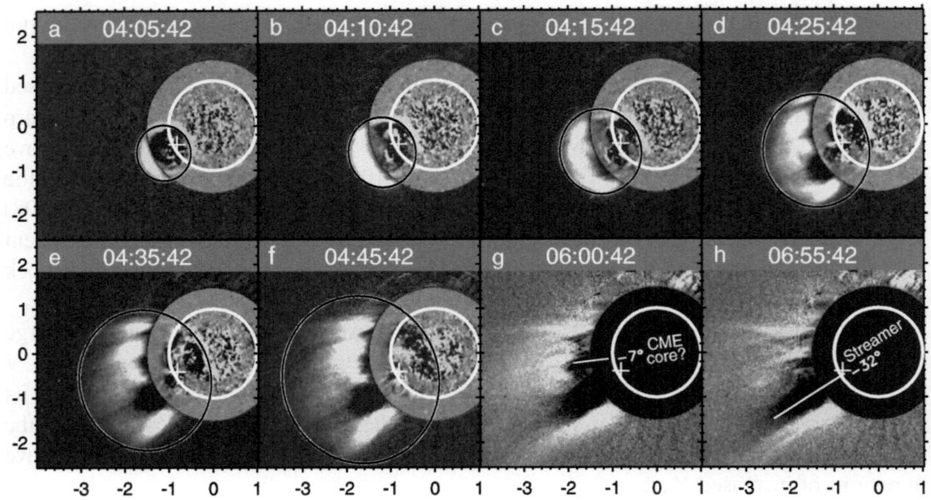

Figure 5 Coronal transient in STEREO-B/COR1 fixed-difference images with inserted EUV wave fronts in running-difference EUVI images (a – f) from Figure 2. Thick white circles denote the solar limb. Black-on-white ovals outline the CME edge according to the shock-PL fit. The cross marks the eruption site. The axes show distances from the solar disk center in solar radii.

(1999). This disturbance moves slowly and stops at a magnetic separatrix. Such influence of the CME lift-off on magnetic fields is expected to diminish at distances well exceeding the initial size of an eruptive magnetic rope suggested by the post-eruptive arcade and the major deep stationary dimmings nearby (if the CME does not involve magnetic fields connected to remote active regions). The second, a faster disturbance is a trail of a coronal shock wave sweeping over the solar surface. With the qualitative and quantitative properties of the outer propagating disturbance revealed by Veronig *et al.* (2010) and in this section, it is difficult to imagine an alternative to its interpretation as a trail of a coronal shock wave. The presence of both predicted wave and non-wave components of the EUV transient in this event offers a promising opportunity to settle debates over the nature of EUV waves.

2.4. White-Light Coronal Transient Observed with COR1 and LASCO/C3

Figure 5 shows eight STEREO-B/COR1 images of a coronal transient. An image observed at 03:50 was subtracted from all others. The ovals outlining the edge of the transient correspond to the shock-PL fit with the same t_0 = 03:47:48. To coordinate the ovals with the non-radially expanding transient (Section 2.2), their centers are increasingly shifted and different expansion factors are used in the radial direction $\delta_{\mathrm{rad}} = 2.80$ and the transversal one $\delta_{\mathrm{trans}} = 2.85$, *i.e.*, the front tended to become oblate. The ovals match the fronts in EUVI images and cling to the outermost edges of the transient. The leading edge decelerated as a freely propagating shock wave. The foremost part looks like a plasma flow streaming along the fan of coronal rays. CME structures are surmised well behind the leading edge. This picture suggests the plasma flow successively involved into the motion by a freely propagating shock front, whose speed was the phase velocity of the involvement. A structure at an angle of $-7°$ (PA = 97°) in Figure 5g, h might be the CME core. The density-depleted major streamer appears in difference images as a wide radial darkening around $-32°$ (PA = 122°).

We consider LASCO/C3 images only (C2 data became available later). Figure 6 presents four of 12 images, in which the CME is detectable. Probable frontal structure and core are

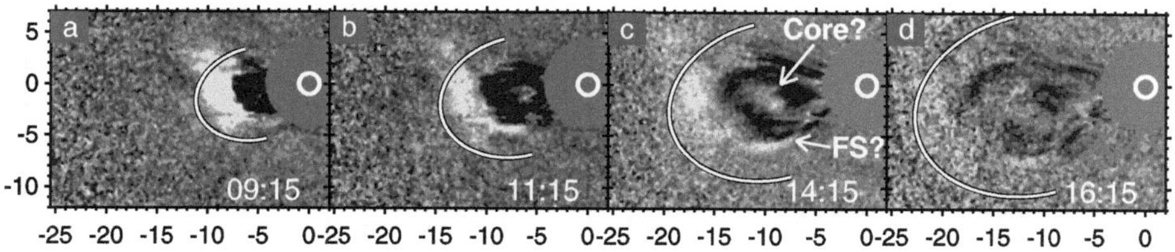

Figure 6 The CME observed with LASCO/C3 (running differences). The thick white circle denotes the solar limb. The white-on-black ovals outline the outermost CME edge according to the shock-PL fit. The axes show distances from the solar disk center in solar radii.

indicated in Figure 6c. A ragged, diffuse presumable plasma flow seems to be present ahead. We outlined the CME with an oval corresponding to expansion of the shock wave. The ovals in Figure 6 calculated according to the shock-PL fit with the same start time $t_0 = 03{:}47{:}48$ and $\delta = 2.80$ embrace the CME except for the fastest feature at PA $\approx 70°$, most likely of a non-wave nature. Rather poor observations and the low CME speed ($< 400\ \mathrm{km\,s^{-1}}$) do not allow us to find out if the change of its shape was due to effects of the shock propagation or acceleration of the solar wind. Nevertheless, the shock-wave kinematics does not contradict even LASCO/C3 observations up to $23 R_\odot$.

2.5. Expansion of the Wave Dome and the Type II Burst Drift Rate

Figure 7a shows a dynamic spectrum combined from HiRAS and Learmonth records. The type II burst had a single band most likely corresponding to the harmonic (twice the fundamental frequency), because the fundamental emission must be strongly refracted due to the far backside location of the eruption site. Veronig *et al.* (2010) came to the same conclusion. The drift of the burst is well outlined with the PL model $n = 5.5 \times 10^8 (h/100\ \mathrm{Mm})^{-2.8}$ and the same wave start time $t_0 = 03{:}47{:}48$. The dashed outline corresponds to a presumable fundamental emission.

We consider the direction $\vartheta = -7°$ matching the sharpest part of the EUV wave front non-radially expanding above the limb and a probable CME core in Figure 5g. Figure 7b shows measurements of the wave dome from EUVI 171 Å and 195 Å images along with a shock-PL fit of the 171 Å data and the frequency drift converted into the height–time plot. Figure 7c shows speed–time plots corresponding to the 171 Å fit and the type II burst. Figure 7d presents our PL density model, the Newkirk (1961) model, and the Saito (1970) model at $\phi \approx |\vartheta| = 7°$. The PL model is close to the 2-fold Newkirk model at the onset of the type II burst and later approaches the Saito model. The difference with the Newkirk model here, unlike the events addressed in Paper I, might be due to the non-radial wave expansion. The arrows show the height ranges corresponding to the type II burst and fields of view of coronagraphs.

Figure 7e shows an overall height–time plot including measurements from the SOHO LASCO CME Catalog (Yashiro *et al.*, 2004, http://cdaw.gsfc.nasa.gov/CME_list/) at PA = 97° up to $24 R_\odot$. To coordinate the measurements at different position angles from two different vantage points of STEREO-B and SOHO, all the distances from the CME Catalog are increased by 4.2%. The shock-PL fit with $\delta = 2.80$ corresponding to the Saito model matches all the data. The boundaries of the gray band in Figure 7b, e covering all the measurements in the Catalog correspond to $x_1 = 20 R_\odot \pm 5\%$, $\delta = 2.80 \pm 0.05$, and $t_0 = 03{:}47{:}48 \pm 30$ s in expression (1). This band presents the scatter of measurements in the CME Catalog in terms of uncertainties of the shock-PL fit.

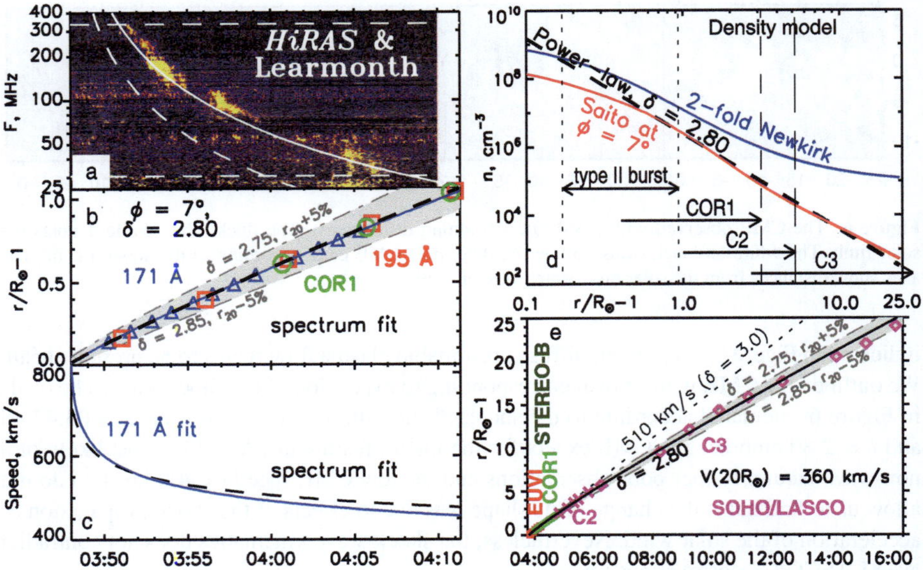

Figure 7 Off-limb expansion of the EUV wave and type II burst. (a) Composite dynamic spectrum. (b) Height–time measurements (symbols) from EUVI at 195 Å (red), 171 Å (blue), and COR1 (green); shock-PL fit (blue line) and fit of the type II burst converted into heights (dashed black). (c) Speed–time plots calculated from the shock-PL fit of the 171 Å data (blue) and from the dynamic spectrum (dashed black). (d) Coronal density models: PL model with $\delta = 2.8$ (dashed black) fitting the dynamic spectrum and models of Newkirk (blue) and Saito for $\phi = 7°$ (red). (e) Overall height–time plot including the CME Catalog data (pink) embraced by the gray band (also shown in panel (b)). The dotted line is a linear fit of data in panel (b).

Deceleration of the wave is not pronounced within $1.1\,R_\odot$ (Figure 7b). This explains why Veronig *et al.* (2010) found a constant wave speed of 650 km s^{-1} in the radial direction ($\vartheta \approx -32°$). The linear-fit speed in the direction $\vartheta \approx -7°$ was 510 km s^{-1}. Deceleration of the wave becomes detectable from COR1 measurements up to $3R_\odot$ from the eruption site. In fact, this determines the measurement accuracy of δ ($\delta = 3.0$ for the constant speed). The dotted line in Figure 7e is a constant-speed plot extrapolating the linear fit in Figure 7b.

The fact that the shock-PL fit matches the expansion of the slow coronal transient up to $24R_\odot$, where its speed became comparable with the solar wind speed, suggests that the leading wave and the trailing piston were not completely independent. Synergy between the piston and wave discussed in Paper I (Section 4.3) is indeed expected to become pronounced at large distances from the Sun.

3. Discussion

The detailed STEREO/SECCHI observations of the EUV wave allow us to compare the results of the shock-PL fit proposed in Paper I with those of the weak shock modeling described in Paper II (hereafter WS model). The EUV wave propagated mainly over quiet-Sun regions without large-scale features except for the polar coronal hole. Since the EUV wave most likely was a near-surface trail of a large-scale coronal MHD wave, its kinematics should not be significantly affected by small-scale inhomogeneities, as the observations confirm (Section 2.3). We describe the global propagation of the EUV wave outside of the active region assuming only radial variations of coronal plasma parameters. The on-disk

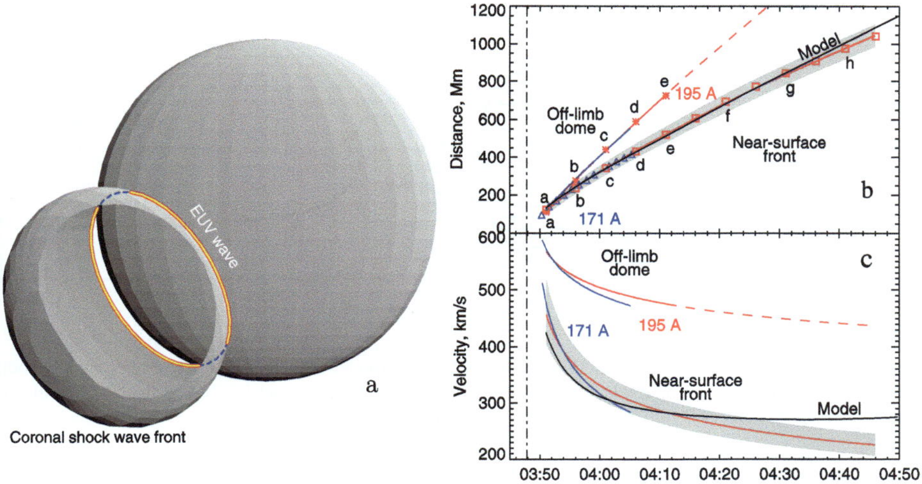

Figure 8 Measurements, fit, and modeling of shock front propagation. (a) The modeled shock front. (b) Distance–time plots of the on-disk wave (measured along the great circle in Figure 2) and the off-limb dome, their shock-PL fit, and a modeled plot (195 Å red, 171 Å blue, model black; the labels a – h denote the corresponding frames in Figure 2). (c) The same for the calculated velocities. The gray bands present extreme uncertainties discussed in Section 2.2.

EUV wave decelerated from $\gtrsim 390$ to $\lesssim 290\ \mathrm{km\,s^{-1}}$ (Section 2.1); Veronig *et al.* (2010) found broadening of the wave profile. These facts along with estimates of V_{fast} in the lower corona above the quiet Sun indicate that the shock was weak to moderate, so that the WS model appears to apply.

The model is not yet able to incorporate coronal magnetic fields extrapolated from real magnetograms. We therefore model kinematics of only an on-disk wave running over the quiet Sun. We use the barometric density falloff of isothermal coronal plasma $n(r) = 4 \times 10^8 \exp\{9.71\,(R_{\odot}/r - 1)\}\ \mathrm{cm^{-3}}$ with coronal temperature $T = 1.5$ MK ($C_{\mathrm{s}} = 180\ \mathrm{km\,s^{-1}}$), and the radial magnetic field model $B_r = 1.35\,(R_{\odot}/r)^2$ G. $V_{\mathrm{A}} = 170\ \mathrm{km\,s^{-1}}$ at 40 Mm and increases upwards. We assume that the wave originates at an initial surface, whose spatial extent is 200 Mm, and inside which the wave source is located. Then an EUV wave front can be observed at 03:51. The model shock wave has an initial length of 80 Mm and an amplitude of $1.5\,V_{\mathrm{fast}\,0}$ ($V_{\mathrm{fast}\,0}$ corresponds to the source height of 80 Mm). We search for EUV signatures of the coronal wave at a height of 40 Mm.

Figure 8 shows some results of the modeling and the measurements. Figure 8a presents the computed 3D shock front. The color rim is the section of the wave dome at 40 Mm. The on-disk EUV front could be partly covered by the dome. Figure 8b shows the distance–time plots of the on-disk EUV front measured at 195 Å (red squares) and 171 Å (blue triangles). The red and blue lines show the corresponding shock-PL fit; the black line presents model results. Figure 8c shows the velocity–time plots obtained by differentiating of the shock-PL curves and the modeled plot. The EUV wave appreciably decelerates due to damping and then slightly accelerates because of an increasing tilt of the front to the solar surface that is discussed in Paper II. The deviation of the speed supplied by the shock-PL fit from the result of the WS modeling does not exceed 15%.

We have also modeled propagation of a shock wave upwards. Active regions determine a V_{A} distribution in their vicinities. To simulate this effect, we have added a radial magnetic dipole into our radial magnetic field model as Warmuth and Mann (2005) did. A 'horizontal'

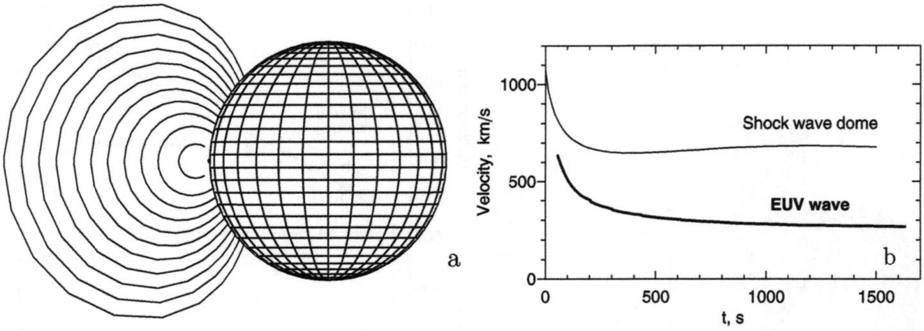

Figure 9 A freely propagating weak shock wave in the WS model containing an active region. (a) Shock fronts separated by 2.5-min intervals. Note the progressive rise of the geometric wave center. (b) Calculated shock front speeds upwards (thin) and along the solar surface (thick).

dipole seems to conform to the active region on 17 January. Embedding such a dipole into the model results in a strongly anisotropic V_{fast} distribution in the corona with a domain of very low $V_{\mathrm{fast}} \approx C_{\mathrm{s}}$ near a null point of the magnetic field as well as a domain of enhanced V_{fast}. This causes asymmetric wave front propagation actually observed in this event. However, the domain of influence of a 'horizontal' dipole is too large, comparable with the solar hemisphere, whereas an estimate from the extrapolated magnetic field shows it to be rather compact along the solar surface ($\lesssim$260 Mm). Therefore, we employ the 'parallel' dipole of Warmuth and Mann (2005), which provides a compact domain of enhanced V_{A}. We adjust the height falloff of the magnetic field above the active region following Gary (2001), but decrease the magnetic field strength to obtain a realistic V_{fast} distribution with model parameters used.

Figure 9 shows model results. The wave source is located above the limb in the equatorial plane. The shock front is oblate in the radial direction presumably due to predominant upwards increase of V_{fast}: ray trajectories are refracted into regions of lower V_{fast}. The effect agrees with the COR1 observations (Figure 5). The speed of the upwards wave expansion is about twice higher than that of the on-disk EUV wave (Figure 9b). This confirms our suggestion in Paper I to overcome the absence of correlation between the speeds of EUV waves and exciters of type II bursts stated by Klassen *et al.* (2000).

The twice-higher upwards speed of the EUV wave relative to the on-disk one prompted Veronig *et al.* (2010) to state that the upward dome expansion was driven all the time by the CME. The authors mentioned that the upward – lateral speed difference could be due to direction-dependent falloffs of V_{fast}, but preferred the CME-driven option seemingly favored by the limited lateral extent of the dimming. However, the latter fact only means that CME-related opening magnetic fields occurred in a limited region and did not involve remote regions. The main expansion of CMEs is radial, but this fact does not guarantee that all CME-associated shocks are driven continuously (see Paper I).

The speed difference in Figure 9 was obtained for a *freely propagating wave* and the direction-dependent V_{fast} above the active region. The front shapes match the observations. The results agree with our considerations and measurements in Paper I and support the scenario of an impulsively generated freely propagating weak shock wave (see also Pomoell, Vainio, and Kissmann, 2008).

The WS modeling explains the disaccord between the EUV wave fronts identified by us and Veronig *et al.* (2010): our red fronts in Figure 2 lag behind the blue ones identified by the authors. The difference is most likely due to a projection effect combined with a different

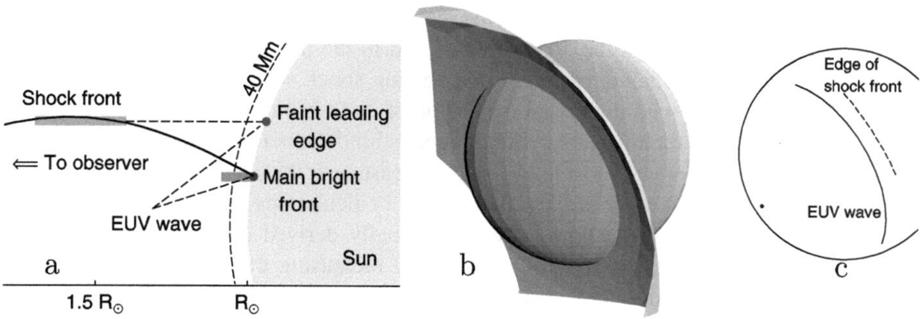

Figure 10 Modeling the situation at about 04:30. (a) The relation between the foremost edge of the EUV wave and its main part. Thick horizontal bar shows cross section of the wave front presumably contributing to the foremost edge of the wave detected by Veronig *et al.* (2010). (b) A portion of the modeled shock front. (c) The lower edge (solid) of the shock front shown in panel (b) and a projection of the faint front's foremost edge (dashed) on the solar surface.

sensitivity of measurements as Figure 10 explains. Plasma is compressed by the shock front over the whole its surface. The largest column emission measure of the compression region is near the solar surface, where the plasma density is higher. Veronig *et al.* (2010) probably detected a high-altitude outermost edge of the convex wave front. Figure 10a demonstrates the calculated 2D cross section of the wave front with its outermost edge at a height of $\approx 0.5 R_{\odot}$. Figure 10b presents a portion of the modeled wave front. Figure 10c shows the calculated on-disk projections of the faint leading edge and the main bright EUV wave front corresponding to about 04:30. The situation resembles the seemingly disaccord between the results of Warmuth *et al.* (2004) and White and Thompson (2005) discussed in Paper I for a similar reason, *i.e.*, the convex shape of the wave front and its tilt toward the solar surface.

4. Summary and Concluding Remarks

Our analysis has confirmed the major conclusion of Veronig *et al.* (2010) that both the on-disk EUV wave and the dome expanding above the limb were due to a coronal shock wave. In addition to the authors' arguments, we have established that

i) the front shape and its changes,
ii) kinematics of both the on-disk front and the off-limb dome up to $24 R_{\odot}$, and even
iii) the difference between our and the authors' measurements all corresponded to the expected propagation of a shock wave.

We have also found that, in agreement with the shock-wave hypothesis, kinematics of the global wave front

iv) corresponded to the drift rate of the type II burst and
v) was controlled by large-scale distribution of the fast-mode speed, while its local inhomogeneities affected the brightness and sharpness of the EUV wave, *e.g.*, it was brightest *in loci* of the fast-mode speed minima.

We do not see any support to the presumption of Veronig *et al.* (2010) that the shock wave was driven by the CME all the time. On the contrary, we consider the shock wave to be excited by an impulsively erupting magnetic rope structure and to be freely propagating afterwards like a decelerating blast wave. This scenario has been argued for and observationally confirmed in Paper I. All the conclusions listed in the preceding paragraph are based on considerations and modeling of freely propagating shock waves. The free wave propagation is also consistent with the fastest expansion of its front in EUVI images in the radial direction, while CME structures apparently lagged behind the wave front.

The shock in this event was most likely weak, at least, near the solar surface, in agreement with the conclusion of Veronig *et al.* (2010). Model calculations for a weak shock match observations. Nevertheless, the power-law fit (formally derived under the assumption of a strong self-similar shock wave with continuously increasing mass) provides reasonable results starting from the early shock appearance up to the latest detectable signatures of the on-disk EUV wave, and even up to distances $> 20 R_{\odot}$ from the Sun, although with somewhat variable parameters.

We have additionally revealed another large-scale EUV brightening, which was quasi-stationary. No manifestations of magnetic field opening were found outside of the eruption region, while the propagating on-disk EUV wave was well visible there. The presence in this event of the two different EUV components predicted by models offers a promising opportunity to reconcile conflicting opinions about the nature of "EUV waves": the propagating EUV wave was of a shock-wave nature for sure, and the quasi-stationary EUV transient was presumably associated with a stretching CME structure.

We specify the conclusion of Veronig *et al.* (2010) that the dome observed in white light was not the CME. Indeed, the leading part most likely was not a magnetoplasma CME component. Coronagraph images, their shock-PL fit, and our considerations indicate that this was a plasma flow successively involved into the motion by the freely propagating shock front. The plasma flow was slower than the shock front, whose speed was the phase velocity of this involvement. Thus, the leading part of the transient was a plasma flow, *i.e.*, a *coronal mass ejection* by definition, but it was a *shock-driven plasma flow*.

Acknowledgements We thank M. Temmer, A. Warmuth, and P.-F. Chen for fruitful discussions and S. Kalashnikov for the assistance in data processing. We thank an anonymous reviewer for useful remarks. We thank the teams operating all instruments whose data are used here for their efforts and open data policies: the ESA & NASA SOHO/EIT & LASCO and STEREO/SECCHI telescopes; the NICT HiRAS (Japan), the IPS Radio and Space Services Learmonth Observatory (Australia), and the USAF RSTN radio telescopes. We appreciatively use the CME catalog generated and maintained at the CDAW Data Center by NASA and the Catholic University of America in cooperation with the Naval Research Laboratory. SOLIS data used here are produced cooperatively by NSF/NSO and NASA/LWS. The research was supported by the Russian Foundation of Basic Research under grant 09-02-00115.

References

Afanasyev, A.N., Uralov, A.M.: 2011, *Solar Phys.* doi:10.1007/s11207-011-9730-9. this issue.
Attrill, G.D.R.: 2010, *Astrophys. J.* **718**, 494.
Attrill, G.D.R., Harra, L.K., van Driel-Gesztelyi, L., Démoulin, P.: 2007, *Astrophys. J. Lett.* **656**, L101.
Biesecker, D.A., Myers, D.C., Thompson, B.J., Hammer, D.M., Vourlidas, A.: 2002, *Astrophys. J.* **569**, 1009.
Brosius, J.W., Landi, E., Cook, J.W., Newmark, J.S., Gopalswamy, N., Lara, A.: 2002, *Astrophys. J.* **574**, 453.
Brueckner, G.E., Howard, R.A., Koomen, M.J., Korendyke, C.M., Michels, D.J., Moses, J.D., *et al.*: 1995, *Solar Phys.* **162**, 357.
Chen, P.F., Fang, C., Shibata, K.: 2005, *Astrophys. J.* **622**, 1202.
Chen, P.F., Wu, S.T., Shibata, K., Fang, C.: 2002, *Astrophys. J. Lett.* **572**, L99.
Cohen, O., Attrill, G.D.R., Manchester, W.B. IV, Wills-Davey, M.J.: 2009, *Astrophys. J.* **705**, 587.
Delaboudinière, J.-P., Artzner, G.E., Brunaud, J., Gabriel, A.H., Hochedez, J.F., Millier, F., *et al.*: 1995, *Solar Phys.* **162**, 291.

Delannée, C., Aulanier, G.: 1999, *Solar Phys.* **190**, 107.
Gallagher, P.T., Long, D.M.: 2010, *Space Sci. Rev.* **127**. doi:10.1007/s11214-010-9710-7.
Gary, G.A.: 2001, *Solar Phys.* **203**, 71.
Gopalswamy, N., Yashiro, S., Temmer, M., Davila, J., Thompson, W.T., Jones, S., McAteer, R.T.J., Wuelser, J.-P., Freeland, S., Howard, R.A.: 2009, *Astrophys. J. Lett.* **691**, L123.
Grechnev, V.V., Uralov, A.M., Slemzin, V.A., Chertok, I.M., Kuzmenko, I.V., Shibasaki, K.: 2008, *Solar Phys.* **253**, 263.
Grechnev, V.V., Uralov, A.M., Chertok, I.M., Kuzmenko, I.V., Afanasyev, A.N., Meshalkina, N.S., Kalashnikov, S.S., Kubo, Y.: 2011, *Solar Phys.* doi:10.1007/s11207-011-9780-z.
Howard, R.A., Moses, J.D., Vourlidas, A., Newmark, J.S., Socker, D.G., Plunkett, S.P., Korendyke, C.M., Cook, J.W., *et al.*: 2008, *Space Sci. Rev.* **136**, 67.
Kaiser, M.L., Kucera, T.A., Davila, J.M., St. Cyr, O.C., Guhathakurta, M., Christian, E.: 2008, *Space Sci. Rev.* **136**, 5.
Khan, J.I., Aurass, H.: 2002, *Astron. Astrophys.* **383**, 1018.
Kienreich, I.W., Temmer, M., Veronig, A.M.: 2009, *Astrophys. J. Lett.* **703**, L118.
Klassen, A., Aurass, H., Mann, G., Thompson, B.J.: 2000, *Astron. Astrophys. Suppl. Ser.* **141**, 357.
Liu, W., Nitta, N.V., Schrijver, C.J., Title, A.M., Tarbell, T.D.: 2010, *Astrophys. J. Lett.* **723**, L53.
Long, D.M., Gallagher, P.T., McAteer, R.T.J., Bloomfield, D.S.: 2008, *Astrophys. J. Lett.* **680**, L81.
Newkirk, G. Jr.: 1961, *Astrophys. J.* **133**, 983.
Patsourakos, S., Vourlidas, A.: 2009, *Astrophys. J. Lett.* **700**, L182.
Patsourakos, S., Vourlidas, A., Wang, Y.M., Stenborg, G., Thernisien, A.: 2009, *Solar Phys.* **259**, 49.
Pomoell, J., Vainio, R., Kissmann, R.: 2008, *Solar Phys.* **253**, 249.
Rudenko, G.V.: 2001, *Solar Phys.* **198**, 5.
Saito, K.: 1970, *Ann. Tokyo Astron. Obs.* **12**, 53.
Thompson, B.J., Plunkett, S.P., Gurman, J.B., Newmark, J.S., St. Cyr, O.C., Michels, D.J.: 1998, *Geophys. Res. Lett.* **25**, 2465.
Thompson, B.J., Gurman, J.B., Neupert, W.M., Newmark, J.S., Delaboudinière, J.-P., St. Cyr, O.C., Stezelberger, S., Dere, K.P., *et al.*: 1999, *Astrophys. J. Lett.* **517**, L151.
Veronig, A.M., Temmer, M., Vršnak, B.: 2008, *Astrophys. J. Lett.* **681**, L113.
Veronig, A.M., Muhr, N., Kienreich, I.W., Temmer, M., Vršnak, B.: 2010, *Astrophys. J. Lett.* **716**, L57.
Warmuth, A., Mann, G.: 2005, *Astron. Astrophys.* **435**, 1123.
Warmuth, A., Vršnak, B., Aurass, H., Hanslmeier, A.: 2001, *Astrophys. J. Lett.* **560**, L105.
Warmuth, A., Vršnak, B., Magdalenić, J., Hanslmeier, A., Otruba, W.: 2004, *Astron. Astrophys.* **418**, 1101.
Wills-Davey, M.J., Attrill, G.D.R.: 2009, *Space Sci. Rev.* **149**, 325.
White, S.M., Thompson, B.J.: 2005, *Astrophys. J. Lett.* **620**, L63.
Yang, H.Q., Chen, P.F.: 2010, *Solar Phys.* **266**, 59.
Yashiro, S., Gopalswamy, N., Michalek, G., St. Cyr, O.C., Plunkett, S.P., Rich, N.B., Howard, R.A.: 2004, *J. Geophys. Res.* **109**, A07105.
Zhukov, A.N., Auchère, F.: 2004, *Astron. Astrophys.* **427**, 705.
Zhukov, A.N., Rodriguez, L., de Patoul, J.: 2009, *Solar Phys.* **259**, 73.

Solar Phys (2011) 273:479–491
DOI 10.1007/s11207-011-9730-9

Coronal Shock Waves, EUV Waves, and Their Relation to CMEs. II. Modeling MHD Shock Wave Propagation Along the Solar Surface, Using Nonlinear Geometrical Acoustics

A.N. Afanasyev · A.M. Uralov

Received: 22 November 2010 / Accepted: 15 February 2011 / Published online: 31 March 2011

Abstract We model the propagation of a coronal shock wave, using nonlinear geometrical acoustics. The method is based on the Wentzel–Kramers–Brillouin (WKB) approach and takes into account the main properties of nonlinear waves: *i*) dependence of the wave front velocity on the wave amplitude, *ii*) nonlinear dissipation of the wave energy, and *iii*) progressive increase in the duration of solitary shock waves. We address the method in detail and present results of the modeling of the propagation of shock-associated extreme-ultraviolet (EUV) waves as well as Moreton waves along the solar surface in the simplest solar corona model. The calculations reveal deceleration and lengthening of the waves. In contrast, waves considered in the linear approximation keep their length unchanged and slightly accelerate.

Keywords Magnetohydrodynamics · Waves, propagation · Waves, shock

1. Introduction

Many solar eruptive events appear to initiate magnetohydrodynamic (MHD) waves in the corona. This conjecture is supported by observations of their manifestations. First, these are waves visible in chromospheric spectral lines. For several decades, Moreton waves have been known (Moreton and Ramsey, 1960) observed in the Hα line. Uchida (1968) proposed that a Moreton wave represented the chromospheric trail of a coronal fast-mode wave. More recent studies show the possible coronal nature of Moreton waves (see, *e.g.*, Balasubramaniam, Pevtsov, and Neidig, 2007). Similar phenomena are propagating waves observed in the He I 10 830 Å line (see, *e.g.*, Vršnak *et al.*, 2002b; Gilbert *et al.*, 2004). Second, type

Energy Storage and Relase through the Solar Activity Cycle – Models Meet Radio Observations
Guest Editors: Christophe Marqué and Alexander Nindos.

A.N. Afanasyev (✉) · A.M. Uralov
Institute of Solar-Terrestrial Physics SB RAS, P.O. Box 291, Lermontov St. 126A, Irkutsk 664033, Russia
e-mail: afa@iszf.irk.ru

A.M. Uralov
e-mail: uralov@iszf.irk.ru

II radio bursts are considered to be signatures of coronal shock waves propagating upwards in the corona (Uchida, 1960). Furthermore, presumable signatures of coronal waves can be observed in soft X-rays (Narukage *et al.*, 2002; Khan and Aurass, 2002; Hudson *et al.*, 2003; Warmuth, Mann, and Aurass, 2005), microwaves (Warmuth *et al.*, 2004a; White and Thompson, 2005), and metric radio-wavelengths (Vršnak *et al.*, 2005).

Other phenomena that can be manifestations of MHD waves are large-scale wave-like disturbances discovered in 1998 with the Extreme-ultraviolet Imaging Telescope (EIT) and referred to as "EIT waves" or "extreme-ultraviolet (EUV) waves". This term appears to include phenomena having different physical nature and therefore various morphological and dynamical properties. For more details, we refer the reader to the Introduction of an accompanying paper (Grechnev *et al.*, 2011b). Note here that describing EUV waves in terms of fast-mode MHD waves seems to be possible and correct for those cases when we really deal with "wave" phenomena. Just this class of EUV transients is the subject of our consideration.

Thus, fast-mode MHD waves are responsible for a number of solar transients and it is important to describe their propagation through the corona. But some questions about modeling the waves remain.

In his original approach, Uchida (1968) modeled a coronal disturbance as a linear and short fast-mode MHD wave propagating from a point source. To calculate a dome-like surface of the wave front and find the position of a Moreton wave (as an intersection line of the dome and the solar surface), he used the linear geometric acoustics (the Wentzel–Kramers–Brillouin (WKB) approach). Some recent studies (see, *e.g.*, Wang, 2000; Patsourakos *et al.*, 2009) also used the same approach under the assumption of a linear disturbance. In this case, the disturbance moves along rays, which curve into regions of a reduced Alfvén speed. This results in the appearance of wave "imprints" (*e.g.*, Moreton waves) running along the spherical solar surface. Note that in the linear approximation the amplitude and duration of the wave do not affect the shape of the wave front and the speed of its motion along the rays. Neither amplitude nor duration of the wave were calculated in the mentioned papers; the geometry of the propagation was their only interest. Henceforth in the paper, we consider the wave amplitude to be the perturbation magnitude of the plasma velocity in the wave.

Uchida's model of a linear MHD wave has demonstrated the possibility to describe Moreton waves in these terms. Later papers by Uchida with collaborators (Uchida, Altschuler, and Newkirk, 1973; Uchida, 1974) and recent papers (Wang, 2000; Patsourakos *et al.*, 2009), in which the coronal magnetic field was calculated from photospheric magnetograms, provided a more accurate quantitative description of Moreton and EUV waves in terms of the linear model.

However, the linear model predicts acceleration of Moreton and EUV waves, whereas observations show their systematic deceleration (Warmuth *et al.* 2001; 2004a). Also, it is often pointed out by many authors that the speeds of coronal waves sometimes well exceed the fast-mode ones (Narukage *et al.* 2002, 2004; Vršnak *et al.*, 2002a; Warmuth *et al.*, 2004b; Muhr *et al.*, 2010). These facts suggest that the linear approximation does not always correctly describe the propagation of those waves. Probably, the disturbance responsible for the transients listed above is nonlinear and most likely is a shock wave.

Studies of the propagation of shock waves meet difficulties due to their nonlinearity. Analytic methods to describe the propagation of shock waves are approximate and often describe the behavior of some extreme classes of nonlinear waves (*e.g.*, very strong self-similar waves, or weak waves, *etc.*). The present paper is devoted to the case of a weak shock wave that appears to be the most acceptable (see, *e.g.*, Warmuth *et al.*, 2004b; Vršnak *et al.*, 2002a). We do not discuss the appearance of a shock wave. Vršnak and Lulić (2000) approximately described this process, based on an analogy with an accelerating flat piston. The cases of cylindrical and spherical pistons were analyzed by Žic *et al.* (2008). Temmer *et al.* (2009) demonstrated how one could describe the kinematics of a Moreton wave by using the solution of a simple wave without a discontinuity. In our study, we assume that a fast-mode shock wave of moderate intensity appears during a solar eruption on the periphery of an active region and decays to a weak shock when traveling in the corona. The wave manifests as a Moreton wave and an EUV wave on the solar disk. We calculate the propagation of the shock wave in terms of the WKB approach taking into account nonlinear effects.

Such a method in its generally accepted variant involves two independent procedures. In the first one, ray trajectories corresponding to the linear approximation as well as cross sections of ray tubes are calculated. So, the influence of the finite wave amplitude on the wave front shape and ray trajectories is ignored because in linear acoustics the propagation velocities of disturbances are equal to the undisturbed sound speed regardless of their amplitude. In the second procedure, the nonlinear variation of the wave amplitude and duration are computed along the linear rays obtained. Such a non-self-consistent approach is fairly useful in some cases, however, nonlinear effects altering the ray pattern and the wave velocity disappear completely. Therefore, we develop another method, which allows us to consider self-consistently wave propagation and the nonlinear variation of wave characteristics.

The method is described in Section 2. In Section 3 we formulate the problem and present results of the analytic propagation model of Moreton and EUV waves along the solar surface. Section 4 contains concluding remarks about the method and the results. We note that those who are not interested in the mathematical details of the method can read Section 3 without going through Section 2.

2. Method

The method of nonlinear geometrical acoustics is based on the method of linear geometrical acoustics and allows one to calculate the propagation of disturbances with small (but finite) amplitudes through an inhomogeneous medium. The linear geometrical acoustics is known to be a method to calculate linear disturbances in the ray approximation (see, *e.g.*, Landau and Lifshitz, 1987). In this approximation, a solution is found in the form of $A(\mathbf{r},t)\mathrm{e}^{\mathrm{i}\Psi(\mathbf{r},t)}$ where $A(\mathbf{r},t)$ is the wave amplitude, and $\Psi(\mathbf{r},t)$ is the eikonal, both depending on coordinates and time. By substituting this representation for wave perturbations into the system of linearized equations of ideal magnetohydrodynamics, one can obtain a Hamilton–Jacobi partial differential equation for the eikonal of fast and slow magnetosonic waves:

$$\frac{\partial\Psi}{\partial t} + (\mathbf{V}\mathrm{grad}\Psi) + a|\mathrm{grad}\Psi| = 0,$$

where $\mathbf{V}$ is the undisturbed plasma flow velocity (*e.g.*, the solar wind velocity), and a is the magnetosonic speed in plasma. Solving the equation with the method of characteristics gives a system of ray equations, which in spherical coordinates (r,θ,φ) takes the form (Uralova and Uralov, 1994)

$$\frac{\mathrm{d}r}{\mathrm{d}t} = V_r + a\frac{k_r}{k} + k\frac{\partial a}{\partial k_r},$$

$$r\frac{\mathrm{d}\theta}{\mathrm{d}t} = a\frac{k_\theta}{k} + k\frac{\partial a}{\partial k_\theta},$$

$$
\begin{aligned}
r\sin\theta\frac{\mathrm{d}\varphi}{\mathrm{d}t} &= a\frac{k_\varphi}{k} + k\frac{\partial a}{\partial k_\varphi}, \\
\frac{\mathrm{d}k_r}{\mathrm{d}t} &= -\frac{\partial V_r}{\partial r}k_r - \frac{\partial a}{\partial r}k + \frac{a}{kr}\left(k_\theta^2 + k_\varphi^2\right), \\
r\frac{\mathrm{d}k_\theta}{\mathrm{d}t} &= -\frac{\partial V_r}{\partial \theta}k_r - \frac{\partial a}{\partial \theta}k + \frac{a}{k}k_\varphi^2\cot\theta - k_\theta\frac{\mathrm{d}r}{\mathrm{d}t}, \\
r\sin\theta\frac{\mathrm{d}k_\varphi}{\mathrm{d}t} &= -\frac{\partial V_r}{\partial \varphi}k_r - \frac{\partial a}{\partial \varphi}k - \sin\theta k_\varphi\frac{\mathrm{d}r}{\mathrm{d}t} - k_\varphi r\cos\theta\frac{\mathrm{d}\theta}{\mathrm{d}t},
\end{aligned}
\tag{1}
$$

where $k_{r,\theta,\varphi}$ are the components of the wave vector $\mathbf{k} = \mathrm{grad}\Psi$, and k is its magnitude. The system of equations (1) corresponds to the case of a medium in steady-state, where only the radial component V_r of the undisturbed plasma flow exists. By integrating the system (1), one can determine the wave front shape.

This approach has been used for modeling coronal fast-mode MHD waves (Uchida, 1968; Wang, 2000), with only the ray pattern being calculated. However, it is essential not only to find the wave geometry, but also to calculate the wave intensity. The geometrical acoustics allows the wave amplitude variation to be calculated.

In the linear geometrical acoustics, the energy flux of a disturbance traveling in a stationary medium with group velocity $\mathbf{q}_0$ is directed along the rays, and its magnitude is conserved within a ray tube (Blokhintsev, 1981), $\mathrm{div}(\Delta\varepsilon\mathbf{q}_0) = 0$, where $\Delta\varepsilon$ is the average density of the disturbance energy. In a medium moving at a velocity of $\mathbf{V}$, we have to take into account the fact that the wave front phase velocity varies as $q_n = V_n + a$, with the n index denoting the projection normal to the front. The conservation law in this case is $\mathrm{div}(\Delta\varepsilon\mathbf{q}\,q_n/a) = 0$ (Uralov, 1982; Barnes, 1992) where $\mathbf{q} = \mathbf{V} + \mathbf{q}_0$ is the group velocity in a moving medium. The average density of the disturbance energy is $\Delta\varepsilon = \rho(u^2 + v^2)$ where ρ is the undisturbed plasma density, and u, v are the plasma velocity components along the normal to the wave front and across it, respectively. Taking into account the relation between the plasma velocity components $\mu = v/u$ (Kulikovsky and Lyubimov, 2005), it is possible to relate the variation of the wave amplitude to the normal cross section $\mathrm{d}S$ of the ray tube formed by a bundle of rays:

$$
\mathrm{d}S\,q\rho u^2\left(1 + \mu^2\right)\frac{q_n}{a} = \mathrm{const.} \tag{2}
$$

Thus, the general approach to determine the wave amplitude in the ray method relies on calculating the cross section of the ray tube.

There are various techniques for calculating cross sections. They are discussed in Kravtsov and Orlov (1990). We calculate cross sections by using the Jacobian of the transformation to ray coordinates. The volume element $\mathrm{d}W$ of the ray tube is expressed in terms of ray coordinates (η_1, η_2, t) as

$$
\mathrm{d}W = \mathrm{d}x\,\mathrm{d}y\,\mathrm{d}z = r^2\sin\theta\,\mathrm{d}r\,\mathrm{d}\theta\,\mathrm{d}\varphi = r^2\sin\theta\,D(t)\,\mathrm{d}\eta_1\,d\eta_2\,\mathrm{d}t,
$$

where $D(t)$ is the Jacobian of the transformation from spherical coordinates to ray ones. Then for the cross section of the ray tube, $\mathrm{d}S$, we have

$$
\mathrm{d}S = \frac{\mathrm{d}W}{\mathrm{d}\sigma} = r^2\sin\theta\frac{D(t)}{q}\,\mathrm{d}\eta_1\,\mathrm{d}\eta_2,
$$

where σ is the ray tube length. Substituting it into Equation (2) yields

$$
D(t)r^2\sin\theta\rho u^2\left(1 + \mu^2\right)\frac{q_n}{a} = \mathrm{const.} \tag{3}
$$

To calculate the Jacobian, we use a method based on numerical integration of a so-called adjoint system of equations. This system consists of differential equations for the derivatives $\partial r/\partial\eta_{1,2}$, $\partial\theta/\partial\eta_{1,2}$, $\partial\varphi/\partial\eta_{1,2}$, $\partial k_r/\partial\eta_{1,2}$, $\partial k_\theta/\partial\eta_{1,2}$, $\partial k_\varphi/\partial\eta_{1,2}$ and is derived from Equation (1) by differentiating the equations with respect to ray coordinates η_1 and η_2. Let the ray coordinates η_1 and η_2 be the angles defining the direction of the outgoing ray from a point source at the initial moment. Note that the ray coordinates do not need to be explicitly determined when the adjoint system is being derived. This becomes essential to specify the initial values for the desired functions. For the case considered, the adjoint system has the following form (in view of the symmetry about η_1 and η_2, instead of 12 equations we give only six for one variable η):

$$
\begin{aligned}
&\frac{\mathrm{d}}{\mathrm{d}t}\left(\frac{\partial r}{\partial\eta}\right)=\frac{\partial V_r}{\partial\eta}+\frac{k_r}{k}\frac{\partial a}{\partial\eta}+\frac{a}{k}\frac{\partial k_r}{\partial\eta}-\frac{ak_r}{k^2}\frac{\partial k}{\partial\eta}+\frac{\partial a}{\partial k_r}\frac{\partial k}{\partial\eta}+k\frac{\partial}{\partial\eta}\left(\frac{\partial a}{\partial k_r}\right),\\
&\frac{\partial r}{\partial\eta}\frac{\mathrm{d}\theta}{\mathrm{d}t}+r\frac{\mathrm{d}}{\mathrm{d}t}\left(\frac{\partial\theta}{\partial\eta}\right)=\frac{k_\theta}{k}\frac{\partial a}{\partial\eta}+\frac{a}{k}\frac{\partial k_\theta}{\partial\eta}-\frac{ak_\theta}{k^2}\frac{\partial k}{\partial\eta}+\frac{\partial a}{\partial k_\theta}\frac{\partial k}{\partial\eta}+k\frac{\partial}{\partial\eta}\left(\frac{\partial a}{\partial k_\theta}\right),\\
&\frac{\partial r}{\partial\eta}\sin\theta\frac{\mathrm{d}\varphi}{\mathrm{d}t}+r\cos\theta\frac{\partial\theta}{\partial\eta}\frac{\mathrm{d}\varphi}{\mathrm{d}t}+r\sin\theta\frac{d}{\mathrm{d}t}\left(\frac{\partial\varphi}{\partial\eta}\right)\\
&\quad=\frac{\partial a}{\partial\eta}\frac{k_\varphi}{k}+\frac{a}{k}\frac{\partial k_\varphi}{\partial\eta}-\frac{ak_\varphi}{k^2}\frac{\partial k}{\partial\eta}+\frac{\partial a}{\partial k_\varphi}\frac{\partial k}{\partial\eta}+k\frac{\partial}{\partial\eta}\left(\frac{\partial a}{\partial k_\varphi}\right),\\
&\frac{\mathrm{d}}{\mathrm{d}t}\left(\frac{\partial k_r}{\partial\eta}\right)=-\frac{\partial V_r}{\partial r}\frac{\partial k_r}{\partial\eta}-k_r\frac{\partial}{\partial\eta}\left(\frac{\partial V_r}{\partial r}\right)-\frac{\partial a}{\partial r}\frac{\partial k}{\partial\eta}-k\frac{\partial}{\partial\eta}\left(\frac{\partial a}{\partial r}\right)+\frac{\partial a}{\partial\eta}\frac{k_\theta^2+k_\varphi^2}{kr}\\
&\qquad-\frac{a}{k^2r}\left(k_\theta^2+k_\varphi^2\right)\frac{\partial k}{\partial\eta}-\frac{a(k_\theta^2+k_\varphi^2)}{kr^2}\frac{\partial r}{\partial\eta}+\frac{a}{kr}\left(2k_\theta\frac{\partial k_\theta}{\partial\eta}+2k_\varphi\frac{\partial k_\varphi}{\partial\eta}\right),\\
&\frac{\partial r}{\partial\eta}\frac{\mathrm{d}k_\theta}{\mathrm{d}t}+r\frac{\mathrm{d}}{\mathrm{d}t}\left(\frac{\partial k_\theta}{\partial\eta}\right)=-\frac{\partial V_r}{\partial\theta}\frac{\partial k_r}{\partial\eta}-k_r\frac{\partial}{\partial\eta}\left(\frac{\partial V_r}{\partial\theta}\right)-\frac{\partial a}{\partial\theta}\frac{\partial k}{\partial\eta}-k\frac{\partial}{\partial\eta}\left(\frac{\partial a}{\partial\theta}\right)\\
&\qquad+\frac{\partial a}{\partial\eta}\frac{k_\varphi^2\cot\theta}{k}-\frac{ak_\varphi^2}{k^2}\cot\theta\frac{\partial k}{\partial\eta}+\frac{2ak_\varphi}{k}\cot\theta\frac{\partial k_\varphi}{\partial\eta}\\
&\qquad-\frac{ak_\varphi^2}{k\sin^2\theta}\frac{\partial\theta}{\partial\eta}-\frac{\partial k_\theta}{\partial\eta}\frac{\mathrm{d}r}{\mathrm{d}t}-k_\theta\frac{\mathrm{d}}{\mathrm{d}t}\left(\frac{\partial r}{\partial\eta}\right),\\
&\frac{\partial r}{\partial\eta}\sin\theta\frac{\mathrm{d}k_\varphi}{\mathrm{d}t}+r\cos\theta\frac{\partial\theta}{\partial\eta}\frac{\mathrm{d}k_\varphi}{\mathrm{d}t}+r\sin\theta\frac{\mathrm{d}}{\mathrm{d}t}\left(\frac{\partial k_\varphi}{\partial\eta}\right)\\
&\quad=-\frac{\partial V_r}{\partial\varphi}\frac{\partial k_r}{\partial\eta}-k_r\frac{\partial}{\partial\eta}\left(\frac{\partial V_r}{\partial\varphi}\right)-\frac{\partial a}{\partial\varphi}\frac{\partial k}{\partial\eta}-k\frac{\partial}{\partial\eta}\left(\frac{\partial a}{\partial\varphi}\right)-\cos\theta\frac{\partial\theta}{\partial\eta}k_\varphi\frac{\mathrm{d}r}{\mathrm{d}t}\\
&\qquad-\sin\theta\frac{\partial k_\varphi}{\partial\eta}\frac{\mathrm{d}r}{\mathrm{d}t}-\sin\theta k_\varphi\frac{\mathrm{d}}{\mathrm{d}t}\left(\frac{\partial r}{\partial\eta}\right)-\frac{\partial k_\varphi}{\partial\eta}r\cos\theta\frac{\mathrm{d}\theta}{\mathrm{d}t}-k_\varphi\frac{\partial r}{\partial\eta}\cos\theta\frac{\mathrm{d}\theta}{\mathrm{d}t}\\
&\qquad+k_\varphi r\sin\theta\frac{\partial\theta}{\partial\eta}\frac{\mathrm{d}\theta}{\mathrm{d}t}-k_\varphi r\cos\theta\frac{\mathrm{d}}{\mathrm{d}t}\left(\frac{\partial\theta}{\partial\eta}\right),
\end{aligned}
\tag{4}
$$

where

$$
\frac{\partial V_r}{\partial\eta}=\sum_\alpha\frac{\partial V_r}{\partial r_\alpha}\frac{\partial r_\alpha}{\partial\eta},\qquad\frac{\partial a}{\partial\eta}=\sum_\alpha\left(\frac{\partial a}{\partial r_\alpha}\frac{\partial r_\alpha}{\partial\eta}+\frac{\partial a}{\partial k_\alpha}\frac{\partial k_\alpha}{\partial\eta}\right),\qquad\frac{\partial k}{\partial\eta}=\frac{1}{k}\sum_\alpha k_\alpha\frac{\partial k_\alpha}{\partial\eta},
$$

$$\frac{\partial}{\partial\eta}\left(\frac{\partial a}{\partial k_\beta}\right)=\sum_\alpha\left(\frac{\partial}{\partial r_\alpha}\left(\frac{\partial a}{\partial k_\beta}\right)\frac{\partial r_\alpha}{\partial\eta}+\frac{\partial}{\partial k_\alpha}\left(\frac{\partial a}{\partial k_\beta}\right)\frac{\partial k_\alpha}{\partial\eta}\right),$$

$$\frac{\partial}{\partial\eta}\left(\frac{\partial a}{\partial r_\beta}\right)=\sum_\alpha\left(\frac{\partial}{\partial r_\alpha}\left(\frac{\partial a}{\partial r_\beta}\right)\frac{\partial r_\alpha}{\partial\eta}+\frac{\partial}{\partial k_\alpha}\left(\frac{\partial a}{\partial r_\beta}\right)\frac{\partial k_\alpha}{\partial\eta}\right),$$

$$\frac{\partial}{\partial\eta}\left(\frac{\partial V_r}{\partial r_\beta}\right)=\sum_\alpha\left(\frac{\partial}{\partial r_\alpha}\left(\frac{\partial V_r}{\partial r_\beta}\right)\frac{\partial r_\alpha}{\partial\eta}\right),\quad r_{\alpha,\beta}=\{r,\theta,\varphi\},\ k_{\alpha,\beta}=\{k_r,k_\theta,k_\varphi\}.$$

Thus, to calculate the propagation of a linear wave and its intensity, at first we have to integrate numerically the ray equations system (1) and adjoint system (4) and then determine the amplitude variations by means of Equation (3).

A nonlinear flat disturbance in an ideal homogeneous medium is described by a simple wave solution and propagates at the supersonic speed determined by its amplitude (Kulikovsky and Lyubimov, 2005). A fast-mode simple-wave element with plasma velocity component u normal to the front moves at $a+\kappa u$, where $\kappa=(1/a)(\mathrm{d}(\rho a)/\mathrm{d}\rho)$ is the numerical coefficient depending both on the plasma beta and the angle between the wave vector and the magnetic field. We do not give here the bulky explicit expression for κ. We only note that values of κ are restricted by the limits: $1/2\le\kappa\le 3/2$. The fact that each simple-wave element travels at its own speed causes wave profile deformation and the appearance of a discontinuity. If a moderate amplitude simple wave has a triangular profile before the discontinuity appears, it will take the shape of the right-angled triangle after the discontinuity forms, with the discontinuity being the leading edge of the disturbance. Note that any nonlinear disturbance profile of a finite duration tends asymptotically to this shape. Let U_{sh} be the jump of the plasma velocity component u in the discontinuity. Then, in the nonlinear geometrical acoustics approximation, the discontinuity moves at a speed of $a+\kappa U_{\mathrm{sh}}/2$. Taking into account this increase in the wave front speed in the ray equations, we are able to correctly describe the propagation of weak shock waves. Then the ray equation system (1) becomes (Uralova and Uralov, 1994)

$$\begin{aligned}
\frac{\mathrm{d}r}{\mathrm{d}t}&=V_r+\left(a+\frac{\kappa U_{\mathrm{sh}}}{2}\right)\frac{k_r}{k}+k\frac{\partial a}{\partial k_r},\\
r\frac{\mathrm{d}\theta}{\mathrm{d}t}&=\left(a+\frac{\kappa U_{\mathrm{sh}}}{2}\right)\frac{k_\theta}{k}+k\frac{\partial a}{\partial k_\theta},\\
r\sin\theta\frac{\mathrm{d}\varphi}{\mathrm{d}t}&=\left(a+\frac{\kappa U_{\mathrm{sh}}}{2}\right)\frac{k_\varphi}{k}+k\frac{\partial a}{\partial k_\varphi},\\
\frac{\mathrm{d}k_r}{\mathrm{d}t}&=-\frac{\partial V_r}{\partial r}k_r-\frac{\partial a}{\partial r}k+\frac{a}{kr}\left(k_\theta^2+k_\varphi^2\right),\\
r\frac{\mathrm{d}k_\theta}{\mathrm{d}t}&=-\frac{\partial V_r}{\partial\theta}k_r-\frac{\partial a}{\partial\theta}k+\frac{a}{k}k_\varphi^2\cot\theta-k_\theta\frac{\mathrm{d}r}{\mathrm{d}t},\\
r\sin\theta\frac{\mathrm{d}k_\varphi}{\mathrm{d}t}&=-\frac{\partial V_r}{\partial\varphi}k_r-\frac{\partial a}{\partial\varphi}k-\sin\theta k_\varphi\frac{\mathrm{d}r}{\mathrm{d}t}-k_\varphi r\cos\theta\frac{\mathrm{d}\theta}{\mathrm{d}t}.
\end{aligned}\tag{5}$$

The generally accepted method of nonlinear geometrical acoustics ignores the additional term $\kappa U_{\mathrm{sh}}/2$ as it is small. However, it is the term that is responsible for wave deceleration due to the amplitude damping. Besides, estimations within the framework of the perturbation theory suggest that the ray pattern variation due to nonlinearity is a correction of the same order of magnitude as the nonlinear variation of the wave amplitude is. It is therefore important to take this into account in the nonlinear geometrical acoustics approximation.

Ray equation system (5) is not closed now because it includes the wave amplitude. In the linear approximation, an amplitude variation can be determined from Equation (3). The nonlinear wave amplitude undergoes additional damping associated with energy dissipation in the discontinuity. As the amplitude, we take the value of the jump U_{sh}. Variations of amplitude U_{sh} and duration T_{sh} of a weak shock wave having a triangular compression phase may be calculated as (Uralov, 1982)

$$U_{\mathrm{sh}} = u_1\left(1+\frac{\tau_1}{T_*}\right)^{-1/2}, \qquad T_{\mathrm{sh}} = T_*\left(1+\frac{\tau_1}{T_*}\right)^{1/2}, \qquad \frac{\mathrm{d}\tau_1}{\mathrm{d}t} = \frac{\kappa u_1}{q_n} \tag{6}$$

where τ_1 is the duration increment of the simple wave with an amplitude of u_1; T_* is the initial duration of the disturbance. Note that the laws (6) of the weak shock wave damping are derived by using values of the amplitude and duration of a simple wave, from which the discontinuity forms. The value of u_1 can be determined from the expression similar to Equation (3).

Thus, solving numerically of 19 ordinary differential equations (5), (4), and (6) enables us to compute propagation of a weak shock wave in an inhomogeneous medium, its amplitude and duration.

3. Analytical Modeling of Wave Propagation

In this section, we employ the nonlinear geometrical acoustics method to describe the propagation of large-scale wave-like transients, namely EUV and Moreton waves. With respect to "EUV wave" phenomena, we address only those disturbances that are associated with a fast-mode MHD shock wave.

To calculate the propagation of a shock wave, we have to specify the solar corona model as well as position and parameters of the shock wave at the initial moment. We use a simple hydrostatic corona model to demonstrate the main particularities of the method and compare results with those obtained in the linear approximation. The corona is considered to be isothermal with temperature $T = 1.5\times10^6$ K and sound speed $c = 181\ \mathrm{km\,s^{-1}}$. The plasma density is distributed in accordance with the barometric law (for details see, *e.g.*, Mann *et al.*, 1999):

$$n(r) = n_0 \exp\left(\frac{R_\odot}{H}\left(\frac{R_\odot}{r}-1\right)\right), \tag{7}$$

where $n_0 = n(R_\odot) = 3\times10^8\ \mathrm{cm^{-3}}$ is the density at the base of the corona, $R_\odot$ the solar radius, $H = 2R_{\mathrm{gas}}T/\widetilde{m}M_{\mathrm{H}}g_\odot \approx 70$ Mm the density scale height, $g_\odot$ the acceleration of gravity on the solar surface, M_{H} the molar mass of hydrogen, $\widetilde{m} = 1.27$ the average atomic weight of an ion, and R_{gas} the gas constant. Let us assume that we have a magnetic field with only a radial component (for details see, *e.g.*, Mann *et al.*, 2003):

$$B_r = \pm B_0\left(\frac{R_\odot}{r}\right)^2, \tag{8}$$

where $B_0 = 2.3$ G is its value at the base of the corona. The sign in Equation (8) depends on the solar hemisphere, but it is not meaningful here. The same model was applied by Uchida (1968).

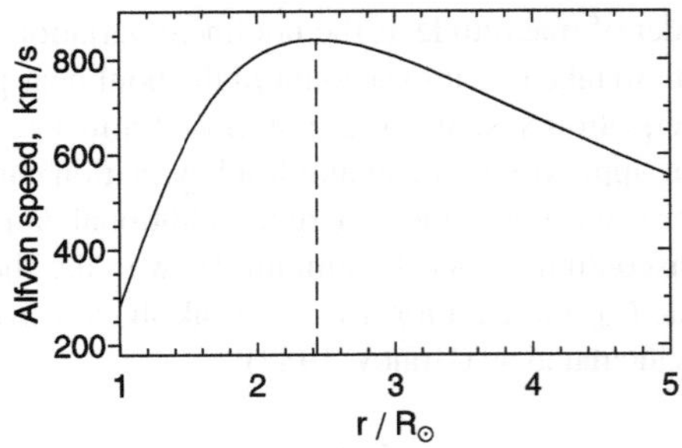

Figure 1 The Alfvén speed distribution in the solar corona model (7), (8).

In the corona model (7) and (8), the Alfvén speed increases with height, peaking at $R_{\odot}^2/4H = 2.43\ R_{\odot}$ (Figure 1). Refraction makes ray trajectories curved towards regions of the lower Alfvén speed. The solar corona has therefore waveguide properties. A portion of the wave energy flux is captured by the coronal waveguide and propagates along the solar surface, giving rise to an EUV and Moreton wave. In the treatment used, the EUV front is observable due to the plasma compression produced by the coronal shock wave. Since the plasma density decreases rapidly with height, a plasma layer near the solar surface contributes substantially to the EUV front emission. The layer thickness is about the density scale height H. So, to estimate the EUV front position, we have to find the intersection line of the calculated shock front and a spherical surface of radius $\lesssim (R_{\odot} + H)$. The Moreton wave corresponds to the chromospheric trail of the coronal shock wave, *i.e.* it can be found at the intersection line of the shock front and the upper chromosphere.

For modeling, it is important to assign the initial characteristics of the shock wave and its start position. We have to specify the initial duration (or length) of the wave and its initial amplitude on some surface. In this study, these values are given according to the strong point-like explosion theory. The wave source located at a height of 80 Mm is characterized by the energy $\tilde{\varepsilon}$ whose release produces the shock wave. When the wave covers a distance of $\Lambda = (\tilde{\varepsilon}/\rho_* a_*^2)^{1/3}$, with ρ_* and a_* being respectively the plasma density and the fast-mode speed at the explosion point, the compression phase profile of the wave is assumed to be triangular. The compression phase length is equal to Λ and the amplitude is χa_*, with χ being a coefficient of the order of unity. In this paper, we employ a value $\chi = 1$ except for the calculations given in Figure 7. The initial duration of the compression phase is supposed to be equal to $T_* = \Lambda(\mathbf{k})/a_*(\mathbf{k})$. We believe that a shock wave arises on the periphery of an active region located within the explosion cavity Λ. This manner to assign initial values does not rely on the specific mechanism of the wave initiation. It is essential only that the energy release producing a shock wave is impulsive. For instance, a wave can be produced by a compact piston acting for a short term (an abruptly accelerating filament and its magnetic envelope).

Figures 2–7 present the results of our modeling. Figure 2 illustrates a 3D image of the shock wave front and the respective 2D section including ray trajectories. The rays go out from the initial surface of size Λ. The wave front gets inclined over the solar surface, with its inclination increasing in time.

Figure 3 shows the distance–time plots of EUV waves (a) as well as the time plots of their velocities (b) along the surface 60 Mm (relative to the solar surface). The curves are given for different values of the wave source energy $\tilde{\varepsilon}$ or different values of the initial shock wave length, as follows from $\Lambda = (\tilde{\varepsilon}/\rho_* a_*^2)^{1/3}$. The EUV wave velocity decreases appreciably due to the nonlinear damping of the coronal wave amplitude. After the amplitude has substantially decreased, the shock wave propagates as a linear one. Having reached its minimum,

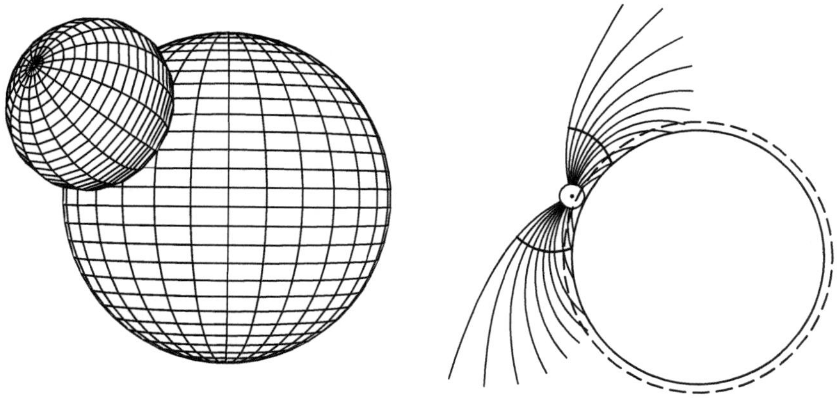

Figure 2 Propagation of a coronal shock wave as seen in a 3D image (left) and a 2D cross-section with ray trajectories (right). The solid arcs drawn across the rays represent shock wave fronts. The dashed line marks a height of 60 Mm above the solar surface. The wave front velocity along this line corresponds to the velocity of an EUV wave. The wave source height is 80 Mm.

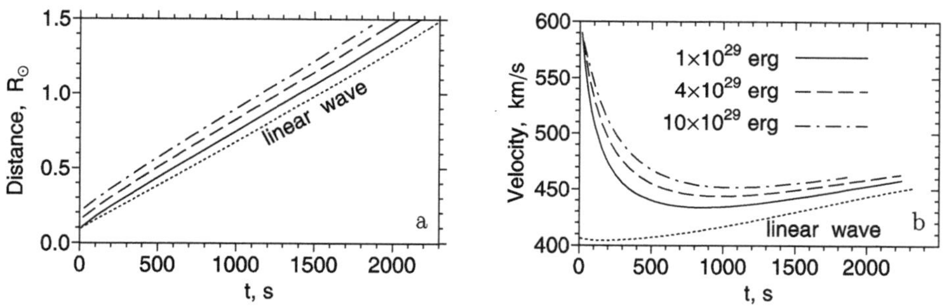

Figure 3 The distance–time plot of EUV waves (a) and the time evolution of their velocities (b) along the surface at 60 Mm height for different energies $\tilde{\varepsilon}$ specified in panel (b). The lowest (dotted) line represents the linear EUV wave.

the wave velocity slightly increases. This is due to the shock front inclination over the solar surface that becomes more and more significant with time and associated with the wave-guide properties of the lower quiet Sun's corona. The larger the wave front inclination, the higher the Moreton and EUV wave velocity. Note that calculated Moreton and EUV wave acceleration must be difficult to observe since it occurs when the wave amplitude becomes low (see Figure 5).

Figure 4 shows positions (a) counted along the surface and velocities (b) of a Moreton wave (thick lines) and an EUV wave (thin lines), which are produced by a single coronal wave. The dotted lines correspond to a linear coronal wave and the solid lines are for a shock one. The linear Moreton wave velocity is lower because the Alfvén speed at the corona base is smaller than that at a height of 60 Mm. Note that a Moreton wave decelerates even in the linear case. This fact is associated with the initial wave source position at a height of 80 Mm above the photosphere. The EUV wave in the linear case does not decelerate since the wave source is located roughly at the same height that the EUV wave is (a little bit higher the dashed line in Figure 2).

The increase in the propagation velocity of the wave front with height results in a front inclination relative to the solar surface. So one can observe the wave signatures correspond-

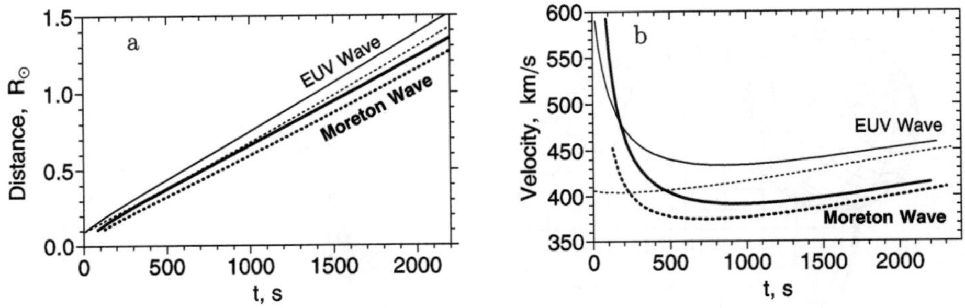

Figure 4 The kinematic plots of a Moreton wave (thick) and an EUV wave (thin), which are produced by a single coronal wave in the linear (dotted) and nonlinear (solid) consideration.

Figure 5 The time evolution of the EUV wave amplitude for different energies $\tilde{\varepsilon}$ of the wave source.

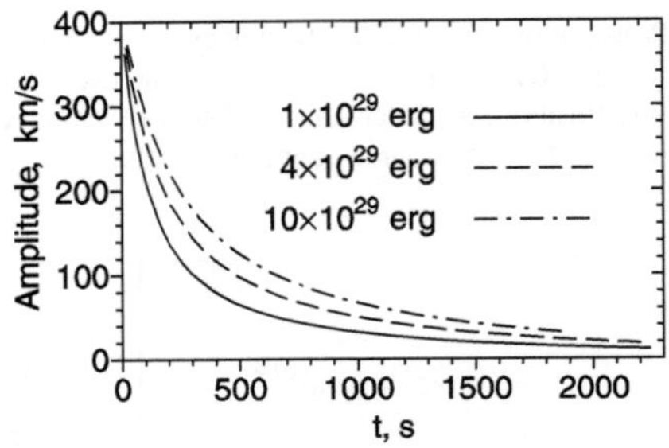

ing to the different heights (*e.g.*, EUV waves and Moreton waves) to be shifted. Increase of the front inclination with time determines a value of this shift. Besides, its time evolution is also determined by the observer position and the sight angle. Such a consideration demonstrates the possibility to explain the offset between Moreton and EUV waves as well as the He I-Hα offset observed by Vršnak *et al.* (2002b) since waves in the He I 10 830 Å line are similar morphologically to EUV ones.

Figure 5 presents time dependence of the EUV wave amplitude for different values of the wave source energy $\tilde{\varepsilon}$. A disturbance having higher $\tilde{\varepsilon}$ and greater length decays more slowly as follows from Equation (6) and as is seen in Figure 5. The Moreton wave amplitude varies in a similar manner, but it has smaller values.

Another effect associated with the nonlinearity of an EUV wave is the increase in its duration T_{sh} and, respectively, length $L = aT_{\mathrm{sh}}$ (also referred to as the wave profile broadening). In the linear case, the wave duration is constant (under the assumption of a steady-state medium). Figure 6 gives the time evolution of the ratio of the EUV wave length to its initial size Λ. If initial amplitudes of disturbances are equal, the relative extension will be faster for disturbances having the lower initial energy (and the shorter initial length).

With respect to the damping of shock waves having the same initial lengths, amplitude decrease is faster for a wave with a higher amplitude. Therefore, shock waves with different initial amplitudes decay to the same level after approximately equal duration (Figure 7, solid curves). For comparison, we also plot the amplitude curves of linear disturbances, which hold an initial ratio of the wave amplitudes throughout propagation. The mentioned property of nonlinear waves (in Figure 7) allows us to be not precise about the value of the initial wave amplitude. Therefore, we can use the value that follows from the strong point-like explosion theory.

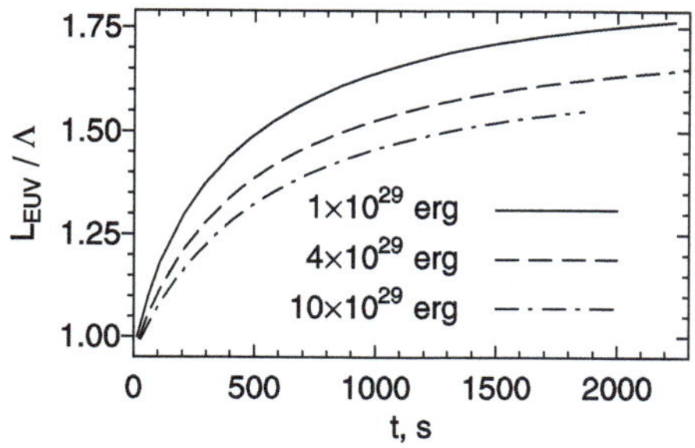

Figure 6 The time dependence of the EUV wave length relative to its initial size for different energies $\tilde{\varepsilon}$.

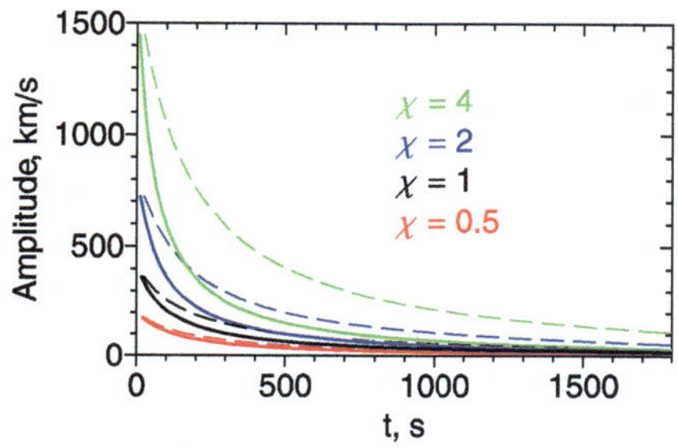

Figure 7 The amplitude damping of shock waves (solid lines) and linear ones (dashed lines) calculated for the wave source energy $\tilde{\varepsilon} = 10^{29}$ erg and different initial amplitudes. The initial amplitude values are specified in the figure as ratios χ of those to a fast-mode speed in the wave source.

4. Discussion and Conclusion

We have modeled the propagation of shock-associated EUV waves and Moreton waves, using the nonlinear geometrical acoustics method. This method takes into account characteristic properties of nonlinear waves:

i) Dependence of the wave velocity on its amplitude.
ii) Wave energy dissipation in the shock front.
iii) Wave duration increase with time.

The method allows one to calculate the nonlinear evolution of the shock wave and its propagation pattern. However, the generally accepted variant of this approach includes nonlinearity effects only for describing the amplitude and duration of a wave, but not for the wave velocity value. So, using such a non-self-consistent approach results in the loss of an important effect concerning wave kinematics.

We have applied another approach, appending an additional term to the ray equations. This has allowed the finite wave amplitude to be taken into account. We have solved self-consistently the modified ray equations and the equations describing the wave amplitude and duration evolution along a ray tube. It is this approach that has been developed in this paper to analyze coronal shock wave propagation along the solar surface.

One of the results of our analysis is the deceleration of EUV and Moreton waves at the initial stage of propagation. Since we use a spherically symmetric and isothermal model of the solar corona, deceleration is a direct consequence of their nonlinearity. Thus, EUV and Moreton waves having a sufficient amplitude (and therefore being observable) have to decelerate in the quiet Sun's regions where average plasma parameters are constant along the solar surface. Note that the large-scale waves under study are registered just in these regions. Calculated wave deceleration is supported by the EUV and Moreton wave observations analyzed by Warmuth *et al.* (2001, 2004a). Also we notice here that EUV waves not associated with a coronal MHD wave show only slight or no deceleration (see, *e.g.*, Wills-Davey and Attrill, 2009).

The simple corona model also let us find other features of wave kinematics, *e.g.*, *i*) the wave source height has effect on the initial portion of the velocity plot, and *ii*) the rate of wave deceleration and damping becomes lower as the wave source energy (or the wave length, respectively) grows. All these findings are applied for modeling EUV wave propagation in the 17 January 2010 event in an accompanying paper (Grechnev *et al.*, 2011a).

Modeling waves in the linear approximation does not reveal their deceleration. On the contrary, linear waves undergo only small acceleration caused by a slightly increasing inclination of the coronal wave front over the solar surface. This effect was first discovered by Uchida (1968). However, because of the error that he made in the expression for the barometric distribution of the coronal plasma density (the scale height was halved), the wave front inclination over the solar surface was very large. As a result, linear waves underwent considerable acceleration in Uchida's original model.

Another important result of our modeling is the duration (and length) increase of Moreton and EUV waves. This effect is also confirmed by observations (see, *e.g.*, Warmuth *et al.*, 2001; Veronig *et al.*, 2010). In contrast, a linear disturbance keeps its duration unchanged in a steady-state medium. Note that in the linear approximation, the wave amplitude and duration also vary due to the viscosity, the thermal conductivity and the finite plasma conductivity, however, these effects are negligible against nonlinear factors.

To summarize, we believe that wave deceleration and its duration increase, both being the attributes of shock wave evolution, point out the crucial role of nonlinearity in the behavior of EUV and Moreton waves (at least, it concerns some of them).

In conclusion, we will briefly discuss the method limitations for solving the shock wave propagation problem. The main limitation is associated with laws (6) of a shock wave damping, which are derived by using the relations for simple flat MHD waves in a homogeneous medium. So, we have to meet two requirements. First, the shock wave length should be smaller than the radius of curvature of the wave front and the smallest medium variation scale. The fulfillment of these conditions also ensures validity of the linear ray approximation (1), which involves actually even less limitations. The smallest variation scale in our modeling is that of plasma density. So, we are aware that our computation lies at the boundary of applicability of nonlinear geometrical acoustics since characteristic shock wave length Λ and density scale $\rho/|\nabla\rho|$ are of the same order of magnitude.

Second, the nonlinear factor U_{sh}/a should be small. Under this condition, damping laws (6) are derived and this very condition ensures a correct calculation of the terms involving U_{sh} in the ray equations system (5). With respect to the limitations of laws (6), the point-like atmospheric explosion theory (Kestenboim, Roslyakov, and Chudov, 1974) suggests that these laws describe satisfactorily spherical shock wave propagation up to $U_{\mathrm{sh}}/a \leq 1$. When we choose the initial value $U_{\mathrm{sh}} = a$ in Section 3, we do not therefore go beyond the scope of the application of relations (6). However, ray equations (5) and (4) at $U_{\mathrm{sh}}/a \approx 1$ are able to yield an error in calculations. Nevertheless, this error is insignificant due to the nearly spherical shape of the wave front at the initial phase of propagation and then it disappears owing to rapid decrease in U_{sh}.

Acknowledgements We thank Dr. V.V. Grechnev for useful discussions and valuable help in preparing the paper. We also thank the anonymous referee and editors for careful reading of the manuscript as well as helpful comments and suggestions. A.A. is very grateful to the scientific organizing committee of the CESRA2010 Workshop for financial support.

The research was supported by the Russian Foundation of Basic Research (Grant No. 10–02–09366) and Siberian Branch of the Russian Academy of Sciences (Lavrentyev Grant 2010 – 2011).

References

Balasubramaniam, K.S., Pevtsov, A.A., Neidig, D.F.: 2007, *Astrophys. J.* **658**, 1372.
Barnes, A.: 1992, *J. Geophys. Res.* **97**, 12105.
Blokhintsev, D.I.: 1981, *Acoustics of an Inhomogeneous Moving Medium*, 2nd edn., Nauka, Moscow (in Russian).
Gilbert, H.R., Holzer, T.E., Thompson, B.J., Burkepile, J.T.: 2004, *Astrophys. J.* **607**, 540.
Grechnev, V.V., Afanasyev, A.N., Uralov, A.M., Chertok, I.M., Eselevich, M.V., Eselevich, V.G., Rudenko, G.V., Kubo, Y.: 2011a, *Solar Phys.* in this issue.
Grechnev, V.V., Uralov, A.M., Chertok, I.M., Kuzmenko, I.V., Afanasyev, A.N., Meshalkina, N.S., Kalashnikov, S.S., Kubo, Y.: 2011b, *Solar Phys.* in this issue.
Hudson, H.S., Khan, J.I., Lemen, J.R., Nitta, N.V., Uchida, Y.: 2003, *Solar Phys.* **212**, 121.
Kestenboim, K.S., Roslyakov, G.S., Chudov, L.A.: 1974, *The Point Explosion*, Nauka, Moscow (in Russian).
Khan, J.I., Aurass, H.: 2002, *Astron. Astrophys.* **383**, 1018.
Kravtsov, Y.A., Orlov, Y.I.: 1990, *Geometrical Optics of Inhomogeneous Media*, Springer, Berlin.
Kulikovsky, A.G., Lyubimov, G.A.: 2005, *Magnetic Hydrodynamics*, 2nd edn., Logos, Moscow (in Russian).
Landau, L.D., Lifshitz, E.M.: 1987, *Fluid Mechanics*, 2nd edn., Pergamon Press, Oxford.
Mann, G., Jansen, F., MacDowall, R.J., Kaiser, M.L., Stone, R.G.: 1999, *Astron. Astrophys.* **348**, 614.
Mann, G., Klassen, A., Aurass, H., Classen, H.-T.: 2003, *Astron. Astrophys.* **400**, 329.
Moreton, G.E., Ramsey, H.E.: 1960, *Publ. Astron. Soc. Japan* **72**, 357.
Muhr, N., Vršnak, B., Temmer, M., Veronig, A.M., Magdalenić, J.: 2010, *Astrophys. J.* **708**, 1639.
Narukage, N., Hudson, H.S., Morimoto, T., Akiyama, S., Kitai, R., Kurokawa, H., Shibata, K.: 2002, *Astrophys. J.* **572**, L109.
Narukage, N., Eto, S., Kadota, M., Kitai, R., Kurokawa, H., Shibata, K.: 2004. In: Stepanov, A.V., Benevolenskaya, E.E., Kosovichev, A.G. (eds.) *Multi-Wavelength Investigations of Solar Activity, Proc. IAU Symp.* **223**, 367.
Patsourakos, S., Vourlidas, A., Wang, Y.M., Stenborg, G., Thernisien, A.: 2009, *Solar Phys.* **259**, 49.
Temmer, M., Vršnak, B., Žic, T., Veronig, A.M.: 2009, *Astrophys. J.* **702**, 1343.
Uchida, Y.: 1960, *Publ. Astron. Soc. Japan* **12**, 376.
Uchida, Y.: 1968, *Solar Phys.* **4**, 30.
Uchida, Y.: 1974, *Solar Phys.* **39**, 431.
Uchida, Y., Altschuler, M.D., Newkirk, G. Jr.: 1973, *Solar Phys.* **28**, 495.
Uralov, A.M.: 1982, *Magn. Gidrodin.* **1**, 45 (in Russian).
Uralova, S.V., Uralov, A.M.: 1994, *Solar Phys.* **152**, 457.
Veronig, A.M., Muhr, N., Kienreich, I.W., Temmer, M., Vršnak, B.: 2010, *Astrophys. J.* **716**, L57.
Vršnak, B., Lulić, S.: 2000, *Solar Phys.* **196**, 157.
Vršnak, B., Magdalenić, J., Aurass, H., Mann, G.: 2002a, *Astron. Astrophys.* **396**, 673.
Vršnak, B., Warmuth, A., Brajša, R., Hanslmeier, A.: 2002b, *Astron. Astrophys.* **394**, 299.
Vršnak, B., Magdalenić, J., Temmer, M., Veronig, A., Warmuth, A., Mann, G., Aurass, H., Otruba, W.: 2005, *Astrophys. J.* **625**, L67.
Wang, Y.-M.: 2000, *Astrophys. J.* **543**, L89.
Warmuth, A., Mann, G., Aurass, H.: 2005, *Astrophys. J.* **626**, L121.
Warmuth, A., Vršnak, B., Aurass, H., Hanslmeier, A.: 2001, *Astrophys. J.* **560**, L105.
Warmuth, A., Vršnak, B., Magdalenić, J., Hanslmeier, A., Otruba, W.: 2004a, *Astron. Astrophys.* **418**, 1101.
Warmuth, A., Vršnak, B., Magdalenić, J., Hanslmeier, A., Otruba, W.: 2004b, *Astron. Astrophys.* **418**, 1117.
White, S.M., Thompson, B.J.: 2005, *Astrophys. J.* **620**, L63.
Wills-Davey, M.J., Attrill, G.D.R.: 2009, *Space Sci. Rev.* **149**, 325.
Žic, T., Vršnak, B., Temmer, M., Jacobs, C.: 2008, *Solar Phys.* **253**, 237.

Solar Phys (2011) 273:493–509
DOI 10.1007/s11207-011-9872-9

The 17 January 2005 Complex Solar Radio Event Associated with Interacting Fast Coronal Mass Ejections

**A. Hillaris · O. Malandraki · K.-L. Klein ·
P. Preka-Papadema · X. Moussas · C. Bouratzis ·
E. Mitsakou · P. Tsitsipis · A. Kontogeorgos**

Received: 31 January 2011 / Accepted: 20 September 2011 / Published online: 8 November 2011

Abstract On 17 January 2005 two fast coronal mass ejections were recorded in close succession during two distinct episodes of a 3B/X3.8 flare. Both were accompanied by metre-

---

Energy Storage and Release through the Solar Activity Cycle – Models Meet Radio Observations
Guest Editors: Christophe Marqué and Alexander Nindos

A. Hillaris (✉) · P. Preka-Papadema · X. Moussas · C. Bouratzis · E. Mitsakou
Section of Astrophysics, Astronomy and Mechanics, Department of Physics, University of Athens, Zografos (Athens), 15783, Greece
e-mail: ahilaris@phys.uoa.gr

P. Preka-Papadema
e-mail: ppreka@phys.uoa.gr

X. Moussas
e-mail: xmoussas@phys.uoa.gr

C. Bouratzis
e-mail: kbouratz@phys.uoa.gr

E. Mitsakou
e-mail: emitsaku@phys.uoa.gr

O. Malandraki
Institute of Astronomy and Astrophysics, National Observatory of Athens, 11810 Athens, Greece
e-mail: omaland@astro.noa.gr

K.-L. Klein
Observatoire de Paris, LESIA-CNRS UMR 8109, Univ. Paris 6 & Paris 7, Observatoire de Meudon, 92195 Meudon, France
e-mail: ludwig.klein@obspm.fr

P. Tsitsipis · A. Kontogeorgos
Technological Educational Institute of Lamia, 35100 Lamia, Greece

P. Tsitsipis
e-mail: tsitsipis@teilam.gr

A. Kontogeorgos
e-mail: akontog@teilam.gr

to-kilometre type-III groups tracing energetic electrons that escape into the interplanetary space and by decametre-to-hectometre type-II bursts attributed to CME-driven shock waves. A peculiar type-III burst group was observed below 600 kHz 1.5 hours after the second type-III group. It occurred without any simultaneous activity at higher frequencies, around the time when the two CMEs were expected to interact. We associate this emission with the interaction of the CMEs at heliocentric distances of about 25 $R_{\odot}$. Near-relativistic electrons observed by the EPAM experiment onboard ACE near 1 AU revealed successive particle releases that can be associated with the two flare/CME events and the low-frequency type-III burst at the time of CME interaction. We compare the pros and cons of shock acceleration and acceleration in the course of magnetic reconnection for the escaping electron beams revealed by the type-III bursts and for the electrons measured *in situ*.

Keywords Radio bursts, meter-wavelengths and longer (m, dkm, hm, km) · Coronal mass ejections · Energetic particles, acceleration · Energetic particles, electrons · Energetic particles, propagation

1. Introduction

The acceleration of charged particles to high energies in the solar corona is related to flares, which reveal the dissipation of magnetically stored energy in complex magnetic-field structures of the low corona, and to coronal mass ejections (CMEs), which are large-scale, complex magnetic-field–plasma structures ejected from the Sun. CMEs can drive bow shocks, and their perturbation of the coronal magnetic field can also give rise to magnetic reconnection, where energy can be released in a similar way as during flares.

When several CMEs are launched along the same path, a faster CME may overtake a slower preceding one, and the two CMEs can merge into a single structure. For this phenomenon Gopalswamy *et al.* (2001) introduced the term *CME Cannibalism.* The CME–CME interaction was found to be associated with a characteristic low-frequency continuum radio emission. Gopalswamy *et al.* (2002) interpreted this type of activity as the radio signature of non-thermal electrons originating either during reconnection between the two CMEs or as the shock of the second, faster CME travels through the body of the first (see Schmidt and Cargill, 2004 for a numerical study of two interacting coronal mass ejections).

In this paper we use radio diagnostics to study electron acceleration during a complex solar event broadly consisting of two stages, each associated with a distinct episode of a flare and with a fast CME, which occurred in close temporal succession on 17 January 2005. The CMEs interacted at a few tens of $R_{\odot}$. Both the flare/CME events and the CME interaction were accompanied by radio emission, which is used here to study electron acceleration scenarios. Energetic electrons in the corona and interplanetary space are traced by their dm-to-km-wave radio emission, mostly excited at or near the electron plasma frequency. The emission provides a diagnostic of the type of the exciter and its path from the low corona (cm–dm wavelengths) to 1 AU (km wavelengths). Radio emissions from exciters moving through the corona appear in dynamic spectra as structures exhibiting a drift in the time–frequency domain. The drift rate depends on their speed and path, resulting in a variety of bursts. Type-III bursts trace the path of supra–thermal electrons guided by magnetic structures. They appear, on dynamic spectra, as short structures (lasting from a fraction of a second at dm-waves to a few tens of minutes at km-waves) with fast negative drift ($\frac{1}{f}\frac{\mathrm{d}f}{\mathrm{d}t} \approx 0.5\ \mathrm{s}^{-1}$; see for example Güdel and Benz, 1988). This corresponds to anti-sunward propagation of the electrons through regions of decreasing ambient density at speeds $\approx c/3$ (*e.g.*, Suzuki and Dulk, 1985). Sunward travelling beams produce reverse drift bursts (RS bursts), and beams propagating in closed loops emit type-U or -J bursts comprising a succession of an initial drift towards lower frequencies and a more or less pronounced RS burst.

Type-II bursts are more slowly drifting bursts ($\frac{1}{f}\frac{\mathrm{d}f}{\mathrm{d}t} \approx 0.001 - 0.01\ \mathrm{s}^{-1}$; see, for example, Table A.1 in Caroubalos *et al.*, 2004) excited by electrons accelerated at travelling shocks and emitting in their upstream region. Finally broadband dm–m wave continuum emission that may last over several minutes or even hours (type-IV burst) is ascribed to electrons confined in closed coronal magnetic structures. The reader is referred to the reviews in McLean and Labrum (1985), Bastian, Benz, and Gary (1998), Nindos *et al.* (2008) and Pick and Vilmer (2008) for more detailed accounts of the radio emission by non-thermal electrons in the corona.

2. Observations and Data Analysis

The 17 January 2005 event consisted of a complex flare, two very fast coronal mass ejections (CMEs), and intense and complex soft X-ray (SXR) and radio emission. In all radiative signatures two successive stages can be distinguished. The CMEs were launched successively from neighbouring regions of the corona and interacted in interplanetary space. The sequence of the observed energetic phenomena is summarized in Table 1 and described, in detail, in the following subsections.

2.1. Optical Observations: Flares and CMEs

Figure 1 displays snapshots in the Hα line obtained from the Kanzelhöhe solar observatory (courtesy M. Temmer; see also Temmer *et al.* (2007), Figure 2, for details on the evolution of the Hα flare ribbons). The only major active region on the disk is NOAA 10720 in the north-western quadrant (N15° W25°). It displayed nearly uninterrupted activity since the early hours of 17 January 2005. The most conspicuous event was a 3B Hα flare reported by Kanzelhöhe 09:16 – 11:57 UT. This flare proceeded successively in two different parts of AR 10720, as shown in the two snapshots of the bottom panel. The first part of the event, referred to as "stage 1" (illustrative snapshot at 09:13 UT), is seen in the eastern part of the active region, close to the sunspots. It is associated with the temporary disappearance or eruption of the filament labelled 'F1' in the upper right panel. Two major flare ribbons are distinguished in the snapshot at 09:13 UT: a narrow band essentially in the east–west direction and a broader north–southward oriented region. The significant offset of the two ribbons with respect to the neutral line shows the eruption of a strongly sheared magnetic field. After about 09:35 UT the brightest emission is seen in the western part of the active region ("stage 2" ; see snapshot at 09:54 UT), together with the eruption of another filament 'F2' (or of a different part of the filament whose northern section erupted before). The brightening consisted of two essentially parallel flare ribbons, which were connected by post flare loops in later snapshots (not shown here). These two stages of the event were also found in the soft X-ray (SXR) and radio emissions, as will be discussed below.

Two CMEs were observed in close succession. A sequence of difference images from the *Large Angle and Spectrometric Coronagraph* (LASCO) aboard the SOHO spacecraft (Brueckner *et al.*, 1995) is displayed in Figure 2: the first CME (henceforth CME_1) is seen in the image at 09:30 UT in the north-western quadrant. While it travelled through the corona, the second, broader CME (CME_2) appeared underneath (image at 09:54 UT). The most

Table 1 Overview of the 17 January 2005 Event and associated activity.

Event	Time UT	Characteristics	Remarks
SXR Start	06:59		AR10720 (N15° W25°)
Type IV	08:40	3.0 – 630 MHz	AR10720
CME_1	09:00		lift-off
SXR Stage 1	09:05		
First cm burst start	09:05		RSTN 15400 MHz
Type III_1	09:07 – 09:28	0.2 – 630 MHz	AR10720
Type II_1	09:11	0.2 – 5 MHz	AR10720
Hα Start	09:13	3B	KANZ, AR10720
CME_1	09:30	2094 $km\,s^{-1}$	On C2
HXR Start	09:35:36		RHESSI Number 5011710
CME_2	09:38		lift-off
SXR Stage 2	09:42		End SXR Stage 1
Second cm burst start	09:43		RSTN 15400 MHz
Type III_2	09:43 – 09:59	0.2 – 630	AR10720
HXR peak	09:49:42	7865 counts s^{-1}	
Type II_2	09:48	0.2 – 8 MHz	AR10720
SXR peak	09:52	X3.8	End SXR Stage 2
CME_2	09:54	2547 $km\,s^{-1}$	On C2
First rise Electron flux	10:00	38 – 315 keV	ACE/EPAM
SXR End	10:07		AR720
HXR End	10:38:52	53 152 112 total counts	RHESSI
Second rise Electron flux	12:00	38 – 315 keV	ACE/EPAM
Type III_3	11:37	0.5 MHz	CME_1, CME_2 merge at 37 $R_\odot$ type II_2 overtakes type II_1
Hα End	11:57		KANZ
Type IV End	15:24	3.0 – 630 MHz	AR10720

conspicuous features of both CMEs are seen above the north-western limb, but both were labelled halo CMEs in the LASCO CME catalog[1] (Yashiro *et al.*, 2004). Speeds of, respectively, 2094 and 2547 $km\,s^{-1}$ were derived from linear fits to the trajectories of their fronts published in the CME catalogue. Formally the CME fronts described by the fits intersected near 12:32 UT at a heliocentric distance of about 38 $R_\odot$. The statistical error of the speeds of the CME fronts and their lift-off times, derived from the above mentioned linear least-squares fit to the measured heliocentric distances, leads to an uncertainty of ±3 h in the time

[1] http://cdaw.gsfc.nasa.gov/CME_list/.

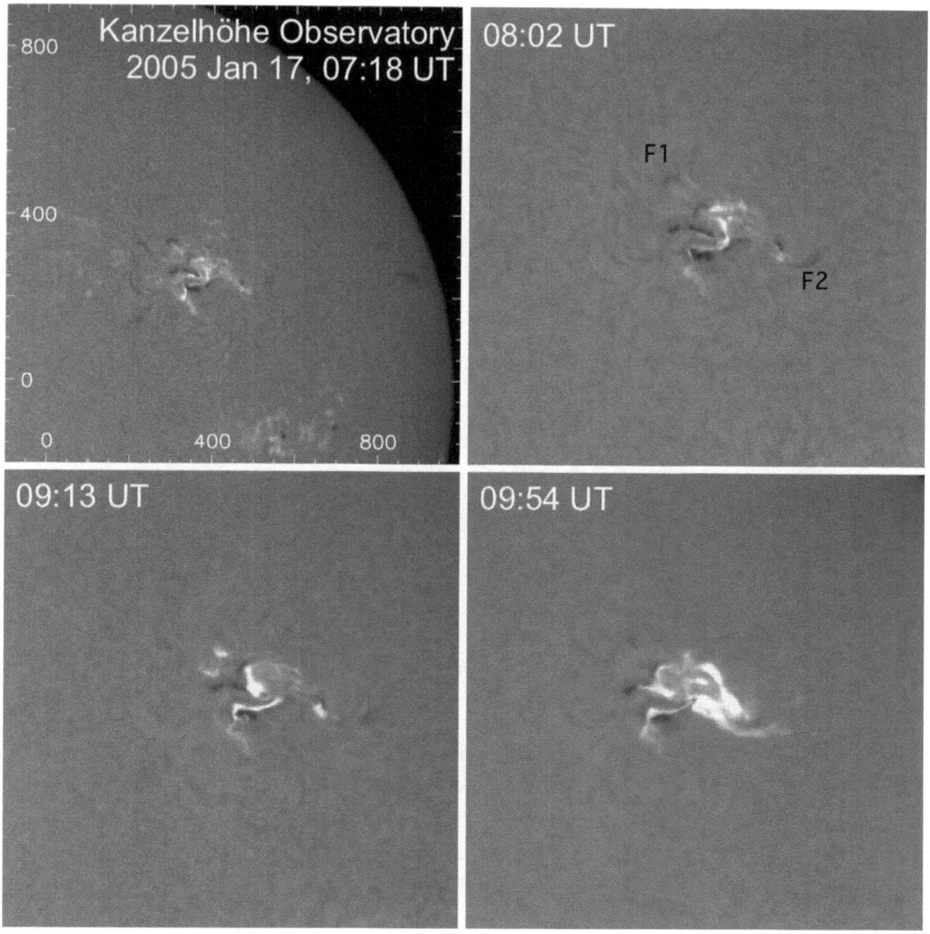

Figure 1 Snapshots of active region NOAA 10720 on 17 January 2005 in Hα line centre (top left) and in the wing, observed at Kanzelhöhe Observatory (courtesy M. Temmer). Solar north is at the top, west on the right. The two snapshots at the top show the active region before the flare under discussion, the two bottom images show two instants during the stages 1 and 2, respectively. These stages were associated with the disappearance of the filaments labelled 'F1' and 'F2'.

of intersection. This uncertainty stems from the fact that the two height–time trajectories are nearly parallel. We will show in Section 2.5 that CME interaction actually occurred well before the formal time of intersection. Of course a single instant of interaction between two complex CMEs is fictitious anyway.

2.2. Soft X-Ray and Radio Emission

An overview of the complex radio event is given in Figure 3. There we present the dynamic flux density spectrum of the radio bursts in the 650 MHz – 20 kHz range (heliocentric distance $\approx 1.1\ R_{\odot}$ to 1 AU) using combined recordings of the *Appareil de Routine pour le Traitement et l'Enregistrement Magnétique de l'Information Spectrale* (ARTEMIS-IV) solar radio-spectrograph (Caroubalos *et al.*, 2001; Kontogeorgos *et al.*, 2006) and the

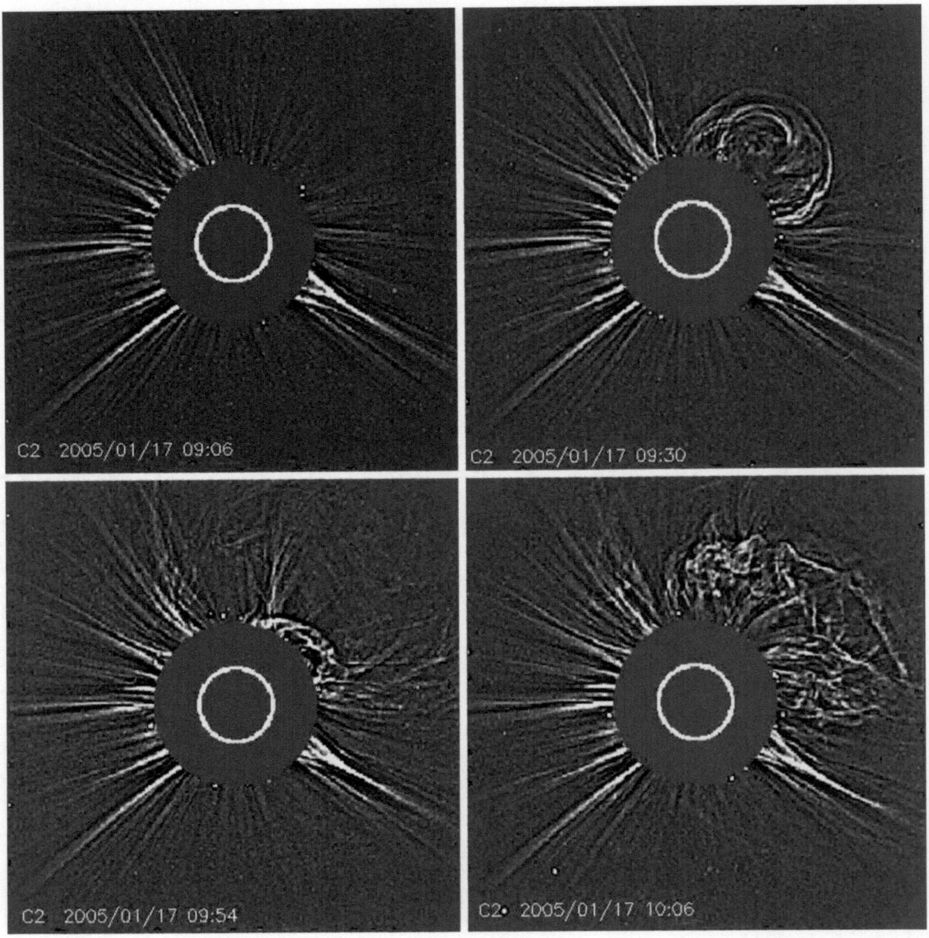

Figure 2 The two LASCO CMEs in close succession; the images have been subjected to high-pass filtering. Top: two frames of the 09:30:05 Halo CME with back-extrapolated lift-off at 09:00:47 UT and plane-of-the-sky speed 2094 $\mathrm{km\,s^{-1}}$. Bottom: two frames of the 09:54:05 Halo CME with back-extrapolated lift-off at 09:38:25 UT and plane-of-the-sky speed 2547 $\mathrm{km\,s^{-1}}$. Solar north is at the top, west on the right.

Wind/WAVES experiment (Bougeret *et al.*, 1995). Several other time histories are superposed on the dynamic spectrum:

- Dashed lines display the approximate frequency–time trajectories of the two CME fronts, using the density model of Vršnak, Magdalenić, and Zlobec (2004), which describes well the coronal density behaviour in the large range of distances from low corona to interplanetary space:

$$\frac{n}{10^8\ \mathrm{cm^{-3}}} = 15.45\left(\frac{R_\odot}{R}\right)^{16} + 3.165\left(\frac{R_\odot}{R}\right)^{6} + 1.0\left(\frac{R_\odot}{R}\right)^{4} + 0.0033\left(\frac{R_\odot}{R}\right)^{2}.$$

The linear fits to the height–time trajectories of the CME fronts in the LASCO images were converted to frequency–time tracks of fundamental (black line) and harmonic (red line) plasma emission.

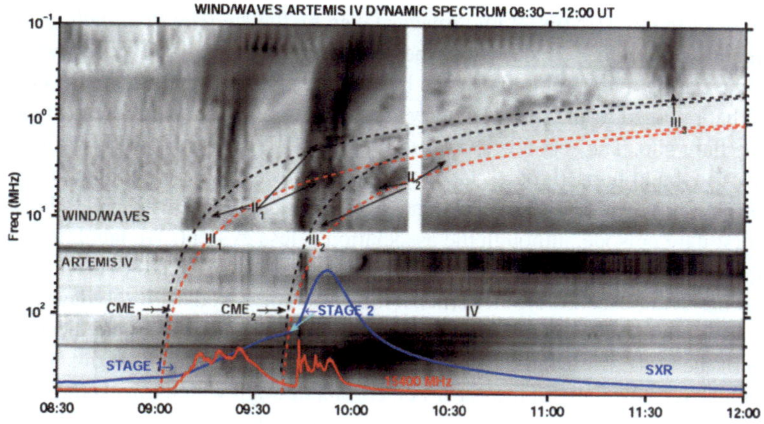

Figure 3 ARTEMIS-IV/*Wind*/WAVES dynamic spectrum (inverse grey scale). Overlays: The profiles of GOES SXR flux (dark blue) and RSTN flux density at 15400 MHz (red) and the frequency–time plots derived from the linear fits to the front trajectories of CME_1 and CME_2 and an empirical density model for fundamental (black dashed curve) and harmonic (red dashed curve) plasma emission. The type-IV continuum, the type-II (II_1 and II_2) and type-III GG bursts (III_1 and III_2), the stages 1 and 2 of the SXR flux rise, and the type-III burst (III_3) around the convergence of the fronts of CME_1 and CME_2 are annotated on the plot.

- The solid blue curve displays the SXR time history (0.1 – 0.8 nm), using GOES on line data (http://www.sel.noaa.gov/ftpmenu/indices), describing thermal emission from the flare-heated plasma.
- The red curve is the microwave time history at 15.4 GHz, produced by non-thermal electrons (energies > 100 keV) in magnetic fields of a few hundred G; these were obtained from the San Vito Solar Observatory of the Radio Solar Telescope Network (RSTN) (Guidice *et al.*, 1981).[2]

The two stages of the flare identified in the Hα observations in Figure 1 correspond to two distinct events of energy release seen in the SXR and microwave time profiles (Figure 3). The SXR time profile had an initial smooth increase between 06:59 UT and 09:05 UT. Subsequently the SXR flux rose slightly faster until 09:45 UT (stage 1), and even faster (stage 2) until the X3.8 peak at 09:52 UT. The gradual rise in stage 1 and the faster rise in stage 2 were each accompanied by strong microwave bursts. The second burst was also observed in hard X-rays by RHESSI[3] (Lin *et al.*, 2002).[4]

2.3. Radio Emission: Decametre-to-Kilometre Waves

The dominant features in the dynamic spectrum observed by *Wind*/WAVES at frequencies below 2 MHz are two groups of type-III bursts, labelled III_1 and III_2. They occurred in association with the SXR and microwave emissions of stages 1 and 2, respectively, and with the two different parts of the Hα flare. The two type-III groups occurred near the extrapolated

[2] http://www.ngdc.noaa.gov/nndc/struts/form?t=102827&s=6&d=8,40,9.

[3] http://hesperia.gsfc.nasa.gov/getfiles/rhessi_data_search.html.

[4] During stage 1 of the SXR flare RHESSI was in the Earth's shadow, so there is no HXR burst associated with the first microwave burst.

lift-off times of the two CMEs. Radio images taken by the *Nançay Radioheliograph* (NRH; Kerdraon and Delouis, 1997) show that the sources are located in the north-western quadrant near the flaring active region.[5] Hence both flare episodes were efficient accelerators of electrons that escaped to the interplanetary space along open magnetic field lines rooted at or near the flare site. The second type-III group (type III_2) was followed by a more slowly drifting narrow-band burst (type II, labelled II_2) produced by a coronal shock wave. Upon closer inspection the spectrum suggests that similar drifting features can also be associated with the first flare episode, although the association is less evident. We label these bursts II_1 in Figure 3. Since the two CMEs are extremely fast, they are expected to drive shock waves in the corona. The observed type-II emission can be compared with the dashed curves in Figure 3, which track fundamental (black) and harmonic (red) emission expected from the trajectory of the CME front and the coronal density model. It is clear that this density model is only indicative, especially in the perturbed corona through which travels the second CME (see discussion in Subsection 2.4). We therefore associate type-II_1 and -II_2 to the bow shocks of the two CMEs, although other interpretations, like shocks on the flanks or shocks from a driver related to the flare, are not excluded.

2.4. Radio Emission: Metre Waves

The dm–m wave emission consisted of a type-IV continuum, the metre wave counterparts of the dekametre–hectometre (DH) type-III groups and of the type-II bursts.

The type-IV continuum started near 08:40 UT during the initial smooth increase of the SXR flux before stage 1. It was first visible as a grey background in the dynamic spectrum, and became progressively more intense. It dominated the metre wave spectrum during and after type III_2, and gradually penetrated to lower frequencies, down to 5 MHz. Images in the EIT 195 Å channel (Delaboudinière *et al.*, 1995) and in the 164 – 432 MHz range taken by the NRH indicate that the thermal (soft X-rays) and non-thermal (radio) emissions all originated near NOAA AR 10720. In the time interval from the start of the type-IV burst to the start of stage 1 a wealth of fine structures was recorded (see Bouratzis *et al.*, 2009). From the high-resolution observations in the 200 – 500 MHz range (see Figures 5, 6 for example) it appears that most bursts are broadband pulsations. Other fine structures of type-IV emission such as spikes, fibre bursts and zebra pattern appear occasionally (see Kuijpers, 1980 for a description of the fine structure of type-IV emission.

During type III_1 the spectral character of the radio emission was clearly different at frequencies below and above the inferred frequency–time track of the CME_1 (see Figure 3). On the low-frequency side of the track strong type-III bursts were prominent after about 09:22 UT. They were preceded by a less regular emission, which Reiner *et al.* (2008) label "complex type-III bursts" because of its varying flux density across the spectrum.

The metre wave counterpart on the high-frequency side of the estimated CME track consisted of a succession of spectral fine structures on the time scale of seconds, with different spectral characteristics superposed on the type-IV continuum, followed after 09:11 UT by the high-frequency extension of the dekametre–hectometre (DH) type-III group III_1. A more detailed view of the difference spectrum is given in the top panel of Figure 4, whilst high-resolution images of the fine structures are in Figure 5. Among these fine structures were broadband pulsations, bursts with ordinary and reverse drift, and fibre bursts due to whistlers travelling upwards in the corona (see Figure 5, *e.g.*, 09:16:20 – 09:16:45 UT). The variety

[5] We have used 164 – 432 MHz contours overlaid on EIT 19.5 Å images.

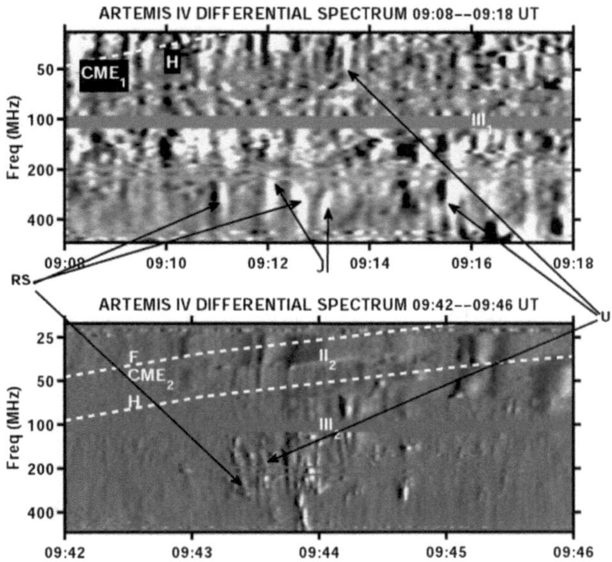

Figure 4 ARTEMIS IV differential spectra in the 20 – 550 MHz range. Top: type III_1 in the 09:08 – 09:18 UT interval. Some bursts of the type-III family (Reverse slope (RS), J-type, U-type) have been annotated. Bottom: type III_2 in the 09:42 – 09:46 UT interval. The frequency–time trajectories of CME_1 and CME_2 (white) are overlaid, assuming fundamental (F) and harmonic plasma emission.

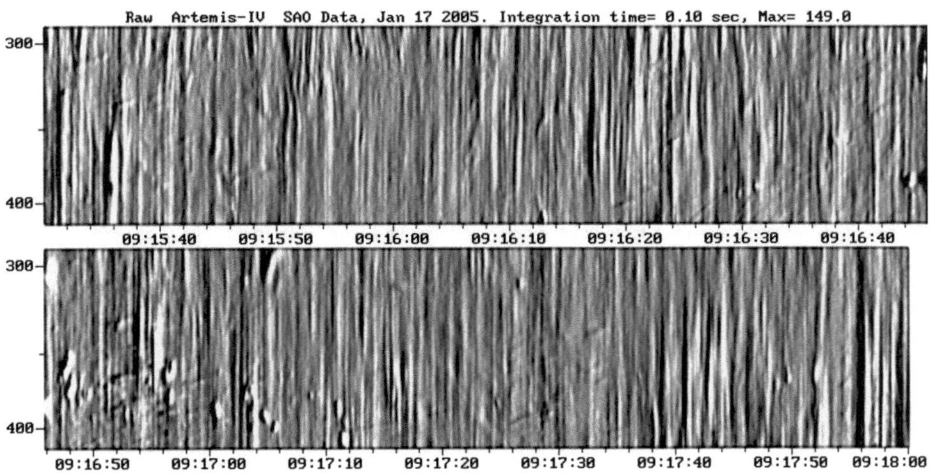

Figure 5 ARTEMIS IV high-resolution (SAO) differential spectra in the 290 – 415 MHz range and the 09:15:30 – 09:18:00 UT interval.

of these bursts shows the acceleration and partial trapping of electron populations in the corona well behind the front of the CME. Indeed, few of the well-identified bursts above 100 MHz seem to continue into the 30 – 70 MHz range. It was only near the end of type III_1 ($\approx$ 08:18 UT) that metre wave type-III bursts appeared as systematic high-frequency extensions of the type-III bursts observed below 2 MHz.

Type III_2 started at 09:43 UT, together with the second microwave burst, near the back-extrapolated lift-off of CME_2 (09:38 UT) and the onset of stage 2 of the SXR burst. A close look at the dynamic spectrum (bottom panel of Figure 4) reveals negative overall drifts below 100 MHz, whilst burst groups with positive overall drift prevailed above 130 MHz. The high-resolution spectrogram in the 300 – 400 MHz range (Figure 6) shows a wealth of individual bursts with different drift rates and zebra pattern. These bursts show again, like in

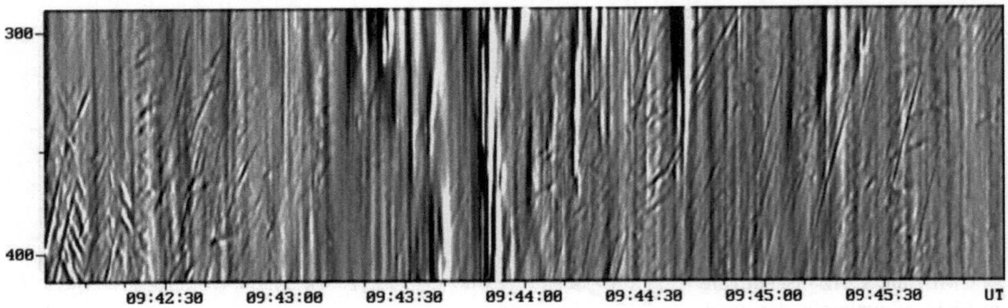

Figure 6 ARTEMIS IV high-resolution (SAO) differential spectra in the 290 – 415 MHz range and the 09:42 – 09:46 UT interval.

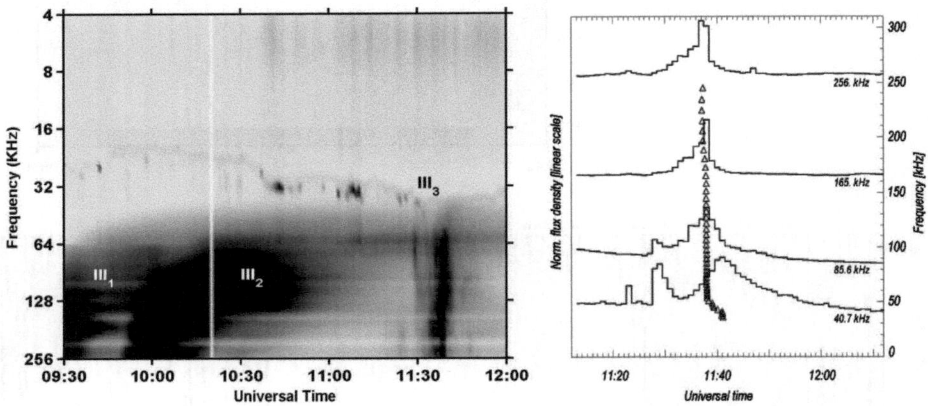

Figure 7 Kilometric radio bursts associated with the flares and with CME interaction. Left: WAVES/TNR dynamic spectrum (inverse grey scale) in the 4 – 256 kHz frequency range from 09:30 to 12:00 UT showing the burst groups type III_1, III_2 and III_3. Right: time histories during type III_3 (each normalized to its maximum) at selected frequencies. The open triangles show the time of maximum of the burst at each frequency in the range 35 – 256 kHz.

stage 1 of the event, that the high-frequency bursts are produced by an accelerator below the CME front, whilst low-frequency bursts show the start of the prominent DH type-III bursts.

2.5. Radio Emission and CME Interaction

Well after the decay of the SXR and microwave emission a third group of bursts (III_3 in Figure 3) is identified (near 11:30 UT), with unusually low starting frequency (0.6 MHz), pointing to an acceleration of the emitting electrons at unusually great height. A more detailed view of the low-frequency radio spectrum of this burst group and the preceding groups III_1 and III_2 is given by the dynamic spectrum as observed by the *Thermal Noise Receiver* (TNR) of *Wind*/WAVES in the left panel of Figure 7. The narrow-band short bursts near 32 kHz are Langmuir wave packets. Together with the fainter continuous band on which they are superposed they indicate how the electron plasma frequency evolves at the *Wind* spacecraft. At the time of III_3 it is about 35 kHz. Using a standard interplanetary density model, where the electron plasma frequency decreases as the inverse of the heliocentric distance R, the starting frequency of III_3 implies $R = 12\ R_\odot$ for fundamental plasma emission, and $R = 25\ R_\odot$ for the harmonic. From the LASCO observations and the uncertainties resulting from the straight-line fits to the CME front trajectories, the heliocentric distances of

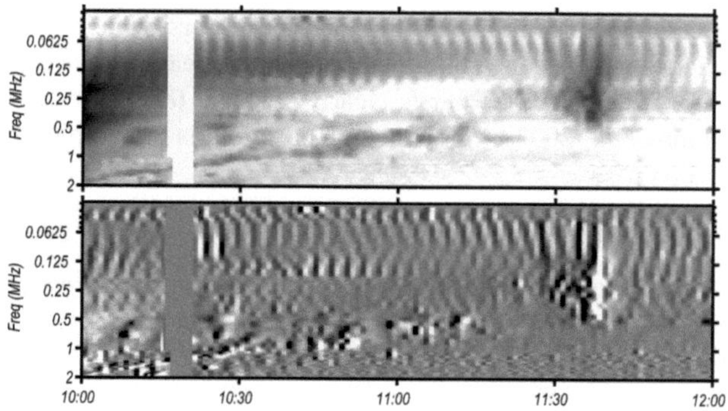

Figure 8 *Wind*/WAVES Spectra in the 2.0 – 0.03 MHz range with details of the merged type-II bursts (10:00 – 11:20 UT) and the low-frequency type III (III_3); about 11:30 – 11:40 UT). Top: intensity, bottom: differential.

the CME fronts at 11:30 UT are, respectively, $(26.9 \pm 4.8)\ R_\odot$ and $(24.5 \pm 5.5)\ R_\odot$. The burst group III_3 is hence consistent with harmonic emission from the vicinity of the CME fronts. This points to a close relationship of this episode of electron acceleration with the interaction of the two CMEs.

Comparison of the three groups of type-III bursts in the TNR spectrum of Figure 7 shows that type III_3 is much shorter than the previous type-III bursts. It has intrinsic structure that indicates a group of bursts. The low-frequency cutoff is near the plasma frequency at the spacecraft at that time. More details are seen in the selected time profiles in the right panel, plotted together with the peak times of the burst at each frequency in the 35 – 256 kHz range (open triangles). The time profiles show that the peak times are not distinguishable over a large part of the frequency spectrum with 1-min integrated data, but that the centre of gravity of the brightest feature shifts to later times at the lower frequencies. We determined the maximum of the burst at each of the TNR frequencies where it is well defined, using a parabolic interpolation between the observed maximum and its two neighbours. It is this interpolated time which is plotted by an open triangle. The peak time spectrum resembles a type-III burst especially at the lower frequencies. The peak time delay is merely 1 min between 250 and 50 kHz, but becomes clear at frequencies below 50 kHz. For comparison, the peak time delay between 50 and 250 kHz is 42 min during the previous burst III_2. The frequency drift rate is hence faster than 3 $kHz\,s^{-1}$ during III_3, as compared to 0.08 $kHz\,s^{-1}$ during III_2. Because of the morphological similarity in the dynamic spectrum, and despite the different drift rates, we assume in the following that the type-III_3 bursts are indeed produced by electron beams travelling in the anti-sunward direction from the acceleration region. Since the emission extends rapidly to the plasma frequency at the spacecraft, we conclude that the electron beams do not travel within the CMEs, but escape rapidly from the acceleration region in the vicinity of the CME fronts to 1 AU. This means that they must travel along pre-existing open solar wind field lines.

To the extent that drift rates reflect the speed of the exciter, the fast frequency drift of the type-III_3 bursts implies that the exciter speed is higher than during the preceding groups III_1 and III_2.

The total and differential radio spectrum observed by *Wind*/WAVES are shown in Figure 8. The spectrum shows a chain of narrow-band emissions with negative frequency drift, indicating the type-II bursts, followed by the high-frequency part of type III_3 between 11:28 and 11:40 UT. The spectrum in Figure 3 leaves it open if this is the continuation of the first type-II burst (II_1), presumably associated with CME_1, or whether it contains contributions from both CMEs. The starting frequency of the type-III bursts is similar to the type-II frequency when extrapolated to the time of the type-III bursts. This is consistent with the type-III electron beams radiating in the upstream region, like the shock-accelerated electrons emitting the type-II burst. One may go one step further and consider this coincidence as a hint that the electron beams are accelerated at the shock, as argued in cases where type-III bursts clearly emanate from type-II lanes (see Bougeret *et al.*, 1998; Mann *et al.*, 2003). We will come back to this problem in the Discussion.

2.6. Solar Wind and Energetic Electrons Near 1 AU

The energetic particle data were obtained from the *Advanced Composition Explorer* (ACE) spacecraft. We use high-resolution intensities of magnetically deflected electrons (DE) in the energy range 38 – 315 keV measured by the B detector of the CA60 telescope of the EPAM experiment (*Electron, Proton and Alpha Monitor*; Gold *et al.*, 1998) on board ACE, and measurements of the angular distributions in the energy range 45 – 312 keV detected by the sunward looking telescope LEFS60. In Figure 9 (top) an overview of the 20-min averaged differential intensities of four channels is presented for the interval 15 – 20 January 2005. AR 10720 produced numerous solar events prior to as well as on 17 January 2005 (Malandraki *et al.*, 2007; Papaioannou *et al.*, 2010); in response to this solar activity, a sequence of energetic electron intensity enhancements was observed. The electron intensities are observed to reach their maximum values during this period following the solar events on 17 January 2005. Figure 9 (bottom) shows 1-min averaged deflected electron intensities (38 – 315 keV) for the time interval 04:00 – 20:00 UT on 17 January 2005. The intensities measured during the time interval 04:00 – 08:00 UT for each electron channel have been averaged to obtain a pre-event background (denoted by horizontal lines in Figure 9). We defined the onset time of the event at ACE for all energy channels as the time when the intensities get $> 2\sigma$ above the background and continue to rise from then on. Using this criterion, we found the first significant rise of the electron intensities to occur at 10:00 UT. No velocity dispersion was observed, probably because the high pre-event ambient intensities (see top panel Figure 9) mask the onset of the electron event (see Malandraki *et al.*, 2005, for a similar case). The spiky increase observed at about 10:40 UT is probably due to X-ray contamination. We found no evidence of a magnetic structure influencing the intensity profiles, which indicates the observed time intensity changes are not due to spatial structures crossing over the spacecraft, but are most likely dominated by temporal effects.

Twenty-minute averaged representative snapshots of pitch angle distributions (PADs) are shown as inserts in the bottom panel of Figure 9. Normalized differential electron intensity is plotted versus the cosine of the pitch angle. Statistically significant PADs are detected first at about 11:00 UT. The PAD snapshot denoted as a in Figure 9 indicates that immediately after the onset of the event unidirectional electron anisotropies are observed.

Based on the observations available we cannot distinguish whether the electrons were directed sunward or anti-sunward, since the magnetic field (not shown) was directed dominantly transverse to the radial direction during this period (see Figure 7 in Lario *et al.*, 2004 for a similar case). However, it is highly likely that the observed electrons are streaming away from the Sun in response to the intense solar activity during this period. Furthermore the type-III bursts clearly indicate that electrons stream away from the Sun (towards regions of lower density). We cannot be certain that the electron population measured at ACE/EPAM is the high-energy counterpart of the electron beams emitting the radio waves, yet the overall timing suggests this.

In the work by Papaioannou *et al.* (2010) a detailed analysis of the plasma and magnetic-field measurements at 1 AU by ACE during the period 16 – 26 January 2005 was carried out. This includes the period under study in the present paper. A forward shock was detected at 07:12 UT on 17 January 2005. We have denoted the arrival time of this shock by a vertical

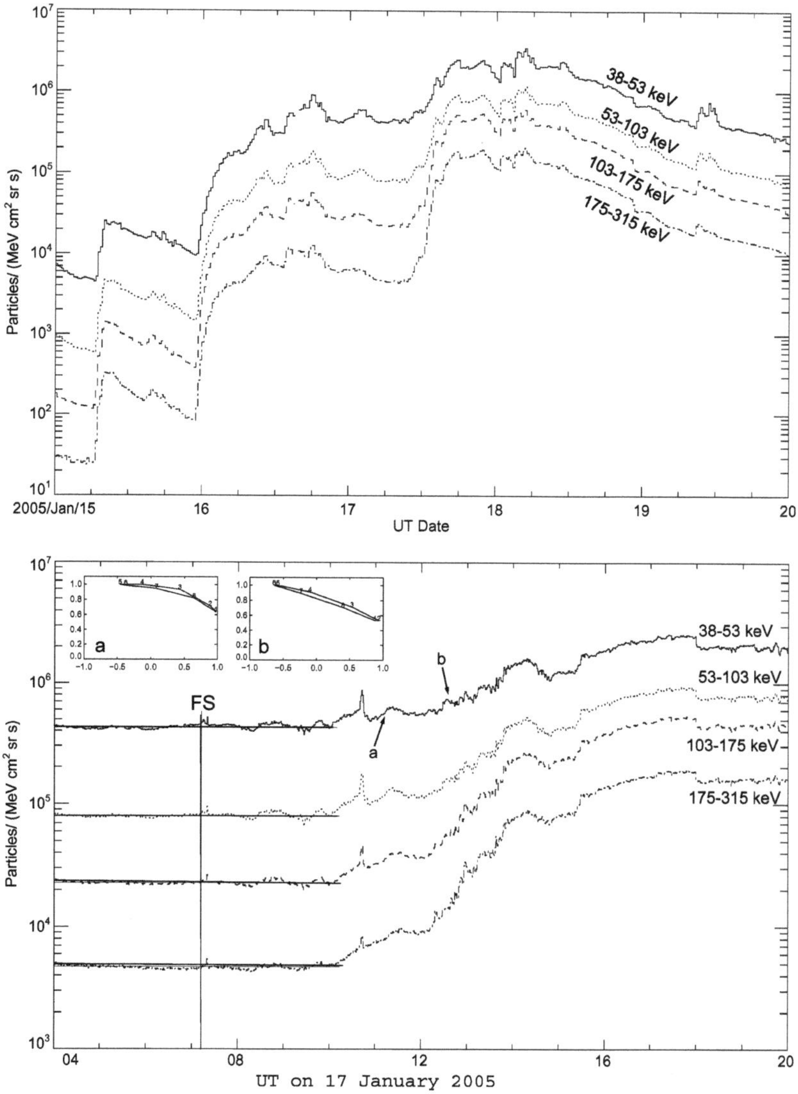

Figure 9 ACE/EPAM energetic electron enhancements in response to the 17 January 2005 solar activity. Top: 20-minute averaged differential intensities for the interval 15 – 20 January 2005. Labels of the abscissa are days. Bottom: 1-min averaged intensities (38 – 315 keV) for the interval 04:00 – 20:00 UT on 17 January 2005 of four magnetically deflected electron channels (38 – 315 keV) from the B detector head of the EPAM/CA60. Labels of the abscissa give the day (17 January) and the hour. The horizontal lines denote pre-event ambient intensity. The solid vertical line marks the arrival of an interplanetary forward shock (FS) at 07:12 UT at ACE. Inserts a and b present pitch angle distribution snapshots at the onset of the event at about 11:00 UT and at about 12:10 UT; normalized differential electron intensity is plotted *versus* the cosine of the pitch angle.

solid line in Figure 9 (bottom). The analysis has shown that after the passage of this shock an unusually extended region exhibiting sheath-like characteristics is observed for ≈ 1.5 day with highly variable magnetic-field magnitude and directions and typical to high proton temperatures (see Figure 3 of Papaioannou *et al.*, 2010 also Ruth Skoug, ACE/SWEPAM

PI team, private communication). This region is probably related to two CMEs ejected in close temporal sequence at the Sun on 15 January (see Figure 2 of Papaioannou *et al.*, 2010). Subsequently, at $\approx$ 23:00 UT on 18 January 2005 the arrival of an ICME at Earth is detected, ending at about 02:30 UT on 20 January.

The energetic electrons observed at 1 AU analyzed in this work were thus detected in the region with disturbed magnetic-field characteristics following the shock on 17 January 2005. For the purposes of this work, as an approximation, we calculated that the nominal Parker spiral for the measured solar wind speed of 620 $\mathrm{km\,s^{-1}}$ (ACE/SWEPAM) at the time of the rise of the electron intensity had a length of about 1.05 AU and was rooted near W 37° on the hypothetical solar wind source surface at 2.5 $R_{\odot}$. This longitude is not contradictory with an active region at W 25°, because non-radial coronal field lines can easily establish a connection (Klein *et al.*, 2008). Supposing that the early rise of the intensities was produced by the faster electrons in an energy channel moving along the interplanetary field line with 0 pitch angle, we estimate a travel time of about 15 min, which indicates the electrons were released from about 09:45 UT at the Sun. This corresponds to a photon arrival time at 09:53 UT. Given that our estimate of the electron rise gives only an upper limit, we consider that this electron release is related to type III_1 and type III_2 (Table 1), but cannot give a more detailed identification. The electron intensities are subsequently observed to exhibit a significant and more abrupt rise at all energies. Extrapolation of this second rise to the pre-event background intensities indicates a start at $\approx$ 12:00 UT. The electron PADs (inset b in Figure 9) indicate stronger unidirectional anisotropies are observed in association with this electron enhancement, which provides evidence for fresh injection of energetic electrons between the Sun and the spacecraft.

The outstanding radio emission near this time is the group of fast type-III bursts during the CME interaction, type III_3. If the electrons are accelerated at a heliocentric distance of about 25 $R_{\odot}$, the path travelled to the spacecraft along the nominal Parker spiral is 0.92 AU for the solar wind speed measured at the time of type III_3 (800 $\mathrm{km\,s^{-1}}$). The inferred upper limit of the solar release time is 11:46 UT for 100 keV electrons. Photons released at that time at 25 $R_{\odot}$ (0.12 AU) will reach the Earth about 7 min later. Since the high background implies that our estimations of the electron rise times are upper limits, we consider that this timing is consistent with the type-III_3 burst group near 11:37 UT (Table 1). This process is hence accompanied by the acceleration of copious amounts of electrons that escape to the vicinity of the Earth.

3. Discussion and Conclusions

On 17 January 2005 two flare/CME events occurred in close temporal succession in the same active region. Both CMEs had very high projected speed, above 2000 $\mathrm{km\,s^{-1}}$, but the second one was faster than the first and eventually overtook it. The CMEs were associated with two successive filament eruptions and SXR enhancements in the same active region. Since the filament eruptions occurred at neighbouring places in the parent active region, the CMEs probably resulted from the eruption of neighbouring parts of the same overall magnetic configuration. The soft X-ray characteristics of the two successive events were different: a slow monotonic rise to moderate flux during the first event, and a more impulsive rise to the X3.8 level ($3.8 \times 10^4\ \mathrm{W\,m^{-2}}$) during the second. Both events had a conspicuous microwave burst, but the first one was stronger than the second, contrary to the soft X-rays. The second burst was also seen in hard X-rays by RHESSI, which was in the Earth's shadow during the first burst.

3.1. Evidence for Evolving Acceleration Regions in the Corona During the Flares

The decametre-to-hectometre wave spectra in the two stages looked similar, with bright groups of type-III bursts signalling the escape of electron beams at heliocentric distances beyond 2 $R_{\odot}$. But radio emission from lower heights shows distinctive differences that point to an evolving acceleration region, *i.e.* either an acceleration region which progresses through the corona or a number of acceleration sites activated in succession.

Shock acceleration was clearly at work during both stages of the flare, as shown by the type-II emission. The strong type-III bursts at the low-frequency side of the estimated spectral track of the CME front could also be ascribed to the acceleration at the type-II shock. The presence of type-III bursts with negative drift at higher frequencies and the fine structures of the type-IV continuum show, however, that at the time when the shock travelled through the high corona, other acceleration regions were active at lower altitude, as is usually the case during complex type-III bursts at decametre and longer waves (see Reiner *et al.*, 2008 and references therein). The type-III bursts (III_1) might then not start near the CME front, but at higher frequencies, and be interrupted by interactions of the electron beams with the turbulence near the front of CME_1. This is a frequently quoted interpretation of complex features in type-III bursts, both at kilometric (MacDowall, 1989) and decametric wavelengths (*e.g.* Reiner and Kaiser, 1999).

In the second type-III group (III_2) the overall frequency drift of the low resolution spectrum (Figure 4) was positive. The persistence of the metric type-IV burst, which suggested acceleration in the lower corona rather than at the shock during the first stage, is again a likely indication of an accelerator that was distinct from the CME shock, and acted in addition to the shock, at lower altitude. This is consistent with an interpretation of the electron acceleration in terms of reconnection in the corona behind the CME (Trottet, 1986; Cliver *et al.*, 1986; Kahler and Hundhausen, 1992). New evidence for this interpretation has recently been provided by Aurass, Landini, and Poletto (2009) using UV and white light coronographic diagnostics along with radio data.

3.2. CME Interaction and Related Radio Emission

Decametric–hectometric radio emission as a signature of CME interaction was discussed in some detail in two event studies (Gopalswamy *et al.*, 2001, 2002). In both events the radio emission had a limited bandwidth, and was referred to as a continuum. In the present case this emission is likely a set of type-III bursts and was therefore labelled type III_3. The starting frequency and the timing of these bursts are consistent with the idea that the electrons are accelerated whilst the faster following CME catches up with the slower preceding one.

The association of CME interaction with particle acceleration has been ascribed by Gopalswamy *et al.* (2001) to acceleration by the shock of the second CME as it traverses the previous one. Problems of this interpretation were discussed by Klein (2006).

Another important question is how the preceding CME could lead to strengthening of the shock of the following one. This problem is still more evident in the 17 January 2005 event, because here the two CMEs are already extremely fast and likely to drive strong shocks even in the ambient solar wind. Their relative speed, however, is rather slow, so that efficient acceleration by the shock of the second CME is not expected in the first CME. Another important feature is that the type-III_3 bursts extend to the plasma frequency at the spacecraft. The electrons hence cannot propagate in closed magnetic structures related to the CMEs. The accelerator must release the electron beams onto open solar wind-type field lines. An alternative scenario to acceleration at the CME shock is again magnetic reconnection. One can surmise that these rapid CMEs were preceded by sheath regions with strong magnetic fields of interplanetary origin, draped around the CME front. These regions are favourable for magnetic reconnection (see the overview by Forsyth *et al.*, 2006 and references therein). The high pressure in the sheath of the second CME will be further enhanced when its progression is slowed down by the previous CME. This makes the configuration favourable to magnetic reconnection involving open solar wind field lines and strong magnetic fields, and allows one to understand qualitatively why accelerated electrons escape immediately towards the outer heliosphere.

In this scenario the type-III emission is expected to start close to the CME, in the upstream region. This can explain why the starting frequency is close to the frequency of the type-II burst, without implying that the electron beams were themselves accelerated at the shock.

3.3. Near-Relativistic Electrons at 1 AU

The flare/CME events under discussion were clearly related with enhanced fluxes of near-relativistic electrons at 1 AU. The peak intensity measured by ACE, of order $10^6\ (\mathrm{MeV\,cm^2\,sr\,s})^{-1}$ in the 38 – 53 keV range, makes the event comparable to the most intense ones of the sample studied by Haggerty and Roelof (2002), as seen in their Figure 3a. The CME speed is well above the speeds of the CMEs identified in that sample (Simnett, Roelof, and Haggerty, 2002).

Since the energetic electrons were observed at 1 AU within the region exhibiting sheath-like characteristics following the shock on 17 January 2005, it is difficult to estimate electron travel times and to relate the *in situ* measurements to solar processes. But the observations strongly suggest that successive intensity increases are first due to the coronal acceleration in the flare/CME event, and then to an episode during the interaction of the two CMEs. The escape of these electrons to ACE confirms the view discussed above that the electrons cannot have been accelerated in the body of the first CME, even if a shock driven by the second one passed through it. Neither can they originate from reconnection between closed magnetic field lines of the two CMEs. The electrons must rather be accelerated in regions from where they have ready access to solar wind magnetic field lines. This is consistent with a common acceleration of the mildly relativistic electrons and the electron beams at lower energies that produce the type-III_3 emission.

Acknowledgements This work was supported in part by the University of Athens Research Center (ELKE/EKPA). The authors appreciate discussions with and assistance of C. Caroubalos, C. Alissandrakis, and S. Hoang. They would also like to thank an anonymous referee for many useful comments on the original manuscript. Hα data were provided by the Kanzelhöhe Observatory, University of Graz, Austria by M. Temmer. The SOHO/LASCO data used here were produced by a consortium of the Naval Research Laboratory (USA), Max-Planck-Institut fuer Aeronomie (Germany), Laboratoire d'Astronomie (France), and the University of Birmingham (UK). The SoHO/LASCO CME catalog is generated and maintained at the CDAW Data Center by NASA and The Catholic University of America in cooperation with the Naval Research Laboratory. SOHO is a project of international cooperation between ESA and NASA. K.L.K. acknowledges the kind hospitality of the solar radio astronomy group at the University of Athens.

References

Aurass, H., Landini, F., Poletto, G.: 2009, *Astron. Astrophys.* **506**, 901.
Bastian, T.S., Benz, A.O., Gary, D.E.: 1998, *Annu. Rev. Astron. Astrophys.* **36**, 131.
Bougeret, J.-L., Kaiser, M.L., Kellogg, P.J., Manning, R., Goetz, K., Monson, S.J. *et al.*: 1995, *Space Sci. Rev.* **71**, 231.

Bougeret, J.-L., Zarka, P., Caroubalos, C., Karlický, M., Leblanc, Y., Maroulis, D. *et al.*: 1998, *Geophys. Res. Lett.* **25**, 2513.
Bouratzis, C., Preka-Papadema, P., Moussas, X., Alissandrakis, C., Hillaris, A.: 2009, *Adv. Space Res.* **43**, 605.
Brueckner, G.E., Howard, R.A., Koomen, M.J., Korendyke, C.M., Michels, D.J., Moses, J.D. *et al.*: 1995, *Solar Phys.* **162**, 357.
Caroubalos, C., Hillaris, A., Bouratzis, C., Alissandrakis, C.E., Preka-Papadema, P., Polygiannakis, J. *et al.*: 2004, *Astron. Astrophys.* **413**, 1125.
Caroubalos, C., Maroulis, D., Patavalis, N., Bougeret, J.L., Dumas, G., Perche, C. *et al.*: 2001, *Exp. Astron.* **11**, 23.
Cliver, E.W., Dennis, B.R., Kiplinger, A.L., Kane, S.R., Neidig, D.F., Sheeley, N.R., Koomen, M.J.: 1986, *Astrophys. J.* **305**, 920.
Delaboudinière, J.P., Artzner, G.E., Brunaud, J., Gabriel, A.H., Hochedez, J.F., Millier, F. *et al.*: 1995, *Solar Phys.* **162**, 291.
Forsyth, R.J., Bothmer, V., Cid, C., Crooker, N.U., Horbury, T.S., Kecskeméty, K., Klecker, B. *et al.*: 2006, *Space Sci. Rev.* **123**, 383.
Gold, R.E., Krimigis, S.M., Hawkins, S.E. III, Haggerty, D.K., Lohr, D.A., Fiore, E., Armstrong, T.P. *et al.*: 1998, *Space Sci. Rev.* **86**, 541.
Gopalswamy, N., Yashiro, S., Kaiser, M.L., Howard, R.A., Bougeret, J.L.: 2001, *Astrophys. J. Lett.* **548**, L91.
Gopalswamy, N., Yashiro, S., Kaiser, M.L., Howard, R.A., Bougeret, J.: 2002, *Geophys. Res. Lett.* **29**(8), 080000.
Güdel, M., Benz, A.O.: 1988, *Astron. Astrophys. Suppl. Ser.* **75**, 243.
Guidice, D.A., Cliver, E.W., Barron, W.R., Kahler, S.: 1981, *Bull. Am. Astron. Soc.* **13**, 553.
Haggerty, D.K., Roelof, E.C.: 2002, *Astrophys. J.* **579**, 841.
Kahler, S.W., Hundhausen, A.J.: 1992, *J. Geophys. Res.* **97**, 1619.
Kerdraon, A., Delouis, J.M.: 1997, In: Trottet, G. (ed.) *Coronal Phys. from Radio and Space Observations, Lecture Notes in Phys.* **483**, Springer, Berlin, 192.
Klein, K.-L.: 2006, In: Gopalswamy, N., Mewaldt, R., Torsti, J. (eds.) *Solar Eruptions and Energetic Particles, AGU Monogr.* **165**, Amer. Geophysical Union, Washington, 233.
Klein, K.-L., Krucker, S., Lointier, G., Kerdraon, A.: 2008, *Astron. Astrophys.* **486**, 589.
Kontogeorgos, A., Tsitsipis, P., Caroubalos, C., Moussas, X., Preka-Papadema, P., Hilaris, A. *et al.*: 2006, *Exp. Astron.* **21**, 41.
Kuijpers, J.: 1980, In: Kundu, M.R., Gergely, T.E. (eds.) *Radio Phys. of the Sun, IAU Symp.* **86**, 341.
Lario, D., Decker, R.B., Roelof, E.C., Reisenfeld, D.B., Sanderson, T.R.: 2004, *J. Geophys. Res.* **109**, 1107.
Lin, R.P., Dennis, B.R., Hurford, G.J., Smith, D.M., Zehnder, A., Harvey, P.R. *et al.*: 2002, *Solar Phys.* **210**, 3.
MacDowall, R.J.: 1989, *Geophys. Res. Lett.* **16**, 923.
Malandraki, O.E., Lario, D., Lanzerotti, L.J., Sarris, E.T., Geranios, A., Tsiropoula, G.: 2005, *J. Geophys. Res.* **110**, 9.
Malandraki, O.E., Marsden, R.G., Tranquille, C., Forsyth, R.J., Elliott, H.A., Lanzerotti, L.J., Geranios, A.: 2007, *J. Geophys. Res.* **112**, 6111.
Mann, G., Klassen, A., Aurass, H., Classen, H.T.: 2003, *Astron. Astrophys.* **400**, 329.
McLean, D.J., Labrum, N.R. (eds.): 1985, *Solar Radiophys.: Studies of Emission from the Sun at Metre Wavelengths.*
Nindos, A., Aurass, H., Klein, K.L., Trottet, G.: 2008, *Solar Phys.* **253**, 3.
Papaioannou, A., Malandraki, O., Belov, A., Skoug, R., Mavromichalaki, H., Eroshenko, E., Abunin, A., Lepri, S.: 2010, *Solar Phys.* **266**, 181.
Pick, M., Vilmer, N.: 2008, *Astron. Astrophys. Rev.* **16**, 1.
Reiner, M.J., Kaiser, M.L.: 1999, *Geophys. Res. Lett.* **26**, 397.
Reiner, M.J., Klein, K., Karlický, M., Jiřička, K., Klassen, A., Kaiser, M.L., Bougeret, J.: 2008, *Solar Phys.* **249**, 337.
Schmidt, J., Cargill, P.: 2004, *Ann. Geophys.* **22**, 2245.
Simnett, G.M., Roelof, E.C., Haggerty, D.K.: 2002, *Astrophys. J.* **579**, 854.
Suzuki, S., Dulk, G.A.: 1985, In: McLean, D., Labrum, N. (eds.) *Solar Radiophys.: Studies of Emission from the Sun at Metre Wavelengths.* Cambridge University Press, Cambridge, 289.
Temmer, M., Veronig, A.M., Vršnak, B., Miklenic, C.: 2007, *Astrophys. J.* **654**, 665.
Trottet, G.: 1986, *Solar Phys.* **104**, 145.
Vršnak, B., Magdalenić, J., Zlobec, P.: 2004, *Astron. Astrophys.* **413**, 753.
Yashiro, S., Gopalswamy, N., Michalek, G., St. Cyr, O.C., Plunkett, S.P., Rich, N.B., Howard, R.A.: 2004, *J. Geophys. Res.* **109**(18), 7105.

Solar Phys (2011) 273:511–523
DOI 10.1007/s11207-011-9770-1

Combined Analysis of Ultraviolet and Radio Observations of the 7 May 2004 CME/Shock Event

Salvatore Mancuso

Received: 12 November 2010 / Accepted: 4 April 2011 / Published online: 11 May 2011

Abstract We report results from the combined analysis of UV and radio observations of a CME-driven shock observed on 7 May 2004 above the southeast limb of the Sun at 1.86 $R_\odot$ with the Ultraviolet Coronagraph Spectrometer (UVCS) on board the *Solar and Heliospheric Observatory* (SOHO). The coronal mass ejection (CME) was first detected in white-light by the SOHO's Large Angle and Spectrometric Coronagraph (LASCO) C2 telescope and shock-associated type II metric emission was recorded simultaneously by ground-based radio spectrographs. The shock speed ($\sim 690\ \mathrm{km\,s^{-1}}$), as deduced from the analysis of the type II emission drift in the radio spectra and the pre-shock local electron density estimated with the diagnostics provided by UVCS observations of the O VI $\lambda\lambda$ 1031.9, 1037.6 doublet line intensities, is just a factor ~ 0.1 higher than the CME speed inferred by means of the white-light (and EUV) data in the middle corona. The local magnetosonic speed, computed from a standard magnetic field model, was estimated as high as $\sim 600\ \mathrm{km\,s^{-1}}$, implying that the CME speed was probably just sufficient to drive a weak fast-mode MHD shock ahead of the front. Simultaneously with the type II radio emission, significant changes in the O VI doublet line intensities and profiles were recorded in the UVCS spectra and found compatible with abrupt post-shock plasma acceleration and modest ion heating. This work provides further evidence for the CME-driven origin of the shocks observed in the middle corona.

Keywords Sun: radio radiation · Sun: coronal mass ejections (CMEs) · Shock waves · Line: profiles

Energy Storage and Release through the Solar Activity Cycle – Models Meet Radio Observations
Guest Editors: Christophe Marqué and Alexander Nindos

S. Mancuso (✉)
Istituto Nazionale di Astrofisica (INAF), Osservatorio Astronomico di Torino, Strada Osservatorio 20, 10025 Pino Torinese (To), Italy
e-mail: mancuso@oato.inaf.it

1. Introduction

The combined analysis of radio, white-light, and ultraviolet spectral signatures detected during coronal transients represents a powerful tool in order to obtain important information about the main coronal plasma physical parameters at distances from the Sun that have not yet been directly accessed by spacecraft. In particular, suitable coronal diagnostics are provided by the analysis of the radiation emitted by the local plasma during the passage of coronal shock waves (see, *e.g.*, Mancuso *et al.*, 2003).

The origin of coronal shock waves is not yet completely understood and has been a long-standing source of controversy in solar physics (see reviews from Nindos *et al.*, 2008; Vršnak and Cliver, 2008). Coronal shocks have been either attributed to flare-ignited blast waves due to a sudden short-range expansion of the flare volume caused by impulsive heating (Uchida, 1974; Vršnak and Lulic, 2000a, 2000b; Hudson *et al.*, 2003; Magdalenić *et al.*, 2008) or, alternatively, driven by flare ejecta (Gopalswamy *et al.*, 1997, 1999; Klein *et al.*, 1999) and/or coronal mass ejections (CMEs) (Gosling *et al.*, 1976; Maxwell, Dryer, and McIntosh, 1985; Cliver, Webb, and Howard, 1999; Mancuso and Raymond, 2004; Gopalswamy, 2006). Even after several decades, the controversy is still open. For example, Liu *et al.* (2009) recently reported evidence of a CME-driven shock based on the analysis of high-cadence observations from STEREO/SECCHI. On the other hand, Magdalenić *et al.* (2010), through the analysis of a small sample of events, claimed that a particular class of flare-generated shock waves may actually exist. In fact, even detailed case studies do not often provide an unambiguous answer to the question whether a given coronal wave was driven by an ejection or ignited by a flare. Especially this is so in the light of the findings of Mancuso and Raymond (2004) who showed, on the basis of the analysis of an extended set of coronal shocks observed in the year 1999, that the apparent lag between CME leading edge and shock source, often invoked as evidence of their blast-wave origin, can be simply attributed to a geometrical effect. The present work is thus aimed to yield further evidence for the CME-driven origin of the shocks observed in the middle corona.

When the difference in speed between an outward propagating CME and the solar wind is larger than the local magnetosonic speed, a forward fast-mode magnetohydrodynamic (MHD) shock is produced ahead of the front (see, *e.g.*, Lin, Mancuso, and Vourlidas, 2006). Although CME-driven shocks may be too faint to be detected by white-light coronagraphs in the visible range, they can be easily identified in radio dynamic spectra as type II radio bursts, that is, slowly drifting, narrow bands of enhanced radio emission near the local electron plasma frequency f_{pe} and/or its harmonics (Wild, 1950; Nelson and Melrose, 1985). The type II emission, generally attributed to locally excited electron plasma (Langmuir) oscillations which in turn non-linearly convert into escaping electromagnetic waves, relates directly to the local electron number density n_e ($f_{pe} \sim 9000\sqrt{n_e\,[\mathrm{cm}^{-3}]}$ Hz) just upstream of the CME-driven shock, and thus to the burst driver's height whenever the coronal electron density radial profile is known (or, at least, can be confidently assumed).

Although observations via ground-based radio spectrographs of coronal type II emissions are very common throughout the solar cycle (see, *e.g.*, Gopalswamy, 2006) and they do represent the clearest remote sensing signature of the presence of a coronal shock, unfortunately they lack to provide spatial information about the actual location of the shock in the corona. In fact, analyses of radio spectra yield only *some* details about the physical properties of the shocked plasma, while their identification with observed features in coronagraphic white-light images is often problematic. Recently, however, Ontiveros and

Vourlidas (2009) demonstrated that careful analysis of well-calibrated base difference images can be profitably used to identify shock signatures in white-light images and even allow, in selected events, to distinguish between shock-compressed plasma and CME material. Notwithstanding the above, both radio and white-light observations have important limitations when crucial topics such as the mechanisms of heating of the shocked coronal ions or the line-of-sight speed of the emitting plasma are involved. It is thus essential for this purpose to use spectroscopic instruments that are sensitive to both kinetic (*via* Doppler shift) and thermal (*via* spectral broadening) effects produced on the coronal plasma by the passage of the MHD shock.

Ultraviolet emission lines present a very important source of diagnostic information about the solar corona, since line intensities and profiles critically depend on the electron density, the electron temperature, the density, the kinetic temperature of the emitting ions, and on the outflow velocity. In the past decade, spectral observations obtained by the UltraViolet Coronograph Spectrometer (UVCS; Kohl *et al.*, 1995) on board the *Solar and Heliospheric Observatory* (SOHO) have been effectively used for shock detection and modeling and coronal diagnostics (see, *e.g.*, Raymond *et al.*, 2000; Mancuso *et al.*, 2002; Raouafi *et al.*, 2004; Mancuso and Raymond, 2004; Mancuso and Abbo, 2004; Ciaravella *et al.*, 2005; Ciaravella, Raymond, and Kahler, 2006; Mancuso, 2007; Mancuso and Avetta, 2008; Bemporad and Mancuso, 2010). Overall, these works have established that when a CME-associated shock reaches the field of view of the UVCS slit (usually in the range between ~ 1.5 and 4 $R_{\odot}$ from the center of the Sun), the post-shock H I Lyα emission fades, the O VI doublet spectral line profiles broaden (and usually dim by $\sim 10\%$) and the emission from heavier ions (such as Si XII) increases by about a factor of two.

In the present work, we analyze EUV, radio, and white-light signatures of a CME/shock event observed on 7 May 2004 at 1.75 $R_{\odot}$ above the southeast limb of the Sun with the UVCS instrument. The above analysis will provide further evidence for the CME-driven origin of the shocks observed in the middle corona. Section 2 describes the white-light, radio, and EUV observations. The data analysis and results are presented in Section 3, and finally a summary and conclusions are given in Section 4.

2. Observations

2.1. White-Light Observations and Flare Activity

On 7 May 2004, a CME event was detected in white-light by the Large Angle Spectrometric Coronagraph (LASCO; Brueckner *et al.*, 1995) C2 telescope on board SOHO. LASCO C2 is a white-light coronal imager designed with a near 360° field of view covering heights from about 2.25 to 6.0 $R_{\odot}$. The CME was associated with a B7.3 class flare observed with the GOES full-Sun X-ray monitors originating at 9:56 UT and located at the East Limb, with peak emission around 10:18 UT. The CME leading edge was first observed in white-light LASCO C2 images at 10:50 UT (Figure 1, left panel). Subsequent images show a bright, irregular front (Figure 1, middle and right panels), expanding eastward with an initial speed around $\sim 600\ \mathrm{km\,s^{-1}}$ (obtained through a second-order polynomial fit to the white-light data) and decelerating away from the Sun at a rate of $\sim -10\ \mathrm{m\,s^{-2}}$ (as reported in the online LASCO CME Catalog; Gopalswamy *et al.*, 2009). In addition, the LASCO images clearly show a streamer being deflected when the expanding northern flank of the CME reaches its location (black arrow in Figure 2, right panel). Streamer deflections (see, *e.g.*, Gosling *et al.*, 1974; Michels *et al.*, 1984; Sheeley, Hakala, and Wang, 2000; Vourlidas *et al.*, 2003)

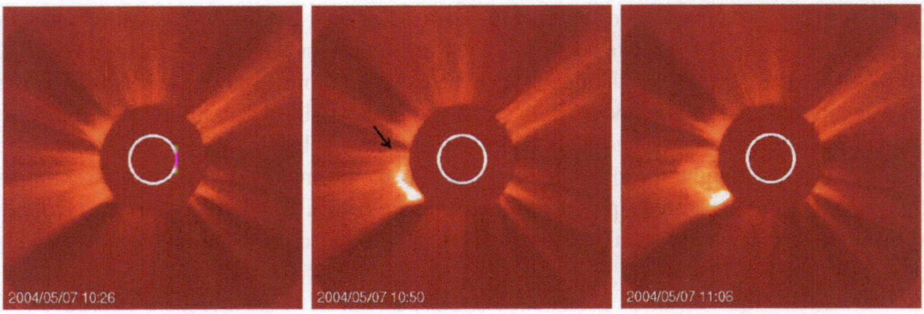

Figure 1 LASCO C2 images at three times during the evolution of the CME which occurred on 7 May 2004. The black arrow points to the deflected streamer.

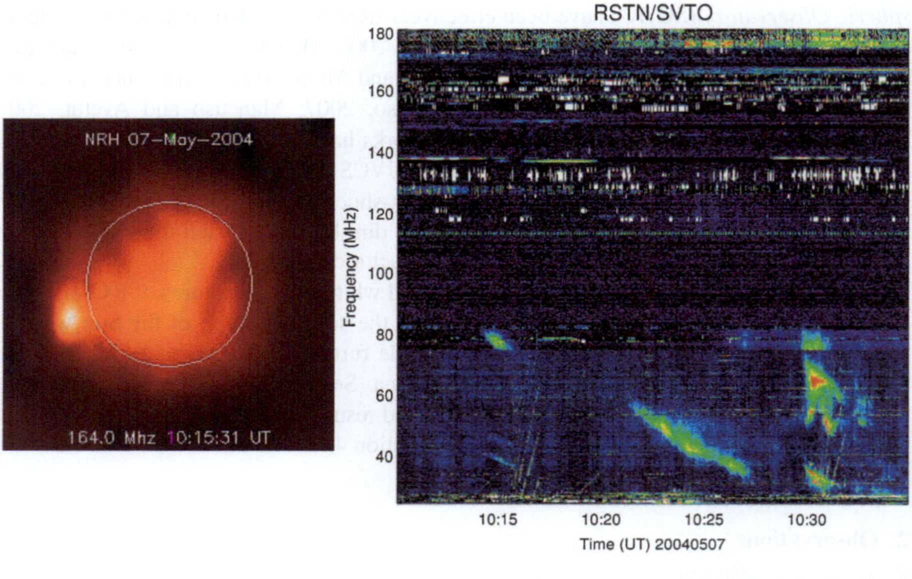

Figure 2 Left panel: Nançay image at 164 MHz showing the emitting region. Right panel: Dynamic spectrum from the San Vito radio spectrometer showing the metric type II radio emissions observed on 7 May 2004.

provide a reliable indication of a CME-driven wave pushing out the streamers, although there remains the question of whether this wave is a shock wave, especially for the cases where the CME speed is not excessively high (see, *e.g.*, Filippov and Srivastava, 2010).

2.2. Radio Observations

Metric type II bursts were reported in the *Solar Geophysical Data* (SGD) by the Izmiran digital radio spectrograph (Gorgutsa *et al.*, 2001) starting at 10:21 UT at 62 MHz and ending at 10:34 UT at 25 MHz. The same emission was also observed at metric wavelengths by the San Vito radio spectrometer (Figure 2, right panel), operated by the United States Air Force Radio Solar Telescope Network (RSTN) and by the radio spectrograph of the Astrophysikalisches Institut Potsdam (AIP; Mann *et al.*, 1992) starting at 10:14.2 UT at

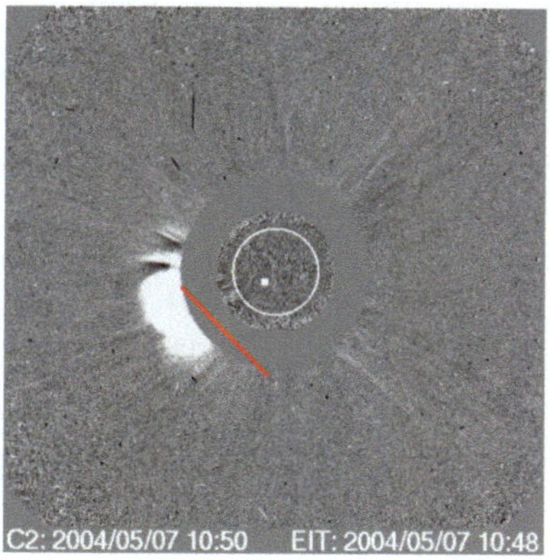

Figure 3 LASCO C2 running difference image of the 7 May 2004 CME imaged at 10:50 UT. The overlaid bar represents the approximate position of the UVCS slit at 1.75 $R_{\odot}$.

85 MHz. Interestingly, strong emission in the 164 MHz band just above the southeast limb (Figure 2, left panel), approximately at the heliographic latitude of the CME/shock emission observed by the UVCS instrument, was also imaged by the Nançay Radioheliograph (NRH; The Radioheliograph Group, 1993) and might be harmonically related to the initial type II emission observed by the AIP radio spectrograph. However, since there are no associated type II signatures at that frequency (164 MHz) in the radio spectra reported by the above mentioned ground-based observatories and since the EUV and radio observations analyzed in this work refer to higher heliocentric distances, we refrain from making further comments on this particular imaging observation and its association with the coronal shock studied in this paper.

2.3. Ultraviolet Observations

During the CME expansion, UVCS was observing above the southeast limb with the entrance slit centered to a mirror height of 1.75 $R_{\odot}$ from the center of the Sun, with the point along the slit closest to the Sun at an angle 135° from the solar North pole (Figure 3). UVCS is a long-slit UV spectrograph consisting of two spectrometric channels for the observation of spectral lines in the EUV range and a visible light channel for polarimetric measurements of the extended solar corona. The 42′ slit is placed parallel to a tangent to the solar limb and can be moved along the radial direction to observe the entire corona at heliocentric distances between 1.4 and 10 $R_{\odot}$. In order to cover all the possible position angles, the slit can be rotated by 360° about an axis pointing to the center of the Sun. The spectral observations considered here were part of a synoptic study and consisted of a set of forty-nine 120 s exposures separated by about 4 s readout intervals (see Table 1) providing the spectrum of the coronal region in the instantaneous field of view determined by the spectrometer entrance slit. The O VI data were taken with a spatial binning of 3 pixels, giving a spatial resolution of 21 arcsec along the slit. Data calibration is described by Gardner *et al.* (2002). The most prominent spectral lines detected during this time interval were the O VI $\lambda\lambda$ 1031.9, 1037.6 doublet, while other lines within the spectrum observed by the UVCS instrument produced only a very weak signal. The slit width was set equal to 200 μm in order to optimize both spectral resolution and photon flux. The four exposures listed in Table 1 are the only ones that were available with the UVCS slit at a mirror height of 1.75 $R_{\odot}$ and an angle 135° from the solar North pole. A $\sim$5′ portion of the slit over which the observed flux was integrated at each exposure (corresponding to an effective height of 1.86 $R_{\odot}$) was chosen in order to match the location of the intersection of the CME/shock with the slit. Other exposures were

Table 1 Summary of UVCS observations.

Exposure	Start (UT)	End (UT)	O VI ratio	$I_{\mathrm{O\,VI}\ 1032}$ (ph s^{-1} cm^{-2} sr^{-1})	FWHM (Å)
43	10 19:56	10:21:56	2.72	4.71×10^9	0.45
44	10 22:00	10:24:00	2.89	5.20×10^9	0.48
45	10 24:04	10:26:04	1.86	4.41×10^9	0.59
46	10 26:08	10:28:08	2.74	6.92×10^9	1.07

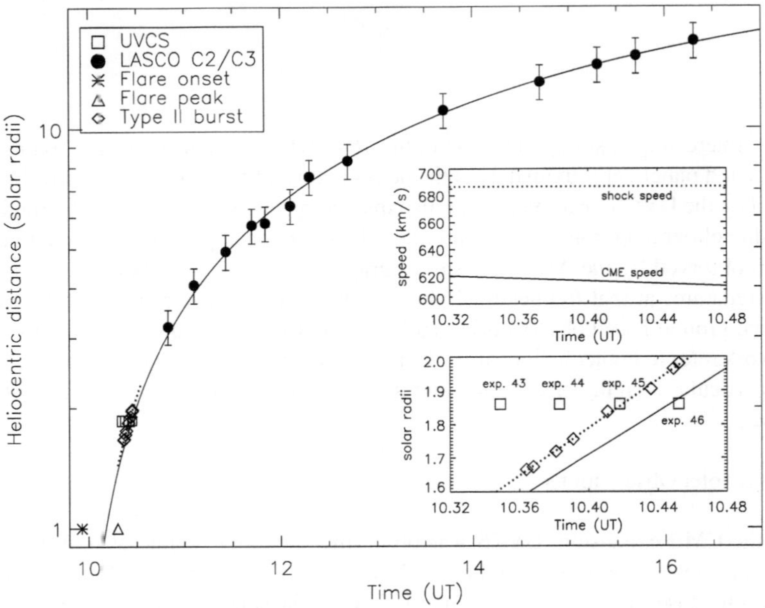

Figure 4 Height–time data for the CME observed by LASCO C2 and C3 from 10:50 to 16:18 UT on 7 May 2004 (filled circles with 10% error bars). The solid line is the best quadratic fit to the measured data. The dotted line shows the height-time shock trajectory as deduced from the drift rates of the type II radio emissions and the electron density estimated with the UVCS diagnostics. The insets show the CME and shock speeds as a function of time and a zoom of the height-time plot during the UVCS observations examined in this work.

also available before and after the CME/shock transit, but at different angles and heights, and therefore they were not used in this study.

3. Analysis and Results

3.1. CME Dynamics

Figure 4 shows the height–time plot (filled circles with approximate error bars estimated as 10% of the measured heights; see Yashiro *et al.*, 2001) derived from determining the location of the CME leading edge through consecutive LASCO coronagraph images from 10:50 to 16:18 UT. The height–time data (comprehensive of the UVCS's CME front detection at

1.86 $R_\odot$ in exp. 46) are well fit by a quadratic function (solid line). The extrapolation to the solar surface of the second-order polynomial fit to the height–time CME trajectory yields an onset time within the interval between the flare's onset and the soft X-ray peak emission. We remark that for this event projection effects should be unimportant due to both the location of the associated flare and the absence of important Doppler shifts in the observed EUV spectra.

3.2. Electron Density

In order to estimate the shock speed from the type II emission frequency drift seen in the radio spectrograph observations presented in the previous section, the relationship between the electron density and the distance r from the Sun's center has to be known or at least assumed. It is well established observationally that interplanetary type II radio emissions occur just upstream of the shock (Cairns, 1986; Reiner *et al.*, 1997; Thejappa, MacDowall, and Vinas, 1997; Bale *et al.*, 1999; Knock *et al.*, 2001). Type II radio emission thus refers to the pre-shocked plasma. Ultraviolet lines in the lower corona form essentially by collisional excitation followed by radiative de-excitation. The intensities of both the radiatively and the collisionally excited lines depend on the electron density n_e. For a nearly static plasma, as we expect above the active region (at least below $\sim 2\ R_\odot$), Doppler dimming is negligible and the coronal electron density can be directly estimated from the observed ratio of the radiative and collisional components of the O VI 1032 Å line (Noci, Kohl, and Withbroe, 1987), given by

$$\frac{I_{\mathrm{rad}}}{I_{\mathrm{col}}} = \frac{5.75 \cdot 10^2 \lambda^2 \mathrm{e}^{E/k_\mathrm{B} T_\mathrm{e}} \sqrt{T_\mathrm{e}} I_{\mathrm{disk}}}{\pi g n_\mathrm{e} \sqrt{\Delta\lambda_{\mathrm{cor}}^2 + \Delta\lambda_{\mathrm{ex}}^2}} \Delta\Omega(r), \tag{1}$$

where $\lambda = 1031.912$ Å, $E = hc/\lambda$ is the energy difference between the levels involved in the transition, g is the effective Gaunt factor, $\Delta\lambda_{\mathrm{ex}}$ is the $1/e$ half-width of the disk line, $\Delta\lambda_{\mathrm{cor}}$ is the $1/e$ half-width of the coronal absorption profile, $\Delta\Omega(r) = 2\pi[1 - \sqrt{1-(R_\odot/r)^2}]$ is the solid angle subtended by the solar disk at a distance r, T_e is the electron temperature, and I_{disk} is the intensity of the line integrated over the disk. Past UVCS observations in streamers around 1.75 $R_\odot$ yield $\log T_\mathrm{e} \sim 6.0-6.3$ range (see, *e.g.*, Ko *et al.*, 2002; Bemporad *et al.*, 2003; Uzzo, Ko, and Raymond, 2004). We thus assume $\log T_\mathrm{e} = 6.15$ (that is, $T_\mathrm{e} = 1.4 \times 10^6$ K). For the O VI λ 1032 disk intensity, we obtained an estimate of 418 erg s^{-1} cm^{-2} sr^{-1} by averaging the values obtained from UVCS disk measurements on October and December 1996 (Zangrilli *et al.*, 2002) and multiplying by a factor of 1.18 corresponding to the parametrized increase of the solar spectral irradiances around the O VI 1032 Å wavelength (see, *e.g.*, Lean *et al.*, 1997, 2005; Lean, 2001) at the time of the CME detection with respect to the above mentioned values. In order to obtain an estimate of n_e at the distance from the Sun corresponding to the UVCS slit's field of view, the two coronal intensity components, I_{rad} and I_{col}, have been derived from the simple relationship between the intensities of the collisionally excited and radiatively excited components of the O VI doublet (Noci, Kohl, and Withbroe, 1987), that is, $\frac{I_{\mathrm{col}}(1032)}{I_{\mathrm{col}}(1037)} = 2$ and $\frac{I_{\mathrm{rad}}(1032)}{I_{\mathrm{rad}}(1037)} = 4$, with the total intensity given by $I_{\mathrm{tot}} = I_{\mathrm{rad}} + I_{\mathrm{col}}$. From the analysis of the pre-shock observed spectra, we thus derived a value of $n_\mathrm{e} = 5.12 \cdot 10^6\ \mathrm{cm}^{-3}$ at 1.86 $R_\odot$ (the distance from the Sun corresponding to the average height of the $\sim 5'$ portion of the UVCS slit where the shock has been detected), that is, 0.58 times a Newkirk (1961) model at that height.

3.3. Shock Speed

In order to extract the frequencies from the San Vito's radio dynamic spectrum of the 7 May 2004 event, we used a software developed by John Kennewell and colleagues at Learmonth Solar Observatory, freely available for download from the NOAA National Geophysical Data Center (NGDC) ftp site.[1] The above website provides access to a number of Space Weather databases, including the RSTN solar radio spectral data used in this work. By using the observed frequency drift rate of the metric type II emission and assuming a $0.58\times$ Newkirk (1961) electron density radial profile, we obtained an average shock speed of $\sim 690\ \mathrm{km\,s^{-1}}$, that is, a factor ~ 1.1 higher than the CME speed ($\sim 620\ \mathrm{km\,s^{-1}}$) as deduced from a quadratic fit to the LASCO data, comprehensive of the UVCS's CME front detection at $1.86\ R_{\odot}$ in exp. 46. In Figure 4, we plot the position of the type II emission as a function of time (diamonds) and a linear fit to the same data (dotted line). In the two insets of the same figure, we also show an enlargement of the CME/shock trajectory and the shock (dotted line) and CME speeds (solid line) during the time of the type II emission.

3.4. The Ultraviolet Spectra

A major indicator of a shock in the UVCS spectra is the simultaneity between the type II radio emission detected by the radio dynamic spectra and the EUV detection. In order to discuss this topic, we need to go into more detail on the mechanism of formation of the O VI spectral lines. In general, the two major features observed in the UVCS spectra after the passage of a shock are the dimming and broadening of the O VI spectral line profiles (see, *e.g.*, Mancuso *et al.*, 2002). As already discussed, the coronal O VI emission is due to both thermal electron impact excitation and resonant scattering of the radiation coming from the disk. At the passage of a CME/shock, because of the sudden plasma acceleration, fewer ions are able to scatter the disk radiation because their coronal absorption profiles are Doppler shifted apart with respect to the disk's radiation. The O VI line intensities are thus expected to drop abruptly, unless the collisional component of the shocked plasma is able to balance such effect. Moreover, the higher electron temperature in the shocked plasma excites O VI ions to higher ionization states, which further reduces the line intensities.

Portions of spectra taken from four successive exposures containing the O VI doublet lines are shown in Figure 5 (left panels), along with their respective Gaussian fits (right panels). The $\sim 5'$ portion of the slit over which the observed flux was integrated at each exposure was chosen in order to match the location of the intersection of the CME/shock with the slit's field of view. At 10:25 UT (exp. 45 in Figure 5), the O VI spectral line suddenly broadened by a factor of ~ 1.3 and dimmed by $\sim 10\%$ with respect to the average intensity in the two previous exposures representing the unshocked plasma (see Table 1). As can be evinced from Figure 4, the timing of the shock passage, as detected by the ground-based radio spectrographs, nicely overlaps the average observation time in exp. 45, thus confirming the identification of the shock feature in the EUV spectrum. Hence, according to our previous discussion, the UVCS data provide a direct signature of shock plasma acceleration.

Another major indicator of a shock is the coincidence between the electron density estimate obtained through the radio data and the one obtained with the EUV data. The frequency of the type II emission at the time when the shock has been observed by the UVCS slit indicates the electron density of the pre-shock plasma. This latter can be directly compared with the pre-shock electron density already determined from the O VI doublet ra-

[1] ftp://ftp.ngdc.noaa.gov/STP/SOLAR_DATA/SOLAR_RADIO/SPECTRAL_RSTN.

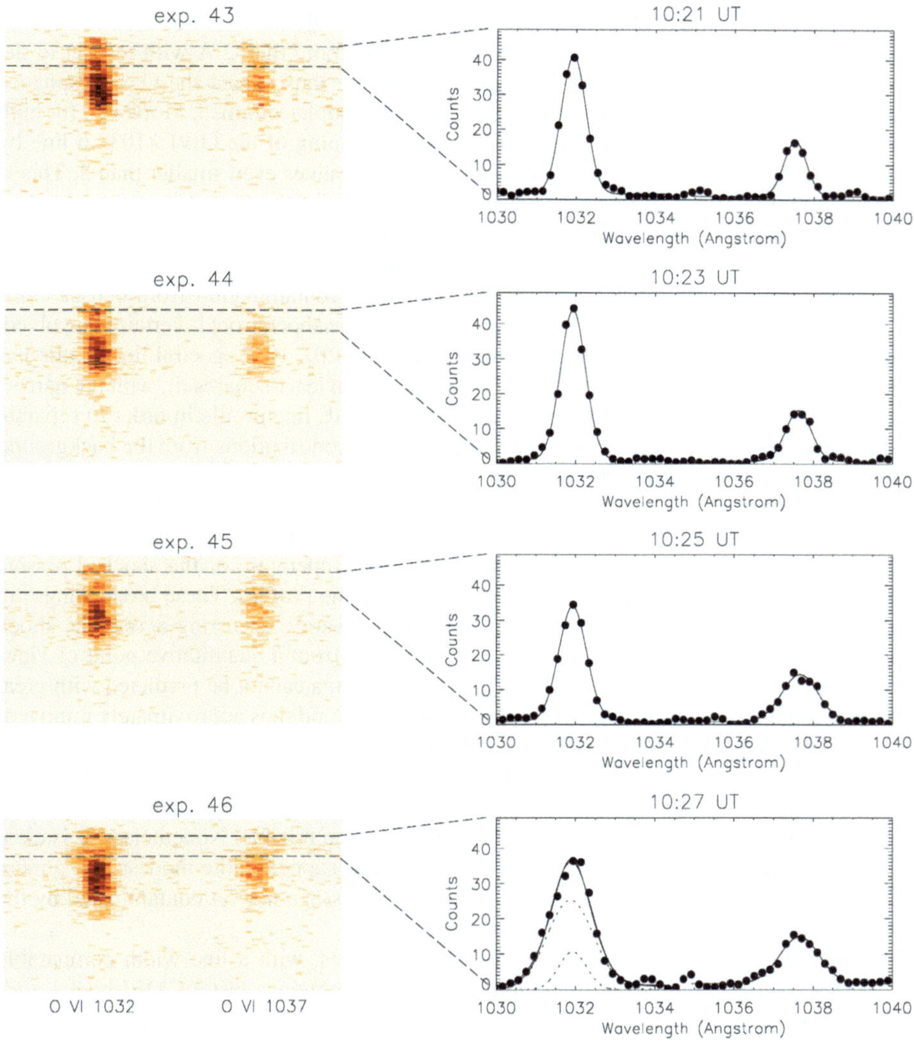

Figure 5 Left panels: Spectra of the 7 May 2004 CME/shock taken at 1.75 $R_\odot$ showing the temporal evolution during the CME/shock passage of the O VI doublet line intensities along the UVCS slit. The vertical axis yields the instantaneous field of view (about 42′), while the horizontal axis is along the spectral dispersion. The dashed lines indicate the $\sim 5'$ portion of the slit where the emission has been spatially integrated (see Table 1). Right panels: Spectral line profiles of the O VI doublet lines averaged over the $\sim 5'$ slit's portion (filled circles) together with the respective Gaussian fits (solid lines).

tio ($n_e = 5.12 \cdot 10^6$ cm^{-3}). In the 120 s interval around the UVCS detection at 10:25 UT, the frequency of the harmonic lane in the type II radio burst (see Figure 2, right panel) is $\sim$38–44 MHz, implying that the plasma traversed by the shock had an electron density within $4.5-6.0 \cdot 10^6$ cm^{-3}. The two independent density estimates are thus in outstanding agreement.

The shock identification is further confirmed by the analysis of the O VI double ratio. At the shock passage, we have seen that the radiative component of the coronal O VI emission is suddenly Doppler-dimmed, so that the $I_{\text{O VI 1032}}/I_{\text{O VI 1037}}$ line ratio (hereafter defined as X) is expected to decrease quite abruptly, eventually reaching a value of two. For a radial velocity $> 600\ \mathrm{km\,s^{-1}}$, the O VI absorption profile is shifted by more than 2 Å with respect to the chromospheric emission profile of the same ion and we thus expect the O VI resonance scattering component of the emission to be completely Doppler-dimmed. Moreover, for high speeds, it is also necessary to take into account the pumping of the O VI λ1037.6 line by the C II $\lambda\lambda$1036.3, 1037.0 lines, which would lead to X values even smaller than 2. This is confirmed by the observed temporal behavior of the ratio of the O VI line intensities (see Table 1): from a pre-shock ratio $X \sim 2.7-2.9$, the post-shock ratio X decreases down to ~ 1.9 in exp. 45, taken at the time of the shock passage. A value of $X \sim 2.7$ is, however, recovered just after the shock transit in exp. 46, when contamination from denser CME material expanding along the line of sight and following the shock front is certainly involved. The contribution from the CME front to the observed O VI λ 1032 spectral line profile has been qualitatively reproduced by means of a two-Gaussian least-squares fit, with the narrow component fixed to the width and centroid of the pre-CME line profile in order to separate and possibly identify, at least qualitatively, the expected contributions from the background (and foreground) corona and the CME front (see, *e.g.*, Ciaravella, Raymond, and Kahler, 2006). As shown in Figure 5, bottom panel, the broad profile of the CME front overlaps the much brighter and narrower profile of the background corona. We remark, however, that both expansion of the emitting volume and enhanced temperature of the shocked plasma may contribute to the observed broadening of the O VI line profiles. These broadenings are actually theoretically expected because of the plasma heating occurring across the shock (see, *e.g.*, Pagano *et al.*, 2008). Unfortunately, however, from a quantitative point of view, the thermal widths of the spectral lines in a shocked plasma cannot be predicted with great certainty and the effect of bulk motions can be estimated (and thus approximately removed) only by assuming a three-dimensional expansion model (see, *e.g.*, Mancuso and Avetta, 2008). However, for the present event, this kind of analysis will not be attempted due to the observed asymmetry and clearly non-hemispherical shape of the expanding CME leading edge that would make such an estimate too uncertain to be reliable. Nonetheless, it is clear that the observed broadening of the O VI line profile does appear quite modest, just a mere $\sim 20\%$ at least in exp. 45, corresponding to the shock passage not yet contaminated by the CME material.

We remark that a strong broadening of the O VI lines, with a line width comparable to the shock speed, as noticed instead in the analysis of previous faster CME/shocks studied by UVCS, was, however, not to be expected for the present event. The shock speed ($\sim 690\ \mathrm{km\,s^{-1}}$), as deduced in the previous section from the analysis of the type II emission drift in the radio spectra, is in fact just a factor ~ 0.1 higher than the CME speed inferred by the white-light (and EUV) data in the middle corona ($\sim 620\ \mathrm{km\,s^{-1}}$). In order to compare this speed with the local magnetosonic speed, we need an estimate of the coronal magnetic field at that height. A direct measurement of the magnetic field in the outer corona remains a very difficult and challenging problem. There are only a few estimates of the magnetic field in the outer corona (see, *e.g.*, Dulk *et al.*, 1976; Akhmedov *et al.*, 1982; Patzold *et al.*, 1987; Mancuso and Spangler, 2000; Mancuso *et al.*, 2003; Ingleby, Spangler, and Whiting, 2007; Mancuso and Garzelli, 2007; Ramesh, Kathiravan, and Sastry, 2010). By using the widely accepted empirical model by Dulk and McLean (1978), we get $B = 0.63$ G at the UVCS slit height, from which we deduce that the magnetosonic speed could be as high as $\sim 600\ \mathrm{km\,s^{-1}}$. It is thus quite evident that the CME on 7 May 2004 was just fast enough to drive a shock ahead in the ambient corona, but that the CME-driven shock was probably too weak in order to be able to heat considerably the coronal plasma, as in fact evinced from the modest O VI line broadening observed by UVCS immediately after the shock front passage.

4. Summary and Conclusions

In this work, we analyzed white-light, radio, and EUV data of a CME-driven shock observed in the solar corona on 7 May 2004. Clear signatures of the transit of a shock associated to the CME observed in white-light by the LASCO coronagraph have been identified in both UVCS and radio spectrograph data. Dynamic spectra from ground-based radio spectrographs display evidence for metric type II radio emission a few minutes after the CME initiation, further associated with a localized radio feature detected by the Nançay Radioheliograph at 164.0 MHz in the inner corona, just above the CME's source, located at the East limb. Although not seen with radio spectrographs, this feature might be attributed to a first stage of type II emission, even though the shock speed at that height might be too slow, with respect to the local magnetosonic speed, in order to drive a shock. A different shock signature is provided by white-light images acquired by the LASCO coronagraph, showing deflection of a coronal streamer during the front expansion. The strongest evidence of the shock passage in the middle corona was, however, obtained through the combined analysis of radio and EUV spectral observations that confirmed both the simultaneity between the type II emission in the radio dynamic spectra and the UVCS detection at 1.86 $R_{\odot}$, and the agreement between the electron density estimate obtained through the radio data and the one retrieved from the observed ratio of the radiative and collisional components of the O VI 1032 Å line in the pre-shock UVCS spectra. The analysis of the O VI doublet intensity ratio further provided evidence of abrupt plasma acceleration due to the shock transit, while the O VI line profiles suggested modest heating of the plasma behind the shock front, although it was not possible, due to the irregular structure of the front, to establish the contribution from both bulk expansion and shock plasma heating. The shock speed ($\sim 690\ \mathrm{km\,s^{-1}}$), as deduced from the type II emission drift in the radio spectra and the pre-shock local electron density, was estimated to be just a factor ~ 0.1 higher than the CME speed inferred by the white-light (and EUV) data in the middle corona ($\sim 620\ \mathrm{km\,s^{-1}}$). By using the Dulk and McLean (1978) magnetic field strength estimate, yielding $B = 0.63$ G at the UVCS slit height, we deduced that the local magnetosonic speed could be as high as $\sim 600\ \mathrm{km\,s^{-1}}$. This implies that the CME was just fast enough to drive a shock ahead in the ambient corona and that the CME-driven shock was too weak in order to heat considerably the plasma, as in fact evidenced from the modest O VI spectral line broadening observed by UVCS immediately after the shock front passage.

In conclusion, this work, besides providing estimates of a few important coronal plasma parameters, yields further evidence for the CME-driven origin of the shocks observed in the middle corona.

Acknowledgements The author thanks Dr. K.L. Klein and the members of the Scientific Organizing Committee for the invitation to hold a review contribution on coronal shocks at the CESRA 2010 workshop. SOHO is a project of international cooperation between ESA and NASA. The LASCO CME Catalog is generated and maintained by NASA and Catholic University of America in cooperation with the Naval Research Laboratory.

References

Akhmedov, S.B., Gelfreikh, G.B., Bogod, V.M., Korzhavin, A.N.: 1982, *Solar Phys.* **79**, 41.

Bale, S.D., Reiner, M.J., Bougeret, J.-L., Kaiser, M.L., Krucker, S., Larson, D.E., *et al.*: 1999, *Geophys. Res. Lett.* **26**, 1573.

Bemporad, A., Mancuso, S.: 2010, *Astrophys. J.* **720**, 130.

Bemporad, A., Poletto, G., Suess, S.T., Ko, Y.K., Parenti, S., Riley, P., *et al.*: 2003, *Astrophys. J.* **593**, 1146.

Brueckner, G.E., Howard, R.A., Koomen, M.J., Korendyke, C.M., Michels, D.J., Moses, J.D., *et al.*: 1995, *Solar Phys.* **162**, 357.
Cairns, I.H.: 1986, *Proc. Astron. Soc. Aust.* **6**, 444.
Ciaravella, A., Raymond, J.C., Kahler, S.W.: 2006, *Astrophys. J.* **652**, 774.
Ciaravella, A., Raymond, J.C., Kahler, S.W., Vourlidas, A., Li, J.: 2005, *Astrophys. J.* **621**, 1121.
Cliver, E.W., Webb, D.F., Howard, R.A.: 1999, *Solar Phys.* **187**, 89.
Dulk, G., McLean, D.J.: 1978, *Solar Phys.* **57**, 279.
Dulk, G.A., Jacques, S., Smerd, S.F., MacQueen, R.M., Gosling, J.T., Steward, R.T., *et al.*: 1976, *Solar Phys.* **49**, 369.
Filippov, B., Srivastava, A.K.: 2010, *Solar Phys.* **266**, 123.
Gardner, L.D., Smith, P.L., Kohl, J.L., Atkins, N., Ciaravella, A., Miralles, M.P., *et al.*: 2002, In: Pauluhn, A., Huber, M.C.E., von Steiger, R. (eds.) *The Radiometric Calibration of SOHO*, ESA, Noordwijk, 161.
Gopalswamy, N.: 2006, *Geophys. Monogr. Ser.* **165**, AGU, Washington, 207.
Gopalswamy, N., Kundu, M.R., Manoharan, P.K., Raoult, A., Nitta, N., Zarka, P.: 1997, *Astrophys. J.* **486**, 1036.
Gopalswamy, N., Nitta, N., Manoharan, P.K., Raoult, A., Pick, M.: 1999, *Astron. Astrophys.* **347**, 684.
Gopalswamy, N., Yashiro, S., Michalek, G., Stenborg, G., Vourlidas, A., Freeland, S., *et al.*: 2009, *Earth Moon, Planets* **104**, 295.
Gorgutsa, R.V., Gnezdilov, A.A., Markeev, A.K., Sobolev, D.E.: 2001, *Astron. Astrophys. Trans.* **20**, 547.
Gosling, J.T., Hildner, E., MacQueen, R.M., Munro, R.H., Poland, A.I., Ross, C.L.: 1974, *J. Geophys. Res.* **79**, 4581.
Gosling, J.T., Hildner, E., MacQueen, R.M., Munro, R.H., Poland, A.I., Ross, C.L.: 1976, *Solar Phys.* **48**, 389.
Hudson, H.S., Khan, J.I., Lemen, J.R., Nitta, N.V., Uchida, Y.: 2003, *Solar Phys.* **212**, 121.
Ingleby, L.D., Spangler, S.R., Whiting, C.A.: 2007, *Astrophys. J.* **668**, 520.
Klein, K.-L., Khan, J.I., Vilmer, N., Delouis, J.-M., Aurass, H.: 1999, *Astron. Astrophys.* **346**, L53.
Knock, S.A., Cairns, I.H., Robinson, P.A., Kuncic, Z.: 2001, *J. Geophys. Res.* **106**, 25041.
Ko, Y.-K., Raymond, J.C., Li, J., Ciaravella, A., Michels, J., Fineschi, S., *et al.*: 2002, *Astrophys. J.* **578**, 979.
Kohl, J.L., Esser, R., Gardner, L.D., Habbal, S., Daigneau, P.S., Dennis, E.F., *et al.*: 1995, *Solar Phys.* **162**, 313.
Lean, J.: 2001, *Geophys. Res. Lett.* **28**, 4119.
Lean, J., Rottman, G.J., Kyle, H.L., Woods, T.N., Hickey, J.R., Puga, L.C.: 1997, *J. Geophys. Res.* **102**, 29939.
Lean, J., Rottman, G.J., Harder, J., Kopp, G.: 2005, *Solar Phys.* **230**, 27.
Lin, J., Mancuso, S., Vourlidas, A.: 2006, *Astrophys. J.* **649**, 1110.
Liu, Y., Luhmann, J.G., Bale, S.D., Lin, R.P.: 2009, *Astrophys. J.* **691**, L151.
Magdalenić, J., Vršnak, B., Pohjolainen, S., Temmer, M., Aurass, H., Lehtinen, N.J.: 2008, *Solar Phys.* **253**, 305.
Magdalenić, J., Marqué, C., Zhukov, A.N., Vršnak, B., Žic, T.: 2010, *Astrophys. J.* **718**, 266.
Mancuso, S.: 2007, *Astron. Astrophys.* **463**, 1137.
Mancuso, S., Abbo, L.: 2004, *Astron. Astrophys.* **415**, L17.
Mancuso, S., Avetta, D.: 2008, *Astrophys. J.* **677**, 683.
Mancuso, S., Garzelli, M.V.: 2007, *Astron. Astrophys.* **466**, L5.
Mancuso, S., Raymond, J.C.: 2004, *Astron. Astrophys.* **413**, 363.
Mancuso, S., Spangler, S.R.: 2000, *Astrophys. J.* **539**, 480.
Mancuso, S., Raymond, J.C., Kohl, J., Ko, Y.-K., Uzzo, M., Wu, R.: 2002, *Astron. Astrophys.* **383**, 267.
Mancuso, S., Raymond, J.C., Kohl, J., Ko, Y.-K., Uzzo, M., Wu, R.: 2003, *Astron. Astrophys.* **400**, 347.
Mann, G., Aurass, H., Voigt, W., Paschke, J.: 1992, In: *Proc. First SOHO Workshop: Coronal Streamers, Coronal Loops, and Coronal and Solar Wind Composition* **SP-348**, ESA, Noordwijk, 129.
Maxwell, A., Dryer, M., McIntosh, P.: 1985, *Solar Phys.* **97**, 401.
Michels, D.J., Sheeley, N.R. Jr., Howard, R.A., Koomen, M.J., Schwenn, R., Mulhauser, K.H., *et al.*: 1984, *Adv. Space Res.* **4**, 311.
Nelson, G.J., Melrose, D.B.: 1985, In: McLean, D.J., Labrum, N.R. (eds.) *Solar Radiophysics*, Cambridge, 333.
Newkirk, G.: 1961, *Astrophys. J.* **133**, 983.
Nindos, A., Aurass, H., Klein, K.-L., Trottet, G.: 2008, *Solar Phys.* **253**, 3.
Noci, G., Kohl, J.L., Withbroe, G.L.: 1987, *Astrophys. J.* **315**, 706.
Ontiveros, V., Vourlidas, A.: 2009, *Astrophys. J.* **693**, 267.
Pagano, P., Raymond, J.C., Reale, F., Orlando, S.: 2008, *Astron. Astrophys.* **481**, 835.
Patzold, M., Bird, M.K., Volland, H., Levy, G.S., Seidel, B.L., Stelzried, C.T.: 1987, *Solar Phys.* **109**, 91.
Ramesh, R., Kathiravan, C., Sastry, C.V.: 2010, *Astrophys. J.* **711**, 1029.

Raouafi, N.-E., Mancuso, S., Solanki, S.K., Inhester, B., Mierla, M., Stenborg, G., *et al.*: 2004, *Astron. Astrophys.* **424**, 1039.
Raymond, J.C., Thompson, B.J., St. Cyr, O.C., Gopalswamy, N., Kahler, S., Kaiser, M., *et al.*: 2000, *Geophys. Res. Lett.* **27**, 1439.
Reiner, M.J., Kasaba, Y., Kaiser, M.L., Matsumoto, H., Nagano, I., Bougeret, J.-L.: 1997, *Geophys. Res. Lett.* **24**, 919.
Sheeley, N.R. Jr., Hakala, W.N., Wang, Y.-M.: 2000, *J. Geophys. Res.* **105**, 5081.
The Radioheliograph Group: 1993, *Adv. Space Res.* **13**(9), 411.
Thejappa, G., MacDowall, R.J., Vinas, A.F.: 1997, In: Wilson, A. (ed.) *31st ESLAB Symp. – Correlated Phenomena at the Sun, in the Heliosphere and in Geospace* **SP-415**, ESA, Noordwijk, 189.
Uchida, Y.: 1974, *Solar Phys.* **39**, 431.
Uzzo, M., Ko, Y.-K., Raymond, J.C.: 2004, *Astrophys. J.* **603**, 760.
Vourlidas, A., Wu, S.T., Wang, A.H., Subramanian, P., Howard, R.A.: 2003, *Astrophys. J.* **598**, 1392.
Vršnak, B., Cliver, E.W.: 2008, *Solar Phys.* **253**, 215.
Vršnak, B., Lulic, S.: 2000a, *Solar Phys.* **196**, 157.
Vršnak, B., Lulic, S.: 2000b, *Solar Phys.* **196**, 181.
Wild, J.P.: 1950, *Aust. J. Sci. Res.* **A3**, 541.
Yashiro, S., Gopalswamy, N., St. Cyr, O.C., Lawrence, G., Michalek, G., Young, C.A., Plunkett, S.P., Howard, R.A.: 2001, AGU Spring Meeting Abstracts, 31.
Zangrilli, L., Poletto, G., Nicolosi, P., Noci, G., Romoli, M.: 2002, *Astrophys. J.* **574**, 477.

Solar Phys (2011) 273:525–536
DOI 10.1007/s11207-011-9870-y

The Connection of Solar Wind Parameters with Radio and UV Emission from Coronal Holes

D.V. Prosovetsky · I.N. Myagkova

Received: 29 November 2010 / Accepted: 20 September 2011 / Published online: 8 November 2011

Abstract This paper presents the results of a comparison between observations of coronal holes in UV (SOHO EIT) and radio emission (17, 5.7 GHz, 327 and 150.9 MHz, from NoRH, SSRT and Nançay radioheliographs), and solar wind parameters, from ACE spacecraft data over the period 12 March – 31 May 2007. The increase in the solar wind velocity up to $\sim 600\ \mathrm{km\,s^{-1}}$ was found to correlate with a decrease in the UV flux in the central parts of the solar disk. A connection between the parameters of the radio emission from three different layers of the solar atmosphere and the solar wind velocity near the Earth's orbit was discovered. Such a connection is suggestive of a common mechanism of solar wind acceleration from chromospheric heights to the upper corona.

Keywords Corona, quiet · Coronal holes · Radio emission, quiet · Solar wind

1. Introduction

The outward expansion of solar plasma referred to as *'solar wind'* by Parker (1958a) has been intensively studied by ground-based and space experiments since the 1950s (*e.g.*, Biermann, 1951, 1952, 1957; Vsehvyatskiy *et al.*, 1955; Parker, 1958a, 1958b; Licht, 1960; Harrison, 1961; Yallop, 1961; Gringauz *et al.*, 1962; Neugebauer and Snyder, 1966). The solar wind consists of three components – the slow, the fast and the sporadic ones. The existence of different components of the solar wind has been revealed through the observations of the *Mariner-2* spacecraft (Neugebauer and Snyder, 1966). According to Sheeley *et al.*

Energy Storage and Release through the Solar Activity Cycle – Models Meet Radio Observations
Guest Editors: Christophe Marqué and Alexander Nindos

D.V. Prosovetsky (✉)
The Institute of Solar-Terrestrial Physics SB RAS, Irkutsk, Russia
e-mail: proso@iszf.irk.ru

I.N. Myagkova
Lomonosov Moscow State University Skobeltsyn Institute of Nuclear Physics, MSU SINP, Moscow, Russia
e-mail: irina@srd.sinp.msu.ru

(1985), the sporadic component is related to coronal mass ejections. Unlike the sporadic component, the slow and fast components are regular: the slow one exists always, and the fast one is observed periodically. Sources of the latter are coronal holes (CHs) – regions of a unipolar magnetic field with an open configuration (Hundhausen, 1972).

The slow and fast SW components do not only differ in plasma velocity (up to 450 and 800 $\mathrm{km\,s^{-1}}$, respectively). According to observations of radio scintillations due to inhomogeneities of the interplanetary medium, one of the main SW parameters – the velocity distribution depending on the distance from the Sun – is different for the slow and fast components. The maximum velocity of the slow component is reached at a distance of more than 10 solar radii (Wang *et al.*, 1998), whereas the fast component reaches its maximum velocity near the Sun, at a distance of one or two solar radii (Grall, Coles, and Klinglesmith, 1996). These findings suggest that the mechanism accelerating SW particles is either completely different or has essentially different parameters for different components. Differences in conditions of the fast SW component acceleration are likely to be found at distances of less than one solar radius above the photosphere.

This assumption is also supported by observations of the non-thermal velocity component made with the use of the SUMER UV spectrograph on board SOHO (Wilhelm *et al.*, 1995). Investigation of the ultraviolet spectrum of the CH atmosphere (Chae, Hle, and Lemaire, 1998) shows that the maximum of the fluctuation velocity (*i.e.*, the non-thermal component δV in $V_{\mathrm{solar}} = \sqrt{\frac{2kT}{M} + \delta V^2}$, where k is the Boltzmann constant, T is the ion temperature and M is the mass of the ion emitting the line) is reached at CH levels in the solar atmosphere with temperatures of about 10^5 K (in the transition region). The non-thermal velocity component specifies wave flux $F_{\mathrm{wave}} = \rho \langle \delta V^2 \rangle V_{\mathrm{A}}$, where ρ is the plasma density, V_{A} is the Alfvén velocity. Alfvén waves are probably responsible for acceleration of SW particles and heating of the corona in CH (Cranmer, 2004).

Consequently, parameters of the fast SW component may be formed in the lower solar atmosphere (chromosphere, transition region and lower corona). Some authors also confirm that SW parameters and the solar emission in the optical and UV wavebands are related (*e.g.*, Vršnak, Temmer, and Veronig, 2007; Obridko *et al.*, 2009). The intensity of areas and ratio of areas of CH emission/absorption lines were proposed in the paper of Stepanian *et al.* (2008), Shugai, Veselovsky, and Trichtchenko (2009), Obridko *et al.* (2009) to predict the SW characteristics. However, these methods are generally based on the emission analysis at a single wavelength (sometimes at two wavelengths); *i.e.*, they do not completely reflect the distribution of the plasma parameters and energy release processes in the solar atmosphere.

Supplementary information about the characteristics of the acceleration of the fast SW component could be gained from a thorough study of the UV and radio emission of CH, obtained with high spatial resolution. However, there are only a few papers (*e.g.*, Chae, Hle, and Lemaire, 1998) devoted to spectral observations of the UV emission obtained from SUMER data. This is because of the limitations of the observing programmes of this instrument. Besides, no studies of the dependence between the high-speed SW characteristic and the radio CH emission have been made so far.

Previously, various scientific groups made observations of the radio CH emission over a wide frequency range. Comparative analysis of such observations shows that the frequency range 6 – 17 GHz is characterised by an increased emission (compared to the quiet Sun). This fact cannot be explained by typical models of the solar atmosphere (Maksimov *et al.*, 2006). The *Nobeyama Radioheliograph* (NoRH, 17 GHz, Nakajima *et al.*, 1994) regularly observes an increased level of brightness temperatures in CH (Gopalswamy *et al.*, 1999; Nindos, Kundu, and White, 1999; Moran *et al.*, 2001). Analysis of simultaneous observations with NoRH and the *Siberian Solar Radio Telescope* (SSRT, 5.7 GHz, Grechnev *et al.*,

2003) revealed a linear anticorrelation between brightness temperatures at 5.7 and 17 GHz in the CH regions (Krissinel *et al.*, 2000). According to the results of observations near the solar limb, carried out at SSRT and NoRH, such regions are radially stratified (Maksimov *et al.*, 2004). Analysing observations of the microwave emission, Maksimov *et al.* (2006) showed that the corona (at least, in some parts of the CH) is heated by the wave flux propagating from the photosphere, and the heights where the increased radio emission is formed correspond to the heights of the chromosphere and transition region. Some researchers (see Cranmer, 2004) consider the same mechanisms for the SW acceleration and coronal heating. Observations of the UV spectrum imply a possible connection between energy release and microwave emission in the lower solar atmosphere.

According to the study of the magnetic field configuration of CH, there is a connection between SW characteristics and the so-called "superradial" divergence of magnetic tubes. "Superradial" divergence is determined by measurements of the photospheric magnetic field and its extrapolation (in the potential approximation) to the source surface (Wang and Sheeley, 1990). There is also a connection between SW characteristics and the unipolar field area at the CH base (Eselevich *et al.*, 2009). Forecasting SW parameters using the findings from these studies does not always yield correct results. Probably the magnetic field configuration cannot completely determine the plasma density and velocity of SW, since the SW plasma carries frozen-in magnetic flux out to the outer solar atmosphere and thus changes its configuration. This restricts the application of the magnetic field extrapolation (both potential and non-potential), which does not take this effect into account when forecasting SW parameters.

The purpose of this paper is to study a possible connection between the SW velocity and emission of the lower solar atmosphere of the CH.

2. Observations

2.1. Data

The period from 12 March to 31 May 2007 (during the last solar minimum) was studied. We chose this interval because of the fact that the main parameters of the fast SW component during a solar minimum are defined by characteristics of the CH. Besides, there was a complete set of experimental data available for this period.

To perform the analysis, we used radio images of the Sun recorded by NoRH, SSRT and *Nançay Radioheliograph* at 17 GHz, 5.7 GHz, 327 MHz and 150.9 MHz, and also SOHO/EIT data at the wavelength of the Fe XII line, $\lambda = 195$ Å. Variations in SW parameters were measured by SWEPAM (*Solar Wind Electron, Proton and Alpha Monitor*) on board the ACE spacecraft (*Advanced Composition Explorer*) orbiting near the libration point L1 in the Sun–Earth system (1.5 million km from the Earth toward the Sun), see http://cdaweb.gsfc.nasa.gov/istp_public.

2.2. Data Processing

We determined the UV emission characteristics of CH using a technique similar to the one developed by Vršnak, Temmer, and Veronig (2007). This method implies the determination of fractional CH areas in central meridional slices of the solar disc. Then, the UV emission flux was calculated in this region. We noticed, however, that the polar regions did not contribute significantly to the SW stream registered near the Earth. Thus the region (unlike in

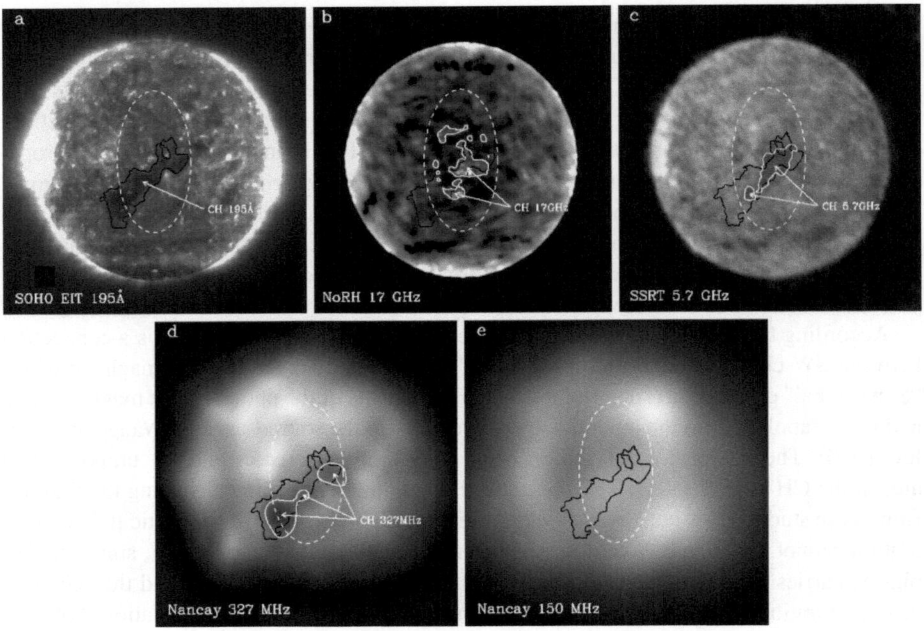

Figure 1 Example of the CH boundary location in the UV and radio emission at four frequencies on 22 March 2007. (a) CH boundaries in the central part of the solar disc with density below $1 \cdot 10^9$ cm^{-3} in the Fe XII band, around $\lambda = 195$ Å. (b – e) Brightness temperatures of the radio emission in the central part of the solar disc (white contour). (b) Above 10^4 K at 17 GHz as deduced from NoRH observations. (c) Below $1.6 \cdot 10^4$ K at 5.7 GHz as deduced from SSRT observations. (d) Below $6.3 \cdot 10^5$ K at 327 MHz as deduced from Nançay radioheliograph observations. (e) The radio emission at 150.9 MHz as deduced from Nançay radioheliograph observations. At this frequency, depression of the radio emission connected with a coronal hole is not observed. CH contours from UV data are overplotted on radio images (see Figure 2a, black contour).

Vršnak, Temmer, and Veronig, 2007) was bounded by the ellipse $R_\odot^2 = X^2 + \frac{Y^2}{1.9}$, where $R_\odot$ is the solar radius. The relation between the semiaxes of the ellipse was determined when comparing positions of CH boundaries and the SW velocity at the point L1 (*i.e.*, we were trying to find out if the high-speed SW stream was registered from a CH that had the latitude and longitude corresponding to the occurrence of a CH boundary on the ellipse boundaries).

The SOHO/EIT data were processed using *SolarSoft* procedures and images from the calibration database of the instrument. The ratio of the emission flux to the CH area (*i.e.*, the average emission flux in the line of Fe XII, $\lambda = 195$ Å) was determined for the CH within the limits of an elliptic central region. The plasma electron density N_e in the CH was deduced from the relationship

$$N_{e,CH} = 8.34 + 0.509 \cdot \log(I_{EIT}) \tag{1}$$

obtained in Brosius *et al.* (2002), where I_{EIT} is the Fe XII $\lambda = 195$ Å line intensity of the calibrated image.

We selected pixels with values of the plasma electron density below $1 \cdot 10^9$ cm^{-3} in the central elliptic area of the solar disc for all images. The specific value of the density was obtained by dividing the plasma electron density sum by the number of such pixels. The procedure allowed us to exclude the points corresponding to the bright features located in the elliptic area. Figure 1a, for example, exhibits images of the equatorial coronal hole in the

Fe XII $\lambda = 195$ Å line acquired on 22 March 2007; their boundaries are shown by contour. A coronal bright point in the northwestern part of the CH with plasma electron density over $1 \cdot 10^9$ cm^{-3} was excluded from the CH boundaries. We thus obtained the variation of the UV emission in the CH with a temporal resolution of about 12 minutes (see Figure 2d).

Figure 1(b–e) presents examples of the CH radio emission at 17, 5.7 GHz, 327 and 150.9 MHz. The contours separate CH boundaries and the quiet Sun. At 17 GHz (Figure 1b), the emission of the coronal hole is brighter; at 5.7 GHz (Figure 1c) and 327 MHz (Figure 1d), the CH is darker than the quiet Sun. Referring to Figure 1(b–d), some parts of the CH within the Fe XII $\lambda = 195$ Å line boundaries produce radio emission at 17 GHz, 5.7 GHz and 327 MHz, which is inhomogeneously distributed on the CH surface. At 150.9 MHz, the coronal hole almost equals the level of the quiet Sun. The radio emission at 150.9 MHz can sometimes be brighter or darker than the quiet Sun in the boundaries of the CH in the line of Fe XII with $\lambda = 195$ Å.

Since the CH radio emission in CH boundaries in 195 Å is irregular, we determined the increment/decrement of the normalised flux of the CH in units of the quiet Sun, relative to the quiet Sun level in the regions inside the ellipse:

$$F_{\mathrm{CH},\nu} = \frac{\sum_{i=1}^{n_{\mathrm{CH}}} (T_{\mathrm{B}\nu,i} - T_{\mathrm{QS}\nu,i})}{\sum_{j=1}^{m} T_{\mathrm{QS}\nu,j}}, \tag{2}$$

where ν is the frequency, $T_{\mathrm{B}\nu,i}$ is the brightness temperature inside a CH (its summation is made over elements S_i of the radio CH surface), $T_{\mathrm{QS}\nu,i}$ and $T_{\mathrm{QS}\nu,i}$ are the brightness temperatures of the quiet Sun. The brightness temperature of the quiet Sun is constant for each frequency, $T_{\mathrm{QS},17} = 10^4$ K (Nakajima *et al.*, 1994), $T_{\mathrm{QS},5.7} = 1.6 \cdot 10^4$ K (Grechnev *et al.*, 2003), $T_{\mathrm{QS},327} = 6.3 \cdot 10^5$ K and $T_{\mathrm{QS},150.9} = 6 \cdot 10^5$ K (Mercier and Chambe, 2009). Summation in the denominator is made throughout the disc (the summation index is j; the number of the entire solar disc pixels is m); its result is constant for each frequency. Summation in the numerator is made for image pixels inside the CH, with emission enhancement at 17 GHz and emission depression at 5.7 GHz, 327 and 150.9 MHz, relative to the quiet Sun level (the summation index is i; the number of CH pixels in radio is n_{CH}). So we obtained data of the normalised radio emission flux of the CH from four atmospheric levels, along with data of the plasma electron density inside the CH.

3. Experimental Data Analysis

3.1. Solar Wind Parameters

The three upper panels (a, b, c) of Figure 2 show SW parameter variations during the period under study: the strength of the B_z component and B of the interplanetary magnetic field (IMF), the SW velocity, V_{SW}, and the proton number density N_{p} in SW, respectively. Referring to Figure 2, ten high-speed SW streams (whose beginning is marked by vertical lines) are evident throughout this period. The first SW stream (with its maximum on 13 March) was excluded from consideration, since this SW stream was generated by a CH that had passed through the central meridian some days before 13 March (when the EIT started its observations after regular maintenance service). The lack of EIT data (139–145 DOY) was caused by SOHO/EIT maintenance operations during these days. In Figure 2, the streams under consideration in this work are numbered from 1 to 9.

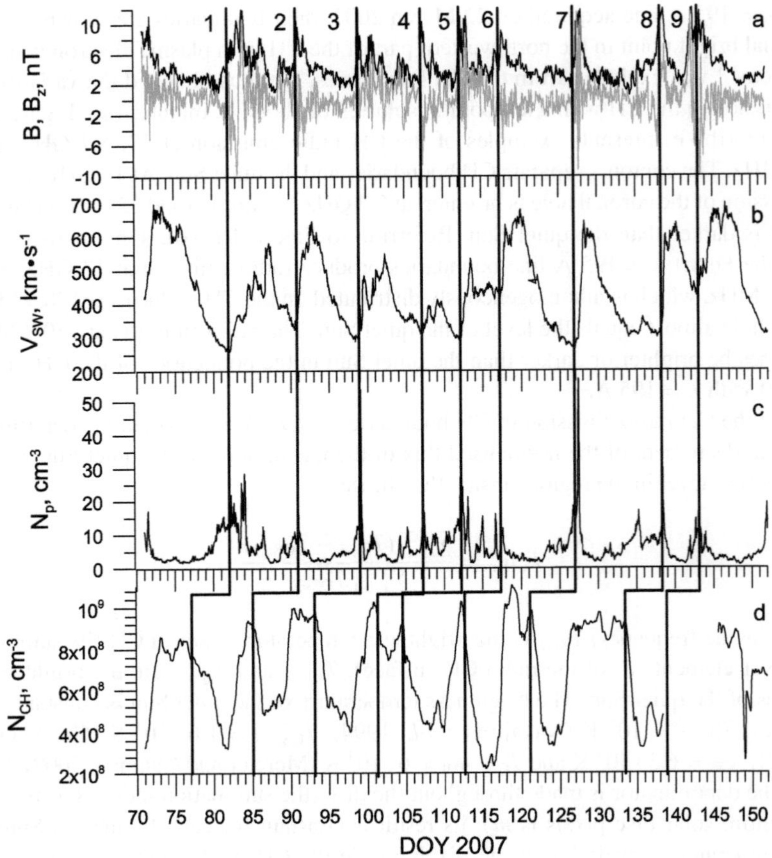

Figure 2 SW parameters measured at L1 in the period 12 March – 31 May 2007: modulus B (black line) and the B_Z component (grey line) of IMF (a); the SW velocity V_{SW} (b); the proton number density N_p (c). Panel (d) shows the variation of the plasma density in the CH, N_{CH} (the curve discontinuity corresponds to the break in the SOHO EIT observations). The x-coordinate is the number of days, starting from 1 January (DOY). Vertical lines correspond to the leading edge of the SW velocity variation and to the leading edge of plasma density decrease due to the CH.

The high-speed streams observed during the period under study have essentially different characteristics: the maximum speed varies from 300 to 700 km s^{-1}, the proton number density at the leading edge of the high-speed stream can reach 40 – 45 cm^{-3} or can be less than 5 cm^{-3} (Figure 2b, c). The variations in the average IMF value B for nine SW streams look similar: we have an abrupt increase at the leading edge of the SW stream and subsequent gradual decrease. During the period under study, B varied in the range from 2 to 10 nT.

3.2. UV Emission of CH

Panel (d) of Figure 2 shows the variation of the plasma density in the CH. It is observed from the figure that the main increases in the SW velocity (1 – 9), taking account of the arrival time at L1, coincide with the plasma density decrease in the CH. The decrease in the UV flux values obtained by the method from Section 2.2 starts when the CH appears at the ellipse boundary. Therefore, the time delay in the arrival of high-speed SW streams is

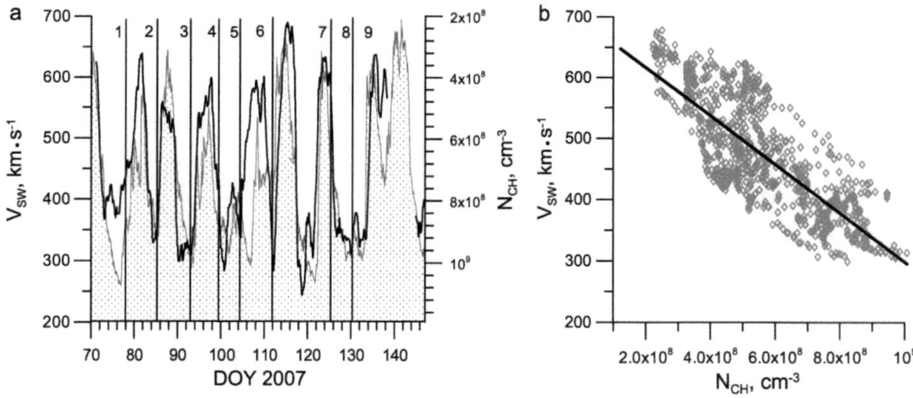

Figure 3 SW velocities at L1 and plasma density in the CH. (a) Superposed variations (shift is 4 days) of density (thick line) and in the SW velocity (filled curve bounded by thin line). The *y*-axis of the plasma density values is reversed. (b) Correlation between density N_{CH} and velocity V_{SW}, taking into account the different time delays of the SW stream arrival at L1. The line shows best fit to the points.

determined from time differences between the beginning of the leading edge of high-speed stream and the start of decrease in the density related to the CH (see Figures 3a and 3d). The delays were determined from the time difference between the leading edge of density variation associated with the CH and the leading edge of the SW speed variation. The time difference varied from 4 to 7 days. A 5-day delay was used for stream 9 with missing UV data. The data gap was due to SOHO maintenance operations.

In Figure 3a, the diagram of SW velocity variations in time is superposed onto the diagram of the variations in the mean density of the CH, taking into consideration the time delay being equal to the time when SW particles with a propagation velocity of 600 km s^{-1} arrive at L1. The y-axis for the density (to the right) is reversed (*i.e.*, higher values are at the bottom). Referring to Figure 3a, SW velocity variations and plasma density in the CH are similar. The discrepancy between the positions of some SW streams and density depressions (*e.g.*, streams 4, 5 and 6) can be explained by the difference in time of the SW particle arrival at L1.

Taking into account the different delays in the SW particle arrival (from 4 to 7 days; see Figure 2) at the point L1, we show the correspondence between density and velocities $V_{\mathrm{SW}} \geq 300$ km s^{-1} (Figure 3b). The slow SW component is thus cast out in the time series where possible. The values N_{CH} and V_{SW} in eight SW streams are approximated by the linear dependence

$$V_{\mathrm{SW}} = -3.97 \cdot 10^{-7} N_{\mathrm{CH}} + 695 \quad (3)$$

with a correlation 0.63. Notice that the parameters of different SW streams significantly differ (especially the delay time between the plasma density decrease in the CH and the arrival of the SW high-speed stream at the Earth). The correlation of each separate SW stream velocity with its corresponding linear dependence reaches 0.8 – 0.9. However, there was no evidence of correlation between the plasma density in the CH and the proton density in the SW.

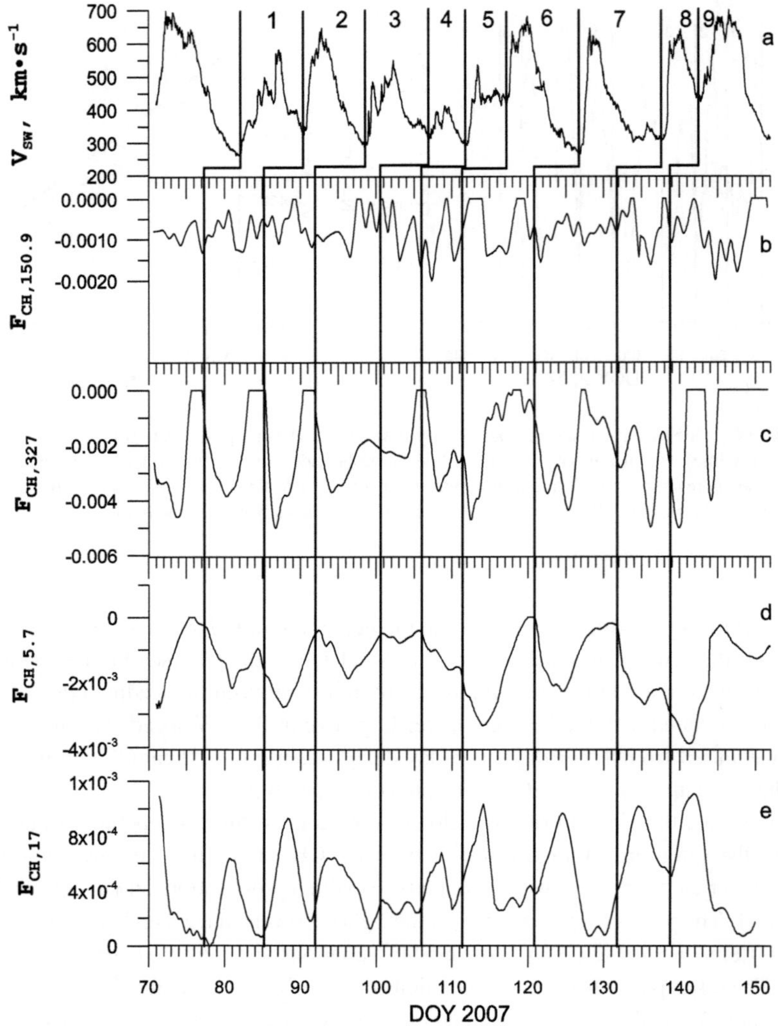

Figure 4 Normalised radio emission fluxes of the CH at 150.9 MHz, 327 MHz, 5.7 and 17 GHz (b, c, d, e) in the quiet Sun units, compared to the SW velocity (a). Vertical lines correspond to the leading edge of SW streams, to the beginnings of CH flux depressions at 150.9 MHz, 327 MHz and 5.7 GHz, and of increases in the microwave flux at 17 GHz.

3.3. Radio Emission of CH

Figure 4(b – e) presents variations in the normalised radio emission flux F_{CH} of the CH at four frequencies, according to data from the NoRH, SSRT and Nançay radioheliographs. The normalised flux was measured with the use of the method described in Section 2. F_{CH} gives the variation of the CH radio flux in units of the quiet Sun, as compared to the flux of the quiet Sun with the same area; *i.e.*, F_{CH} represents a linear function of the CH radio flux with no contribution from the quiet Sun. The normalised radio flux reflects the change in the CH area and the brightness temperature distribution. Negative and positive sign of the F_{CH} imply depression and enhancement of the radio flux (relative to the quiet Sun level),

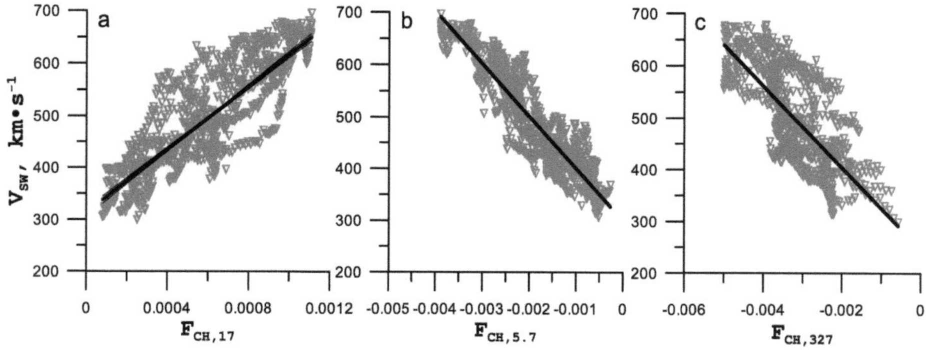

Figure 5 Correlations between the normalised radio flux of CHs at 17 GHz (a), 5.7 GHz (b) and 327 MHz (c) and the SW velocity at L1. Lines show the best fit for the corresponding dataset.

respectively. Vertical lines in Figure 4a mark the beginning of the leading edges of SW streams; those in Figure 4(b – e) the beginning of leading edges of increments/decrements of the normalised radio flux at four frequencies.

The increased radio emission flux (as compared to the quiet Sun) is related to the CH at 17 GHz, whereas the decreased radio emission flux is related to it at 5.7 GHz and 327 MHz. At 150.9 MHz, there is no evidence of an increased nor decreased radio emission flux. Its magnitude is comparable to the flux of neighbouring regions of the solar atmosphere (Figure 4b). Many authors (Dulk and Sheridan, 1974; Trottet and Lantos, 1978; Kosugi, Ishiguro, and Shibasaki, 1986; Borovik *et al.*, 1990; Chiuderi-Drago *et al.*, 1999; Gopalswamy *et al.*, 1999; Nindos, Kundu, and White, 1999; Krissinel *et al.*, 2000; Moran *et al.*, 2001; Maksimov *et al.*, 2006) indicated the increase of the microwave emission at frequencies near 17 GHz, the decrease of the radio emission at frequencies below 5 GHz and above 150.9 MHz, and the drastic decrease in the CH contrast relative to the quiet Sun at frequencies below 150.9 MHz.

Nine CHs were registered during the period under study; time of the SW stream arrival at the point L1 was taken into consideration (Figure 4), and normalised radio emission fluxes with SW velocities $V_{\mathrm{SW}} \geq 300\ \mathrm{km\,s^{-1}}$ were compared at four frequencies (Figure 5a – c). The following relations were obtained from the linear approximation of the experimental data:

$$V_{\mathrm{SW}} = 3.04 \cdot 10^5 F_{\mathrm{CH},17} + 313, \tag{4}$$

$$V_{\mathrm{SW}} = -1.00 \cdot 10^5 F_{\mathrm{CH},5.7} + 300, \tag{5}$$

$$V_{\mathrm{SW}} = -7.85 \cdot 10^4 F_{\mathrm{CH},327} + 247. \tag{6}$$

Observational data correspond to these relations with a correlation of 0.69 at 17 GHz, 0.84 at 5.7 GHz and 0.60 at 327 GHz. No reliable dependence was found for 150.9 MHz. Except for a linear approximation, we also tried other ones, however, the linear fit provided the best correlation. The scatter of flux values could be caused by the noise of antennas and receivers of the radio telescopes, and also by different methods for restoring radio data and the technique for defining UV and radio flux. The difference in methods for processing and restoring radio data has probably caused a difference in correlation coefficients for data obtained at 17 GHz, 5.7 GHz and 327 MHz.

4. Discussion

When analysing experimental data in the previous section, we also examined a probable connection between the other SW parameters (B, B_Z, N_p) and the CH atmosphere characteristics (the radio emission flux F at four frequencies, the UV intensity and its related plasma electron density N_{CH}). No reliable relations were revealed. Probably these SW parameters obtain their terminal values at other atmospheric levels of the CH or in interplanetary space.

Among the SW parameters under study, we found dependencies only between its velocity at L1 and characteristics of the solar atmosphere emission in the UV and radio bands at three frequencies. Some authors (*e.g.*, Vršnak, Temmer, and Veronig, 2007; Obridko *et al.*, 2009) have already called attention to the relation between SW stream speeds and some characteristics of the UV emission (for instance, the area) in the CH. In this study, we established the dependence of the SW velocity on the mean plasma density in the CH and the radio flux at three frequencies.

According to the linear dependencies $V_{SW}(N_{CH})$ (3), $V_{SW}(F_{CH,17})$ (4), $V_{SW}(F_{CH,5.7})$ (5) and $V_{SW}(F_{CH,327})$ (6), all CH characteristics (N_{CH}, $F_{CH,\nu}$) and the SW velocity V_{SW} are interrelated. Actually, the dependencies $N_{CH}(V_{SW})$, $F_{CH,17}(V_{SW})$, $F_{CH,5.2}(V_{SW})$ and $F_{CH,327}(V_{SW})$ obtained from Equations (3), (4), (5) and (6) lead to the following relations:

$$\begin{cases} N_{CH} = -2.52 \cdot 10^6 V_{SW} + 1.75 \cdot 10^9, \\ F_{CH,17} = 3.29 \cdot 10^{-6} V_{SW} - 1.03 \cdot 10^{-3}, \\ F_{CH,5.7} = -1.00 \cdot 10^{-5} V_{SW} + 3.00 \cdot 10^{-3}, \\ F_{CH,327} = -1.27 \cdot 10^{-5} V_{SW} + 3.15 \cdot 10^{-3} \end{cases} \quad (7)$$

and all variables of this system are also linearly dependent. This result confirms a linear dependence between brightness temperatures at 5.7 and 17 GHz in the CH, revealed by Maksimov *et al.* (2006). Notice that no dependence between the SW velocity and the radio emission flux was found for 150.9 MHz at four frequencies. The radio emission flux at this frequency is therefore linearly independent of the emission at the other three frequencies and of the plasma density determined from the Fe XII $\lambda = 195$ Å line emission.

The exact height of the Fe XII $\lambda = 195$ Å line formation in the CH is unknown. Though this line is considered to be coronal, CH limb observations show (according to SOHO EIT data) that the line intensity (and comparable values of plasma density in the CH within the range from 10^7 cm^{-3} to 10^9 cm^{-3}) is observed at heights less than one solar radius. This suggests that plasma density values in the CH obtained in this study correspond to the range of heights where the radio emission is formed at 17 GHz (chromosphere), 5.7 GHz (low corona) and 327 MHz (corona). So we can expect that there will be a frequency or a range of frequencies whose formation heights would correspond to the Fe XII $\lambda = 195$ Å line emission height, and there may be a linear dependence for these frequencies between the emission flux from the CH and the SW velocity. We can also expect that the height of one out of three frequencies under consideration is close to the Fe XII $\lambda = 195$ Å line formation height.

Observations of the CH at 17 GHz (Gopalswamy *et al.*, 1999; Nindos, Kundu, and White, 1999), where the brightness temperature of the CH is higher than that of surrounding quiet regions, confirm energy release in the upper chromosphere and transition region. Besides, observations of the CH at frequencies near 17 GHz also demonstrate an increased brightness temperature in the CH (Gopalswamy *et al.*, 1999; Nindos, Kundu, and White, 1999; Krissinel *et al.*, 2000; Moran *et al.*, 2001; Maksimov *et al.*, 2006).

The connection between the solar atmosphere parameters and high-speed solar wind is evident at heights less than one solar radius; at heights more than one solar radius, no dependence is observed. This is confirmed by the absence of a connection between the radio emission flux in the CH at 150.9 MHz and the velocity of the high-speed SW, whereas such a connection is evident at higher frequencies. The papers of Dulk and Sheridan (1974), Chiuderi-Drago (1974), Trottet and Lantos (1978), Chiuderi-Drago *et al.* (1999) also show that the CH at $\sim$150.9 MHz and below may be invisible on the quiet Sun background, or their brightness temperature may slightly differ from the temperature of the quiet Sun.

The established connection between SW parameters and radio emission established in this work provides a good reason to believe that the acceleration process of the high-speed SW and characteristics of the solar atmosphere are related at heights less than one solar radius. Some authors (Hollweg, 1978; Tu, Zhou, and Marsch, 1995; Tu *et al.*, 2005; Suzuki and Inutsuka, 2006) think that the acceleration of SW particles is caused by the energy transfer from the wave flux to them (for instance, due to the wave – particle resonance), and wave sources are in the lower solar atmosphere. According to data from the SUMER instrument on board SOHO (Chae, Hle, and Lemaire, 1998), the maximum value related to the wave flux is in the transition region. Measured wave velocities in the polar CH are about 130 – 160 $\mathrm{km\,s^{-1}}$ (already in the chromosphere and transition region) and up to 330 $\mathrm{km\,s^{-1}}$ in the lower corona, according to data from the instruments of *Hinode*/EIS and SOHO/SUMER instruments (Gupta *et al.*, 2010). This implies that the particle acceleration in the high-speed SW related to the CH is observed already in the lower solar atmosphere, and within this framework, the connection of the SW velocity near the Earth's orbit with the emission at these heights becomes clear.

Measurements of plasma velocity with UV coronal spectrographs may reveal whether SW particles at coronal heights are accelerated or not. However, no studies on this issue have been published so far. We hope that this issue will be addressed with data from new space observatories (*e.g.*, Solar C).

5. Summary

From the analysis of available observational data, the following conclusions have been drawn.

i) The velocity of the high-speed SW during the period under study is related to the mean emission intensity of the Fe XII line with $\lambda = 195$ Å (and the mean plasma density N_{CH} determined from it) in the CH.

ii) The density has been established between velocity of the high-speed SW and CH radio emission flux in the chromosphere and lower corona during the period under investigation.

Acknowledgements We are grateful to the team of observatories SOHO (instrument EIT), SSRT, Nobeyama and Nançay for providing free access to the data, which enabled this work.

References

Biermann, L.: 1951, *Z. Astrophys.* **29**, 274.
Biermann, L.: 1952, *Z. Naturforsch.* **7a**, 127.
Biermann, L.: 1957, *Observatory* **77**, 109.
Borovik, V.N., Kurbanov, M.S., Livshits, M.A., Ryabov, B.I.: 1990, *Sov. Astron.* **34**(5) 522.

Brosius, J.W., Landi, E., Cook, J.W., Newmark, J.S., Gopalswamy, N., Lara, A.: 2002, *Astrophys. J.* **574**, 453.
Chae, J., Hle, U.S., Lemaire, P.: 1998, *Astrophys. J.* **505**, 957.
Cranmer, S.R.: 2004, In: Walsh, R.W., Ireland, J., Danesy, D., Fleck, B. (eds.) *Proceedings of the SOHO 15 Workshop – Coronal Heating* **SP-575**, European Space Agency, Paris, 154.
Chiuderi-Drago, F.: 1974, In: G. Righini (ed.) *Skylab Solar Workshop: Osserv. Mem. Oss. Astrofis. Arcetri* **164**, 242.
Chiuderi-Drago, F., Landi, E., Fludra, A., Kerdraon, A.: 1999, *Space Sci. Rev.* **87**(1–2), 141.
Dulk, G.A., Sheridan, K.V.: 1974, *Solar Phys.* **36**, 191.
Eselevich, V.G., Fainshtein, V.G., Rudenko, G.V., Eselevich, M.V., Kashapova, L.K.: 2009, *Cosm. Res.* **47**, 95.
Grall, R.R., Coles, W.A., Klinglesmith, M.T.: 1996, In: *Proc. of the 8th International Solar Wind Conference: Solar Wind Eight, AIP Conf. Proc.* **382**, 108.
Gopalswamy, N., Shibasaki, K., Thompson, B.J., Gurman, J., DeForest, C.: 1999, *J. Geophys. Res.* **104**(A5), 9767.
Grechnev, V.V., Lesovoi, S.V., Smolkov, G.Y., *et al.*: 2003, *Solar Phys.* **216**(1), 239.
Gringauz, K.I., Bezrukikh, V.V., Ozerov, V.D., Rybchinskii, R.E.: 1962, *Planet. Space Sci.* **9**(3), 103 (First published: 1960, *Dokl. Acad. Sci. USSR* **131**, 1301).
Gupta, G.R., Banerjee, D., Teriaca, L., Imada, S., Solanki, S.: 2010, *Astrophys. J.* **718**, 11.
Harrison, E.R.: 1961, *Nature* **189**, 993.
Hollweg, J.V.: 1978, *Rev. Geophys. Space Phys.* **16**, 689.
Hundhausen, A.J.: 1972, *Coronal Expansion and Solar Wind*, Springer, Berlin.
Kosugi, T., Ishiguro, M., Shibasaki, K.: 1986, *Publ. Astron. Soc. Japan* **38**(1), 1.
Krissinel, B.B., Kuznetsova, S.M., Maksimov, V.P., Prosovetsky, D.V., Grechnev, V.V., Stepanov, A.P., Shishko, L.F.: 2000, *Publ. Astron. Soc. Japan* **52**, 909.
Licht, A.L.: 1960, *J. Geophys. Res.* **65**(5), 1397.
Maksimov, V.P., Prosovetsky, D.V., Kuznetsova, S.M., Obukhov, A.G.: 2004, *Solar-Terr. Phys.* **6**, 80.
Maksimov, V.P., Prosovetsky, D.V., Grechnev, V.V., Krissinel, B.B., Shibasaki, K.: 2006, *Publ. Astron. Soc. Japan* **58**(1), 1.
Mercier, C., Chambe, G.: 2009, *Astrophys. J.* **700**, L137.
Moran, T., Gopalswamy, N., Dammasch, I.E., Wilhelm, K.A.: 2001, *Astron. Astrophys.* **378**, 1037.
Nakajima, H., Nishio, M., Enome, S., Shibasaki, K., Takano, T., Hanaoka, Y., Torii, C., Sekiguchi, H., *et al.*: 1994, *Proc. IEEE* **82**(5), 705.
Neugebauer, M., Snyder, C.W.: 1966, *J. Geophys. Res.* **71**, 4469.
Nindos, A., Kundu, M.R., White, S.M.: 1999, *Astrophys. J.* **527**(1), 415.
Obridko, V.N., Shelting, B.D., Livshits, I.M., Asgarov, A.B.: 2009, *Solar Phys.* **260**, 191.
Parker, E.: 1958a, *Astrophys. J.* **128**, 664.
Parker, E.: 1958b, *Astrophys. J.* **128**, 677.
Sheeley, N.R. Jr., Howard, R.A., Michels, D.J., Koomen, M.J.: 1985, *J. Geophys. Res.* **90**, 163.
Shugai, Yu.S., Veselovsky, I.S., Trichtchenko, L.D.: 2009, *Geomagn. Aeron.* **49**(4), 41.
Stepanian, N.N., Kuzin, S.V., Fainshtein, V.G., Rudenko, G.V., Malashchuk, V.M., Perebeinos, V.A., Shtertser, N.I., Zhigalkin, R.K., *et al.*: 2008, *Solar Syst. Res.* **42**(1), 83.
Suzuki, T., Inutsuka, S.: 2006, *Astrophys. J.* **632**(1), L49.
Trottet, G., Lantos, P.: 1978, *Astron. Astrophys.* **70**, 245.
Tu, C.-Y., Zhou, C., Marsch, E.: 1995, *Space Sci. Rev.* **73**, 1.
Tu, C.-Y., Zhou, C., Marsch, E., Xia, L.-D., Zhao, L., Wang, J.-X., Wilhelm, K.: 2005, *Science* **308**(5721), 519.
Vršnak, B., Temmer, M., Veronig, A.M.: 2007, *Solar Phys.* **240**(2), 315.
Vsehvyatskiy, S.K., Nikol'skiy, G.M., Ponomarev, E.A., Cherednichenko, V.I.: 1955, *Astron. Zh.* **32**(2), 165.
Wang, Y.-M., Sheeley, N.R. Jr.: 1990, *Astrophys. J.* **355**, 726.
Wang, Y.-M., Sheeley, N.R., Walters, J.H. Jr., Brueckner, G.E., Howard, R.A., Michels, D.J., Lamy, P.L., Schwenn, R., Simnett, G.M.: 1998, *Astrophys. J.* **498**, L165.
Wilhelm, K., Curdt, W., Marsch, E., Schuhle, U., Lemaire, P., Gabriel, A., Vial, J.-C., Grewing, M., *et al.*: 1995, *Solar Phys.* **162**, 189.
Yallop, B.D.: 1961, *Observatory* **81**, 235.

MIX
Papier aus verantwortungsvollen Quellen
Paper from responsible sources
FSC® C105338

If you have any concerns about our products,
you can contact us on
ProductSafety@springernature.com

In case Publisher is established outside the EU,
the EU authorized representative is:
Springer Nature Customer Service Center GmbH
Europaplatz 3, 69115 Heidelberg, Germany

Printed by Libri Plureos GmbH
in Hamburg, Germany